Applied Probability and Statistics

BAILEY · The Elements of Stochastic Processes with Applications to the Natural Sciences

BAILEY · Mathematics, Statistics and Systems for Health

BARTHOLOMEW · Stochastic Models for Social Processes, *Second Edition*

BECK and ARNOLD · Parameter Estimation in Engineering and Science

BENNETT and FRANKLIN · Statistical Analysis in Chemistry and the Chemical Industry

BHAT · Elements of Applied Stochastic Processes

BLOOMFIELD · Fourier Analysis of Time Series: An Introduction

BOX · R. A. Fisher, The Life of a Scientist

BOX [illegible] cal Method fo[illegible]

DATE DUE

BOX, HUNTER, and HUNTER · Statistics for Experimenters: An Introduction to Design, Data Analysis, and Model Building

BROWN and HOLLANDER · Statistics: A Biomedical Introduction

BROWNLEE · Statistical Theory and Methodology in Science and Engineering, *Second Edition*

BURY · Statistical Models in Applied Science

CHAMBERS · Computational Methods for Data Analysis

CHATTERJEE and PRICE · Regression Analysis by Example

CHERNOFF and MOSES · Elementary Decision Theory

CHOW · Analysis and Control of Dynamic Economic Systems

CLELLAND, deCANI, BROWN, BURSK, and MURRAY · Basic Statistics with Business Applications, *Second Edition*

COCHRAN · Sampling Techniques, *Third Edition*

COCHRAN and COX · Experimental Designs, *Second Edition*

COX · Planning of Experiments

COX and MILLER · The Theory of Stochastic Processes, *Second Edition*

DANIEL · Application of Statistics to Industrial Experimentation

DANIEL · Biostatistics: A Foundation for Analysis in the Health Sciences

DANIEL and WOOD · Fitting Equations to Data

DAVID · Order Statistics

DEMING · Sample Design in Business Research

DODGE and ROMIG · Sampling Inspection Tables, *Second Edition*

DRAPER and SMITH · Applied Regression Analysis

DUNN · Basic Statistics: A Primer for the Biomedical Sciences, *Second Edition*

DUNN and CLARK · Applied Statistics: Analysis of Variance and Regression

ELANDT-JOHNSON · Probability Models and Statistical Methods in Genetics

FLEISS · Statistical Methods for Rates and Proportions

GALAMBOS · The Asymptotic Theory of Extreme Order Statistics

GIBBONS, OLKIN, and SOBEL · Selecting and Ordering Populations: A New Statistical Methodology

GNANADESIKAN · Methods for Statistical Data Analysis of Multivariate Observations

GOLDBERGER · Econometric Theory

GOLDSTEIN and DILLON · Discrete Discriminant Analysis

continued on back

D1564478

SPECIFICATION SEARCHES

Ad Hoc Inference with Nonexperimental Data

EDWARD E. LEAMER

University of California, Los Angeles

RECEIVED

JUL 26 1978

MANKATO STATE UNIVERSITY
MEMORIAL LIBRARY
MANKATO, MN

A Wiley-Interscience Publication

JOHN WILEY & SONS

New York · Chichester · Brisbane · Toronto

QA
276
.L337

Copyright © 1978 by John Wiley & Sons, Inc.

All rights reserved. Published simultaneously in Canada.

Reproduction or translation of any part of this work beyond that permitted by Sections 107 or 108 of the 1976 United States Copyright Act without the permission of the copyright owner is unlawful. Requests for permission or further information should be addressed to the Permissions Department, John Wiley & Sons, Inc.

Library of Congress Cataloging in Publication Data

Leamer, Edward E.
Specification searches.

(Wiley series in probability and mathematical statistics)
Bibliography: p. 350
1. Mathematical statistics. I. Title.

QA276.L337 519.5 77-26855
ISBN 0-471-01520-2

Printed in the United States of America

10 9 8 7 6 5 4 3 2 1

To my Parents and to Marjorie

416302

PREFACE

This book could be classified under the heading of "metastatistics." Statistics is the theory of inferences *ideally* drawn from data. Metastatistics is the theory of inferences *actually* drawn from data. Statistical theory takes as given the model, the data, and the purity of the researcher's motives. Metastatistics analyzes how the researcher's motives and opinions influence his choice of model and his choice of data. Metastatistics includes the study of memory and computing failures; it also deals with the social mechanism by which information is transmitted among individuals.

Most of this book deals with a special topic of metastatistics—specification searches. Traditional statistical theory assumes that the statistical model is given. By definition, nonexperimental inference cannot make this assumption, and the usefulness of the traditional theory is rendered doubtful. Moreover, once the model is known, the inferential puzzles that remain are trivial in comparison with the puzzles that arise in the specification of a model. The latter puzzles form the subject of "specimetrics." Specimetrics describes the processes by which a researcher is led to choose one specification of the model rather than another; furthermore, it attempts to identify the inferences that may be properly drawn from a data set when the data-generating mechanism is ambiguous.

My interest in metastatistics stems from my observations of economists at work. The opinion that econometric theory is largely irrelevant is held by an embarrassingly large share of the economics profession. The wide gap between econometric theory and econometric practice might be expected to cause professional tension. In fact, a calm equilibrium permeates our journals and our meetings. We comfortably divide ourselves into a celibate priesthood of statistical theorists, on the one hand, and a legion of inveterate sinner-data analysts, on the other. The priests are empowered to draw up lists of sins and are revered for the special talents they display. Sinners are not expected to avoid sins; they need only confess their errors openly.

In this book I discard the elitism of the statistical priesthood and proceed under the assumption that unavoided sins cannot be sins at all. For several years I have observed colleagues who analyze data. I report herein my understanding of what it is they are doing. The language I use to describe their behavior is the language of the statistical priesthood, and this book may serve partially to bridge the gap between the priests and the sinners. The principal outcome of this effort is an appreciation of why certain sins are unavoidable, even desirable. A new sin does emerge, however. It is a sin not to know why you are sinning. Pointless sin must be avoided.

I began thinking about these problems when I was a graduate student in economics at the University of Michigan, 1966–1970. At that time there was a very active group building an econometric model of the United States. As it happens, the econometric modeling was done in the basement of the building and the econometric theory courses were taught on the top floor (the third). I was perplexed by the fact that the same language was used in both places. Even more amazing was the transmogrification of particular individuals who wantonly sinned in the basement and metamorphosed into the highest of high priests as they ascended to the third floor.

As I struggled against the schizophrenia, I found comforting the Bayesian theory of inference that I was studying in the math department. This truly seemed to be the key toward understanding the difference between the basement and the third floor. Perhaps a complete reconciliation could be achieved.

I am now less optimistic. I am confident that the Bayesian approach helps us understand nonexperimental inference. It helps also to avoid certain errors. But I do not think it can truly solve all the problems. Nor do I foresee developments on the horizon that will make any mathematical theory of inference fully applicable. For better or for worse, real inference will remain a highly complicated, poorly understood phenomenon.

In trying to give a theory of nonexperimental inference, I have accepted an enormous task. The phenomenon of inference is within the purview of statisticians, philosophers, psychologists, and historians. I have found it impossible to master all four fields, though I have tried to understand the central ideas of each. My greatest familiarity is with the theories of statistical inference, but even in that area I am certain that there is much material I have overlooked. My knowledge of philosophy, psychology, and the history of science is spotty, but you will find occasional reference to it. Once I thought I should try to master that literature, but, frankly, I found it inaccessible. Mastering it would have put off the completion of this work for several years. It seemed to be more efficient to say what I have to say, and to be prepared to explore other literature in response to critical comments. Thus I shall welcome the criticism that I am unfamiliar with various materials. Believe me, I have learned much since this project commenced and I do not regard publication as a termination of the process. On the contrary, I feel I had better get it published before I change my mind.

This book is logically divided into three parts. The core of the book consists of Chapters 4 through 9, which analyze six different kinds of specification searches. The first three chapters are introductory, and the last chapter constitutes a footnote describing the inability of anyone actually to behave as described in the first nine chapters.

Chapters 2 and 3 make the book relatively self-contained. It is not necessary that the reader be familiar with Bayesian inference, since a complete (though opinionated) introduction is given in these chapters. Familiarity with classical inference, particularly the normal linear regression model, is assumed, but a review of the standard propositions is provided. Mathematical sophistication is not required to read this book. Matrix algebra is used extensively, but graphs are used whenever possible to clarify the logic. The problems in Appendix 4 may also be helpful.

I can hope, but it seems unlikely, that this book will be used as a text at many universities. I myself have used it as a text for special topics courses both at Harvard and at UCLA. Students in these courses had already been through the usual sequence in graduate econometrics. Many had some first-hand knowledge of econometric practice and were happy to learn that someone else thought econometric theory to be rather remote.

Although it may not be used as a principal text, this book would usefully supplement the fare offered in the standard texts. Some of the material from Chapter 5 on interpretive searches has already filtered into the econometrics courses at the better universities. Chapter 7 on proxy searches could supplement the traditional, rather meager treatment of errors-in-variables problems. Chapter 4 contains two easily read sections on choosing the significance level of a test. Chapter 8, on data selection, has several sections that ought to be valuable reading—especially the unified treatment of multivariate regression, error-components models, and Kalman filtering. The introduction to Chapter 9 on data-instigated models makes informally the important point that there is a fundamental difference between statistical inference and "Sherlock Holmes" inference. What in this book, I ask myself rhetorically, would not make profitable reading?

Which chapters are my favorites? Clearly, the greatest conceptual contribution is the recognition of the difference between statistical inference and "Sherlock Holmes" inference described in Chapter 9. One may doubt whether the discipline suggested there is practical, however. Is the Bode's law example serious? As for practical advice, turn to Chapter 5 on interpretive searches. It includes what I believe to be a correct statement of the problem of collinearity. I also think the distinction between interpretive searches and simplification searches is important. Chapter 10, on judgmental errors (largely a review and interpretation of psychological literature), was the most amusing to write. Chapter 7 was personally useful to me since I learned in writing it that the metaphysical distinction between endogenous and exogenous variables has little to do with the inferences it is proper to draw from a given body of data.

The impossible task of properly listing my debts is now at hand. As far as I can remember, the first time I did any independent thinking was in writing my undergraduate thesis at Princeton under the direction of John Hartigan. It may be interesting to note that I first read Bayes' *Essay* in his course. It made no particular impression on me at the time. At the University of Michigan the statisticians William Ericson and Bruce Hill taught me Bayesian inference at a perfect time in my life. Were it not for that contact, this book would never have been written. Nor would it have been written had I been unable to observe econometricians at work. At the University of Michigan, Daniel Suits, Saul Hymans, and Harold Shapiro were the (unknowing?) rats in my laboratory. Later, at Harvard, Martin Feldstein, Dale Jorgenson, and Zvi Griliches passed under my microscope. My friend Richard Freeman was a constant source of ideas and encouragement. Also at Harvard, I had the good fortune to spend several years with Gary Chamberlain, who served as a student, a collaborator, and a friend. More recently, Herman Leonard has been a stimulating coworker.

Many of my ideas have been influenced by attendees at the semi-annual Seminars in Bayesian Inference in Econometrics. Arnold Zellner especially should be mentioned. He and Jacques Drezé were the first to carry the Bayesian fasces into the econometrics arena. Another attendee at these seminars (and a Harvard colleague), John Pratt, has had a significant influence on my thoughts. Though they may wish to deny it, I have discovered kindred souls in the form of James Dickey and Thomas Rothenberg.

There is a long list of statisticians whose written contributions have influenced me. That debt is acknowledged in the text with references to their work. Parts of Chapters 2 and 4 were originally prepared jointly with Howard Raiffa for an introductory book that remains unpublished. Earlier versions of this book were read in part and commented on by colleagues at UCLA: Mike Darby, John Riley, Bob Clower, and Jack Carr. Walter Vandaele provided useful detailed comments on several chapters. Lynn Shisido and Tom Means checked the entire manuscript and uncovered numerous errors.

I acknowledge also my debt to the National Science Foundation, Social Science Division, headed by Dr. James Blackman, for grants GS-31929 and SOC 76-08863, without which this book would not have been prepared.

Lastly I would like to bring to the attention of interested parties the existence of *SEARCH*, a computer package designed to implement some of the ideas in this book.

EDWARD E. LEAMER

Los Angeles, California
February 1978

CONTENTS

CHAPTER 1

INTRODUCTION

"Data mining," "fishing," "grubbing," "number crunching." These are the value-laden terms we use to disparage each other's empirical work with the linear regression model. A less provocative description would be "specification searching," and a catch-all definition is "the data-dependent process of selecting a statistical model." This definition encompasses both the estimation of different regression equations with different sets of explanatory variables and also the estimation of a single equation using different subsets of the data.

The fact that specification searching invalidates the traditional models of statistical inference is implicit in the pejorative content of the word "fishing," but the industrious implication of the word "mining" suggests that the activity may, in fact, be productive.[1] Although "fishing" too might seem to be a productive activity, the term is usually used in a derogatory way to indicate both the fisherman's great uncertainty over the quantity and quality of fish that might appear in his net and his willingness to accept anything that shows up. Mining, in contrast, is an activity intended to bring to the surface a specific valuable commodity whose existence is likely to be relatively well established before mining commences.[2]

[1]Computer programs for data analysis are given names that reflect this use and abuse of the power of the computer: RAPFE (regression analysis program for economists), ESP (econometric software program), TROLL (time-shared reactive on-line laboratory).

[2]Commercial fishing that involves greatly reduced uncertainty is sometimes called "mining the sea."

This book is about "data mining." It describes how specification searches can be legitimately used to bring to the surface the nuggets of truth that may be buried in a data set. The essential ingredients are judgment and purpose, which jointly determine where in a data set one ought to be digging and also which stones are gems and which are rocks. Without judgment and purpose, a specification search is merely a fishing expedition, and the product of the search will have a value that is difficult or impossible to assess.

The subtitle of this book, "Ad Hoc Inference with Nonexperimental Data," was chosen to suggest that the phenomenon of specification searching is an order of magnitude more common in nonexperimental inference. This can be made definitionally true by asserting that an experiment defines a model. When a specification search occurs, the researcher reveals that he does not think an experiment was conducted. Given this definition, I offer both a descriptive and a prescriptive theory of nonexperimental inference. My observations of economists have led me to the conclusion that there are six logically distinct varieties of specification searches, and each is discussed in this book. The resultant theory is descriptive, in the sense that it springs from observation of nonexperimental scientists at work, but it is also prescriptive, in that it offers alternatives to what seems to be going on now.

A Bayesian approach is used almost exclusively. Anyone who is familiar with the extent to which judgment is used in the analysis of nonexperimental data should have no difficulty in accepting the Bayesian, personal view of inference that is espoused here. Arguments concerning Bayesian versus classical inference are implicit in much of this book, but the battle over the proper philosophical foundations for inference is largely ignored. That battle is intellectually stimulating, and, as far as I am concerned, decidedly one-sided. But it is a battle evidently of little interest to analyzers of real data, perhaps because the practical consequences of accepting the Bayesian view are either ambiguous or minor.

I offer here a different argument in favor of the Bayesian position. The phenomenon of specification searches completely invalidates the traditional models of inference, both Bayesian and classical. But the Bayesian approach is sufficiently flexible that, with suitable alterations, specification searches can be made legitimate, or at least understandable. This does not seem to be the case with the classical model of inference. I am definitely not arguing that one must be a formal Bayesian. I am only claiming that the Bayesian view yields insights. A formal Bayesian encounters insurmountable difficulties in constructing meaningful prior distributions. Thus, most uncertain judgments elude precise quantification. But a way to deal with the fuzziness of quantified probability judgments is to explore the implications of many different, precisely described judgments, a procedure which seems to me to be better than the other approaches that compound the judgment fuzziness with methodological fuzziness. The myth that inference with nonexperimental data (or any data) could be judgment-free creates an insidious and a counter productive goal.

To the extent that I have been successful in identifying all the reasons for specification searches, this book offers a nearly complete normative theory of personal learning with the linear regression model. It parallels to a great degree the commonsense "ad hoceries" that are characteristic of nonexperimental inference. In this book there are, however, several aspects of learning that are either not mentioned or incompletely discussed. First, no mention is made of the simultaneous-equations problem that plagues nonexperimental inference. The simultaneous equations model does bring up the interesting problem of inferring causality, but from the standpoint of specification searches, it is a formal variant of the simple linear model and therefore implies no interesting methodological issues that are not discussed herein. A second shortcoming is the brief treatment of memory failures. The shortcomings of memory seem quite important for any positive theory of personal inference, although a normative theory may proceed usefully with a perfect memory assumption. The usual Bayesian model implicitly does make this assumption, and memory failures may cast doubt on any of its implications. A third shortcoming of this book is its neglect of social learning. It is obvious that the accumulation of opinion is partly, if not largely, a social phenomenon. Unfortunately, the currently available mathematical models of social learning are primitive and are hardly worth discussing, except that they, rightly, remind us of the social-learning phenomenon. It is useful to observe that the social-learning problem is a special memory problem. Social memory is simply the accumulated set of experiences of *all* individuals, and your access to the totality of experiences depends on your contact and communication with the people who had or who heard of the particular experiences. There are, of course, various distortions for various reasons in the communication of these experiences, just as there are features of personal memory that make some events more memorable than others. Thus the significant shortcoming of this book is its inadequate treatment of memory problems, personal and social.

1.1 The Axiom of Specification

In searching for a model of nonexperimental inference, we may easily discard the textbook version of classical inference. It makes implicit use of the following unacceptable specification axiom.

The Axiom of Correct Specification

(a) The set of explanatory variables that are thought to determine (linearly) the dependent variable must be
  (1) unique,
  (2) complete,
  (3) small in number, and
  (4) observable.
(b) Other determinants of the dependent variable must have a probability distribution with at most a few unknown parameters.
(c) All unknown parameters must be constant.

If this axiom were, in fact, accepted, we would find one equation estimated for every phenomenon, and we would have books that compiled these estimates published with the same scientific fanfare that accompanies estimates of the speed of light or the gravitational constant. Quite the contrary, we are literally deluged with regression equations, all offering to "explain" the same event, and instead of a book of findings we have volumes of competing estimates.

The phenomenon of specification searches thus represents an unambiguous rejection of the axiom of specification and literally pulls the foundation from under classical inference. This book presents an alternative theory of inference that either formally allows specification searches or suggests alternatives. The theory rests on the firm (but fuzzy) foundation of probabilistic judgments. It makes use of formal decision theory in those cases in which a specification search seems to be solving a decision problem.

I am certainly not the first to notice the discrepancy between inference as it is described in the textbooks and inference as it is practiced at the computer center. There is a wide spectrum of opinions concerning the effect of specification searches on inference. "Believers" use ad hoc techniques to search for specification, throwing out insignificant variables here and there, for example, but they continue to regard the end result of such a methodology to be identical to the end result obtained in the experimental sciences (or at least cynically to act that way). Believers report the summary statistics from the nth equation as if the other $n-1$ were not tried, as if the nth equation defined a controlled experiment.

At the other extreme are the agnostics, who gladly admit the irrelevance of classical inference. They argue that a nonexperimental scientist is merely identifying relationships that exist in the historical data. He is describing the salient features of the data accurately but economically. Ideally, the data analysis generates hypotheses that need new data to be tested. Agnostics may thus discount any statistical result until it has been employed in a prediction outside the data period. We might interpret such statements in a statistical context as the absence of information concerning the standard errors. A point estimate without an associated standard error does not imply an hypothesis test, nor can it determine unambiguous inferences.

Somewhere between these two extremes is a group of pragmatists. They feel that the believers' contentment stems only from ignorance but that the agnostics have gone too far. This group argues that estimated standard errors are properly enlarged by a specification search but not to the extent that they become infinite. Theil (1961), for example, writes:

> The obvious result is that, if a 'maintained' hypothesis [a specification, in our terms] gives unsatisfactory results, it is not maintained but rejected, and replaced by another 'maintained' hypothesis; etc. It is hardly reasonable to say that this kind of experimentation is incorrect, even if it affects the superstructure built on such 'maintained' hypotheses. [In a footnote, he explains that he is referring specifically to the standard errors calculated by classical formulae.] It is especially unreasonable to reject such an experimental approach, because... the statistical theory which forbids the rejection of a 'maintained' hypothesis is not fully satisfactory either in view of the difficulty of its application.
>
> What *is* incorrect, however, is to act as if the final hypothesis presented is the first one, whereas in fact it is the result of much experimentation.

Although Theil is rejecting classical inference as unworkable and berating the naïveté of the believers, he does not offer a procedure that would allow valid inferences in the context of a specification search. By how much are the standard errors to be enlarged? And which of the many estimates are we to choose? A theory of specification searches is needed to answer these important questions.

1.2 The Six Varieties of Specification Searches

A theory of specification searches can be constructed first by identifying the reason a researcher engages in a search, and second by building formal inferential models that properly carry out his legitimate intentions. By observation of economists analyzing data, I have come to the conclusion that there are six different reasons for specification searches. Each is discussed in a separate chapter of this book. The six searches are listed with chapter references in Table 1.1.

For illustrative purposes, imagine a researcher interested in exploring empirically the theory of demand. In its simplest form the theory may be stated as follows: "*Ceteris paribus*, an individual's purchases of some commodity depends on his income and on the price of the commodity." The problem of the empirical worker is to translate this theoretical assertion into a statement about observable phenomena. He must identify the

Table 1.1

Specification Searches

Name of Search	Designed to	Chapter
Hypothesis-testing search	choose a "true model"	4
Interpretative search	interpret multidimensional evidence	5
Simplification search	construct a "fruitful model"	6
Proxy search	find a quantitative facsimile	7
Data-selection search	select a data set	8
Postdata model construction	improve an existing model	9

observable counterparts of the theoretical variables, he must select other variables that may significantly affect purchases, he must choose a particular functional relationship between the variables, and he must decide which individuals are actually to be observed. Because he cannot make these decisions with complete confidence, the researcher is willing to change his mind if his original choices seem not to work out as well as he might have liked. He does so by changing the specification of his statistical model. He may include more explanatory variables; he may omit certain variables; he may substitute one variable for another; he may discard observations, or he may include new observations.

Suppose the initial model is $\log D_i = \alpha + \beta \log Y_i + \gamma \log P_i + u_i$, where D_i is the purchases of oranges by household i, Y_i is monetery income, P_i is the the price of the commodity, and u_i is a "random disturbance" assumed to be normally distributed, independent of u_j, for $i \neq j$. The variables are observed by asking a random selection of heads of households, "How much did you earn last month, how many oranges did you purchase, and how much did they cost?" Using the replies of 150 households, the following regression equation is estimated:

$$\underset{}{\log D_i} = \underset{(1.1)}{6.2} + \underset{(.21)}{.85} \log Y_i - \underset{(.13)}{.67} \log P_i, \qquad R^2 = .15,$$

with standard errors in parentheses. For a variety of reasons, it is likely that other equations would be estimated with the same data set. Without endorsing the procedures, I now describe a typical search program.

Of special interest is the hypothesis that the fraction of income spent on oranges is not a function of price, $\gamma = -1$. To test this hypothesis, the equation is reestimated with the constraint applied:

$$\log D_i + \log P_i = \underset{(1.0)}{7.2} + \underset{(.20)}{.96} \log Y_i, \quad R^2 = .14.$$

Using a standard F test, this hypothesis is rejected at the .05 level, and it is inferred that the data cast doubt on the hypothesis $\gamma = -1$. This is an example of an hypothesis testing search in which different specifications describe different hypotheses about the phenomenon.

The theory of demand describes the behavior of a single individual, but this sample varies across individuals. The nutritional importance of oranges is greatest in areas with the least sunlight, and it may be inappropriate to treat southerners as if they were identical to northerners in their taste for oranges. Separate regressions are therefore computed for southerners and northerners:

$$\log D_i^N = \underset{(1.9)}{7.3} + \underset{(.41)}{.89}\log Y_i^N - \underset{(.25)}{.60}\log P_i^N \qquad R^2 = .18,$$

$$\log D_i^S = \underset{(2.2)}{7.0} + \underset{(.31)}{.82}\log Y_i^S - \underset{(.26)}{1.10}\log P_i^S \qquad R^2 = .19.$$

These regressions suggest that in the North, income is the relatively more important variable and price the relatively less important variable, but the hypothesis that the coefficients are different is not rejected at the .05 level. This is an example of a *data-selection search*. The same theoretical hypothesis underlies all three specifications: the one estimated with all the data and the pair estimated with subsets. The specifications differ in their choice of data sets.

Next it must be observed that the answer to the income question may be a very poor measurement of the household's true income. As it turns out, households were asked to report their expenditures on a fairly inclusive list of other commodities, and it may be that their total expenditures E_i is a better measurement of income than Y_i. The variable E_i is substituted for Y_i, and the estimated equation becomes

$$\log D_i = \underset{(1.0)}{5.2} + \underset{(.18)}{1.1}\log E_i - \underset{(.16)}{.45}\log P_i, \qquad R^2 = .18.$$

The R^2 has increased, and the coefficient on the income variable has become more significant, which suggests that E_i is the better measurement of income. This is a *proxy variable search*. Competing specifications in a proxy variable search all derive from the same underlying hypothesis. Different estimated regressions reflect different ways of measuring a common set of hypothetical variables.

The R^2s in all these equations are unhappily low. Perhaps there are other variables that might be added to the specification to improve the fit. After all, the theory makes use of the Latin phrase, *ceteris paribus*, other things constant, yet it is the nature of nonexperimental research that other things are not held constant. Although I prefer oranges, if grapefruit are on sale, I will sometimes buy them instead. Adding the price of grapefruit π_i to the equations yields the result

$$\log D_i = \underset{(1.0)}{3.1} + \underset{(.20)}{.83}\log E_i + \underset{(.15)}{.01}\log P_i - \underset{(.60)}{.56}\log \pi_i \qquad R^2 = .20.$$

This specification represents the broader theory: "*Ceteris paribus*, an individual's purchases of some commodity depends on his income, on the price of the commodity, and on the price of 'similar' commodities." The process of revising the underlying theory in response to the data evidence is called *post data model construction*, and the resulting hypothesis is called a *data-instigated* hypothesis. Whereas all other specifications are implicit in the original theoretical statement, a data-instigated hypothesis is not.

In the regression last reported, the coefficients on the price variables are insignificant and of the "wrong" sign. Furthermore, the sum of the coefficients ($.83 + .01 - .56 = .28$) is rather far from zero. The presumption that these coefficients sum to zero derives from the homogeneity postulate that asserts the following. "There is no money illusion: if money income and all prices are multiplied by the same constant, purchases will not change." Applying this homogeneity constraint yields the regression

$$\log D_i = \underset{(.9)}{4.2} + \underset{(.19)}{.52}\log E_i - \underset{(.14)}{.61}\log P_i + \underset{(.31)}{.09}\log \pi_i, \qquad R^2 = .19.$$

The R^2 has fallen only slightly, and the coefficients all have the right sign, two of them significantly so. Thus the constraint seems to improve the specification. This is an example of an *interpretive search*. The underlying hypothesis is taken as given. Restrictions are imposed in the hopes that the estimates may be "improved."

The regression equation now includes three variables, one with a very small coefficient and the other two with coefficients approximately the same size in absolute value. A simple equation would result if π were omitted and the other two coefficients set equal to each other (but opposite in sign):

$$\log D_i = \underset{(.8)}{3.7} + \underset{(.18).}{.58}\log(E_i/P_i) \qquad R^2 = .18.$$

The R^2 is only slightly smaller, and this simple equation is selected. This sixth and final search is a *simplification search*, the function of which is to find a simple but useful model.

The six kinds of specification searches may not yet be clearly different in your mind. In practice, there is little effort made to distinguish one from the other, and it is unsurprising that at first consideration it is difficult to discern the real differences. Moreover, since the searches differ sometimes only in the intent of the researcher and not in his actions, it may be difficult to infer which kind of search actually occurred. By this I do not mean to imply that it is *unimportant* to identify the type of search. Quite the contrary, the effectiveness of a search must be evaluated in terms of its intentions. An apparently successful simplification search may be judged completely unsuccessful as an interpretive search, and so forth.

It is always possible for a researcher to know what kind of search he is employing, and it is absolutely essential for him to communicate that information to the readers of his report. The differences in the searches will perhaps be most clear after this book is read in its entirety, but more may be said in this introductory chapter. Hypothesis-testing searches involve alternative models that have "truth value." It is difficult to find non-Bayesian language that can make such a statement less ambiguous, but in the Bayesian language, hypothesis-testing searches make use of alternative specifications that are assigned positive subjective prior probability. This can be contrasted with an interpretive search, in which only the most general specification is assigned positive probability. In an interpretive search an hypothesis, say, "$\gamma=0$," is thought to be false with probability one, but the hypothesis, "γ is close to zero," is thought to be quite likely. That is to say, the prior distribution for γ concentrates the probability mass in the neighborhood of $\gamma=0$ but assigns zero probability to zero. Given any such probability distribution, it is always possible to find a good approximation to it that does allocate positive probability to $\gamma=0$, and the distinction between hypothesis-testing searches and interpretive searches is thereby blurred. But in a large sample the value $\gamma=0$ almost certainly becomes uninteresting unless it is allocated a positive prior probability. Thus one practical difference between interpretive searches and hypothesis testing searches is that the former are strictly small-sample phenomena. To put this in the language of classical hypothesis testing, the significance level of a test should be a decreasing function of sample size in an hypothesis-testing search but should be relatively constant in an interpretive search. Incidentally, real examples of hypothesis-testing searches are extremely rare. The most general models used in nonexperimental inference are themselves not regarded to be complete descriptions of the phenomena under study. Restrictions on these "false" models could hardly lead to potentially true models. Hypothesis-testing searches are discussed first in this book only because the formal theory of hypothesis testing is most familiar, not because it solves an important problem.

A simple formal example contrasts three of the searches. A researcher may estimate the pair of equations $Y=x\beta+z\gamma+u$ and $Y=x\beta+u$ where β and γ are uncertain parameters, Y and x are observable variables, and u is the "residual error." From this fact alone it is impossible to determine whether he has engaged in an hypothesis-testing search, an interpretive search, or a simplification search. In an hypothesis-testing search, the hypothesis $\gamma=0$ means "the model $Y=\beta x+u$ is true." In an interpretive search the same hypothesis implies only that "it is better to estimate β acting as if $Y=\beta x+u$ were the true model than to estimate β using the more complete model." An analogous simplification hypothesis might be "if prediction of Y is the goal, the value of γ is usefully set to zero."

The motivation for hypothesis-testing searches and interpretive searches is prior information. Hypotheses are conjectured to be true or to be approximately true. The motivation for simplification searches is a loss function that penalizes complexity. Hypotheses are not conjectured to be true, or even approximately true. It is only hoped that a simple model would turn out to be adequate.

The data-selection search described above apparently is also an interpretive search, or possibly postdata model construction. In that example, the data set was split into two subsets, and separate regressions were estimated for each. An interpretive search might make use of a general model with two sets of parameters and might test the hypothesis that better estimates would result if the parameters in the two regimes were treated as if they were identical. But since the more general model was not explicitly stated in the beginning, it might be better to think of the search as post data model construction. Data-selection searches could thus be treated as special cases of these and possibly other searches, but the category is nonetheless useful. A theory rarely indicates an experiment that could be used to test it or to estimate its uncertain parameters. A researcher must construct his own experiment, or in the nonexperimental sciences, he must select observations from the set of recorded nonexperiments. The problems he confronts in doing so lead to a data-selection search, even though these problems may be formally similar to the problems associated with other searches.

To understand this more clearly, consider again the theoretical statement, "Y depends linearly on x and z: $Y=\alpha+x\beta+z\gamma$." To estimate this model, a researcher must select a data set over which the parameters α, β, and γ can be thought to be constant, or he must append to this model some description of how the parameters change from observation to observation. In practice, he will often treat the slope parameters β and γ as constants and try several different probabilistic descriptions of the variability of the level α from observation to observation. By definition, a data-selection search deals with the variability of unobservables (parameters and "error" term). An interpretive search introduces prior information about the means of the unobservables. Postdata model construction adds new unobservables to the model.

There is little difficulty in identifying a proxy variable search, but there is great difficulty in determining the inferences that may be legitimately made in the context of a proxy search. At one extreme, the theory is taken as given, and the data are used to construct a quantitative facsimile of the unquestioned theory. The evidence is completely spent to select a proxy, and no evidence is left over for inference about the theoretical parameters. At the other extreme, perfect measurement is assumed, and none of the evidence is spent to select a proxy. Real proxy searches lie somewhere between, with the evidence partly spent to estimate the theoretical parameters. It is difficult to position a search very precisely between the two extremes.

The last search is what I have called postdata model construction. I also like to call it "Sherlock Holmes inference." Sherlock solves the case by weaving together all the bits of evidence into a plausible story. He would think it indeed preposterous if anyone suggested that he should construct a function indicating the probability of all possible configurations of evidence for all possible hypotheses about the crime. In response to a question from Dr. Watson concerning the likely perpetrators of the crime, Holmes replied, "No data yet.... It is a capital mistake to theorize before you have all the evidence. It biases the judgments."[3]

Sherlock avoids formulating the hypotheses because the set of viable alternatives is immense and any attempt to formulate it completely will involve intolerable costs. If an incomplete set of hypotheses is formulated before the data are observed, there is a great risk of not realizing that the data favor some yet unspecified hypothesis. Instead, evidence is used to direct the construction of a set of "empirically relevant" hypotheses, thereby reducing both the cost of formulating hypotheses and the risk of not identifying the "best" hypothesis. There is, unfortunately, an opportunity cost to this process: the data may not also be used in any obvious way to discriminate among the data-instigated hypotheses. This dilemma is most excruciating when the data set is strictly limited, as in astronomy.

Because statistical inference requires a well-specified theory in advance of the data, Sherlock regards statistical inference to be a "capital mistake." Unlike most of us, however, Sherlock has the luxury of the ultimate extra bit of data—the confession. Even under the greatest coercion, our data sets usually resist our efforts to force a confession from them. Without the confession, it is no longer possible to be confident that any inferences are legitimate.

A solution to this dilemma is to act as if Sherlock Holmes inference solved a certain statistical decision theory problem. Given the model

[3]Doyle (1888), *A Study in Scarlet*.

$Y = x\beta + z\gamma + u$, it is possible to determine before seeing any data whether it is necessary to observe z. The variable z may be thought to be uncorrelated with x, or γ may be thought to be small. Then inferences about β may be made without observing z, in the context of the model $Y = x\beta + u$. If the resulting estimate of β is the wrong sign, or if the pattern of estimated residuals is peculiar, one may legitimately change his mind and observe z.

This formal decision theory problem mimics Sherlock Holmes inference in that the data may induce the use of a more general model, but there is a very important difference. In the decision theory problem, the second model must have been explicitly defined before the data were observed. In sharp contrast, Sherlock Holmes admonishes Dr. Watson against formulating models too completely: "It biases the judgments." Although Sherlock Holmes inference is not, and cannot be, a formal statistical decision theory problem, it is nonetheless desirable to act as if Sherlock were solving the decision theory problem, since legitimate statistical inferences are then implied by a Sherlock Holmes procedure.

Consider again the example of postdata model construction. After getting a low R^2 in a regression of demand for oranges on the price of oranges and monetary income, the price of grapefruit is added to the equation. Any economist will explain that the price of close substitutes surely influences purchases of a commodity, and the use of the price of grapefruit does not reflect a new theory but only a more complete version of the theory that was available all along. Excluding the price of grapefruit cannot be sensible theoretically, although it may be desirable practically. This sounds just like the formal decision theory problem, in which it was first determined that observation of the price of grapefruit was unnecessary. Although the researcher did not explicitly solve this problem, I think he did so implicitly. As a result, it is possible to broaden statistical inference to encompass Sherlock Holmes inference.

In general, the consequence of a specification search is what you might expect. There is greater uncertainty over the parameters than is suggested by the final specification. The data evidence is spent partly to specify the model, and only a part of it is left over to estimate parameters or to discriminate among competing models. The one exception to this rule is a simplification search. Simplification is a decision problem that properly occurs after inferences have already been drawn. It is not necessary to discount the evidence because of the search, but it is quite important to understand that the simplified specification is a tool for some anticipated decision problem and is not a model for inference with the given data set.

With the exception of postdata model construction, the other searches produce an equation that tends to understate the uncertainty, because the equation is estimated as if some parameter were known with certainty, when in fact the parameter remains uncertain. The equation is estimated as if the specification were given, whereas the very fact that a search occurred reveals that there is uncertainty over the specification. Loosely speaking, the apparent statistical evidence implied by the final equation must be discounted; the greater the range of search, the greater must be the discount.

The discount applying to a data-instigated model is somewhat different. A data-instigated model is treated as if it were the model the researcher always believed in. As a result, the final specification is certainly better than the original specification, and there can be no discounting because of uncertainty in the specification. A discount nonetheless applies to the final specification. In estimating the original specification the researcher reveals something about his prior information. He thinks that the variables he has omitted are not important. When he decides to add them to the specification, he is obligated to retain his original prior. This prior tends to adjust his estimates back toward the estimates obtained with the simple model. In that sense, the evidence implied by the final specification is discounted.

The various specification searches can be connected with the axiom of correct specification described in Section 1.1. When the set of explanatory variables is not unique, an hypothesis-testing search occurs. The incompleteness of the list of variables leads to postdata model construction. When the list of variables is excessively long, interpretive and simplification searches may be used. Unobservable variables imply proxy variable searches. Finally, data-selection searches are a response to the researcher's uncertainty over the choice of error distribution or to his concern that parameters may have shifted.

1.3 Data in Economics

There is a growing cynicism among economists toward empirical work. Regression equations are regarded by many to be merely stylistic devices, not unlike footnotes referencing obscure scholarly papers. It is the phenomenon of specification searches that has made the profession uneasy, and a theory of specification searches may help.

Distinguishing the various kinds of specification searches is a step in the right direction. Researchers currently do not distinguish one kind of search from another. Casual examination of papers in economics suggests that interpretive searches are the most prevalent, although these are hardly distinguishable from simplification searches. Hypothesis-testing searches are certainly the least common. Regardless of the type, the rules of search are informal and rarely stated explicitly. As a consequence, there is considerable doubt whether the average nonexperimental scientist is getting what he wants from his techniques. There is even some doubt that he knows what he wants.

Readers of this book are strongly urged not to conclude either that real learning processes could be fully mathematized and therefore trivialized, or that actual learning should be altered to meet fully the mechanical features of any mathematical model. To paraphrase an analogy of Polanyi's (1964), this is a book about violin playing, and while mastery of the technical/mechanical aspects of violin playing is essential, no one would suggest that studying a book alone would lead to great artistry; nor must a great artist completely conform to mechanical standards, the functions of which are primarily to improve the performances of the great mass of lesser artists.

One mathematical model of learning can be discarded, but it is not clear that the other should be retained. I refer, respectively, to classical and Bayesian inferences. Classical inference apparently allows judgments that are either completely certain or "completely uncertain." We are asked to be certain about the parameter spaces but peculiarly uncertain about the choice of parameters within those spaces. Typically, when selecting a parameter space economists also formulate judgments about the likelihood of various values of the parameters within that space. If they do have these uncertain judgments, they may want to make use of Bayesian tools.

It is perhaps more accurate to describe the classical judgmental inputs relative to the strength of the sample evidence, rather than in an absolute sense. The judgments are not absolutely certain or absolutely uncertain; rather they either overwhelm or are overwhelmed by the sample evidence. The process of learning is a "herky-jerky" reaction to the sample evidence, consisting of phases of complete disregard of sample evidence (failure to reject a null hypothesis?) and phases of complete disregard of nonsample evidence (rejection and discarding of the null hypothesis?). A Bayesian approach can obviously deal with the trivial cases of overwhelming sample evidence or overwhelming nonsample evidence, but it considers also the nontrivial problem of mixing two sources of information.[4]

No one who has worked with economic data or who has watched others work with it could retain the notion that economists have either overwhelming sample or overwhelming nonsample information. If this were so, economic research would often involve fitting a single regression equation. No reference to the process that generated the data would be made, no discussion of peculiar coefficients. There would be no collinearity problem, no proxy variables.

[4]The adjective trivial applies to the data-interpretation problem, obviously not to the mathematics that has been developed to solve these problems.

Necessarily, practicing economists have discarded the formal constraints of classical inference, and they have added the essential bits of subjective uncertain information through ad hoc specification searches. This involves trying not two or three different equations but literally thousands. Curiously, they retain a verbal commitment to classical inference, talking about such irrelevant things as "best linear unbiased estimators," "*t*-ratios," and the like.

When these specification searches are most effective, the final result may be an appropriate mixture of sample and nonsample information, essentially a posterior distribution. This process may be accurately captured by a Bayesian learning model, according to which we begin with a well-specified set of certain and uncertain judgments and enlist the data to encourage or discourage subsets of those judgments. (Another possibility, discussed below, is that economists are not doing statistical inference at all.)

It is highly unlikely that an ad hoc specification search could be as effective in implementing quantifiable uncertain judgments as the Bayesian approach. Even if the two techniques were to yield identical descriptions of the postdata uncertainty, the specification search approach has the critical defect that it cannot clearly distinguish sample from nonsample information, and the researcher thus has no way of effectively communicating the judgments that were required to analyze the data. It is then impossible for a reader to evaluate the reported results.

The inferential problems of nonexperimental scientists thus seem to be especially well suited to Bayesian inference. It is apparently astounding that Bayesian theory, which has been available in rudimentary forms for two centuries and in highly developed forms for several decades, has had so little impact on real data analysis. Can it be that the Bayesian philosophy attracts poor salesmen? Or is the product better in theory than in practice? I'm afraid it may be the latter.

In practice the Bayesian model has two defects. The first is that it requires the researcher actually to select a prior probability function. There is no doubt in my mind that uncertain prior information is used to analyze nonexperimental data. But there is also no doubt in my mind that uncertain prior information is impossible to quantify precisely. Ad hoc procedures may, in fact, be efficient methods of using imprecisely defined priors. I like to comment on this suggestion with a slogan: "The mapping is the message." The meaning of a data set is that it changes opinions. It takes particular prior opinions into particular posterior opinions. A data set may thus be fully described in terms of the mapping that it implies from prior distributions into posterior distributions. It is not necessary and it is even undesirable for a researcher to select a particular prior distribution. That task properly belongs to the reader of the report. A researcher should instead describe as completely as possible the mapping from priors into posteriors. He may properly recommend a particular prior, but he has no business forcing it on his reader.

An interesting Bayesian analysis of data, therefore, need not use a single, precisely specified prior distribution, and this hurdle need not be surmounted. But in deflecting our course from this one hurdle, we are forced to surmount another: How can the mapping of priors into posteriors be economically analyzed? This is a question that has not been asked often; it seems to me to be the major issue involved in practical use of the Bayesian tools.

There is, unfortunately, a second, potentially insurmountable difficulty with the Bayesian approach. Whereas I am confident that economists interested in statistical inference should be Bayesians, I question whether they should be interested in statistical inference at all. If, instead, they are doing Sherlock Holmes inference, the choice between a distorted Bayesian and a distorted classical approach is ambiguous. The Bayesian approach encourages more careful formulation of the model space, and to the extent that this is the right direction for the profession to move, the approach seems desirable. But Sherlock warns us against excessive theoretical development before seeing the facts. The process of assigning probabilities to models tends to make a researcher believe and cling to his original set of hypotheses. This straitjackets his Sherlock Holmes instincts, and he may ignore important evidence simply because the relevant hypothesis is outside his immediate field of vision.

I hope that this book will make clear the contribution Bayesian inference can make toward understanding the processes of research with nonexperimental data. In many cases it is enough that we understand what we are doing. In several cases specific alternatives are suggested—alternatives that are unambiguously superior to current procedures. Also, a long chapter is devoted to Sherlock Holmes inference, and it is hoped that the reader will understand the importance of this problem in real research as well as its implications for models of inference.

1.4 A Schematic Model of Inference

The model of inference that is being suggested in this book is indicated schematically in Figure 1.1. Inputs into the inferential process occur at ovals 1, 2, and 3. Major elements of the data analysis are indicated in rectangles 4 to 8. The specification search decisions to redo the analysis with a different set of models or propositions are indicated in diamonds 9 and 10. Solid line linkages may be discussed as problems in statistical

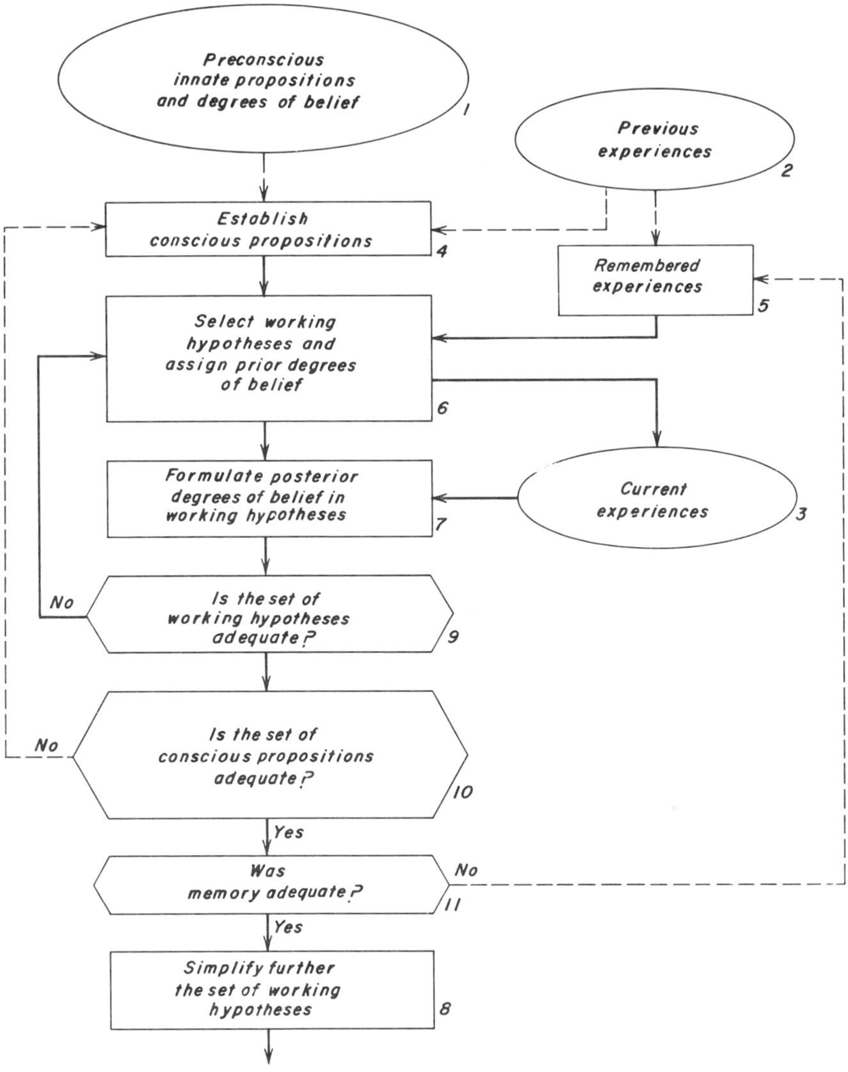

Fig. 1.1 A schematic diagram of inference.

inference as it is currently conceived. Dotted line linkages are philosophically outside the scope of statistical theory.

An individual is thought to have an enormous set of innate but preconscious propositions (oval 1), with innate degrees of belief assigned to them. These are determined, for example, by inherited sensory apparatus. Propositions about heat, hardness, taste are not learned but rather are built into the nervous system. The experiences indicated in oval 2 are used to select from this set of propositions a relatively tiny set of conscious beliefs (rectangle 4). Remembered experiences (rectangle 5) may then be used to determine which of the still large set of conscious propositions are to be used as a basis for a data analysis. The resulting set of working hypotheses (rectangle 6) is only remotely connected to the set of innate propositions, and the degrees of belief assigned to these propositions must be at best crude approximations to the degrees of belief of a Bayesian with unlimited memory and cognitive skills.

The preobservation "theoretical" work terminates temporarily when the set of working hypotheses is established and degrees of belief are assigned to them. Data are then observed (oval 3). The line linking the working hypothesis rectangle to the current experiences oval allows the choice of data to be a function of the working hypotheses. The data and the prior are mixed in rectangle seven to form tentative posterior degrees of belief.

Peculiarities in the data may then force a reconsideration of the several decisions that were explicitly or implicitly made earlier. The data may suggest that one of the excluded conscious hypotheses should now be included in the set of working hypotheses (diamond 9); or the data may induce the researcher to think up "new" hypotheses (diamond 10); or memory may be searched again (diamond 11). If the researcher changes his mind about one or more of his decisions, he will reanalyze the same data set with different working hypotheses or different priors.

Having satisfied himself with his analysis of the given data set, he must either make use of his newly formed opinions for the imminent decision toward which his efforts had been aimed, or he must make ready for future inference and decision problems as yet ill-defined. In either case he may wish to simplify the set of working hypotheses (rectangle 8).

Most treatments of statistical inference deal with a much more restricted description of learning. The set of hypotheses is ordinarily treated as if it

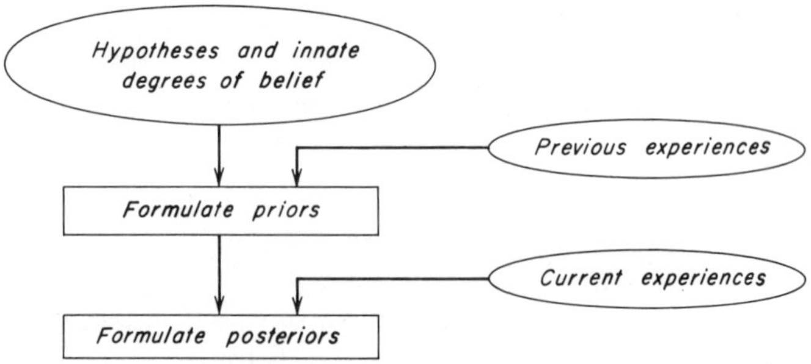

Fig. 1.2 Simple statistical inference.

were complete; that is, the set of working hypotheses, the set of conscious propositions, and the set of preconscious propositions are implicitly treated as if they were identical. Furthermore, the problem of fallible memory is ignored, and if prior information is used, it is assumed to represent accurately all previous experiences. The postdata simplification problem (rectangle 8) is also not discussed. The result is Figure 1.2.

It is possible to extend the logic of statistical decision theory to include both predata and postdata simplification problems. In the case of predata simplification we may formally ask the question: given the costs associated with working with a complete set of hypotheses, is it not better to use a restricted set of hypotheses? Given current experiences, it is logically proper to re-evaluate that decision and therefore to redo the analysis with an enlarged set of hypotheses. The result is Figure 1.3.

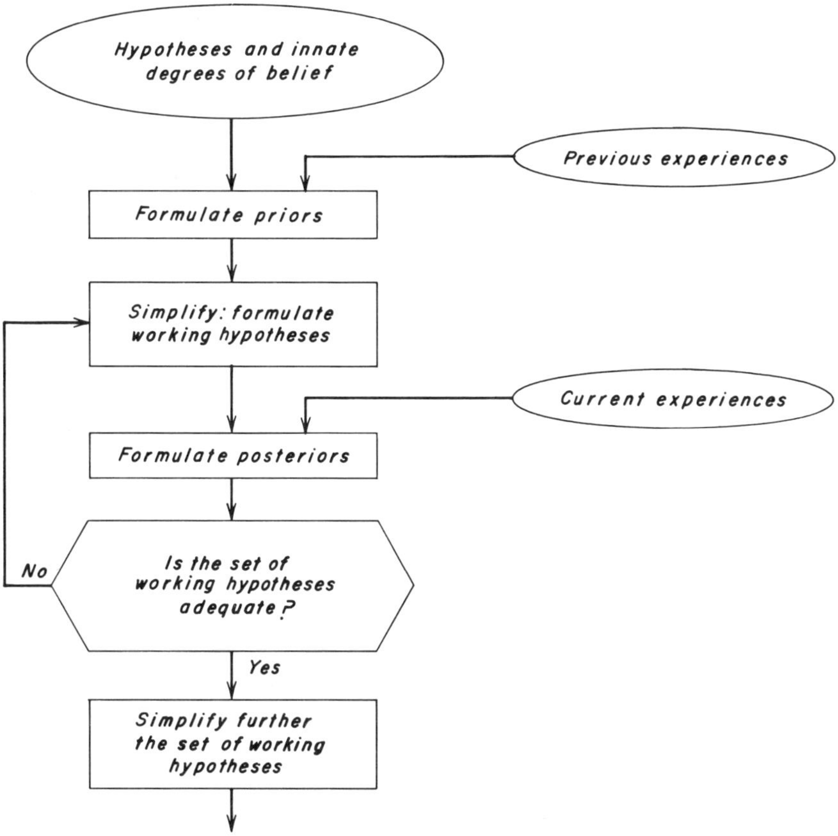

Fig. 1.3 Complete statistical inference.

That part of Figure 1.1 that is not also part of Figure 1.3 is philosophically outside the range of statistical inference. There is, first of all, the problem of memory failure associated with rectangle 5 and diamond 11. Second, there is the problem that the set of conscious hypotheses or propositions is a small subset of the complete set of propositions. This leads to rectangle 4 and diamond 10, which deal with the elicitation of hypotheses from the enormous file of innate propositions.

Four of the six kinds of specification searches lie within the framework of simple statistical inference: interpretive searches, hypothesis-testing searches, proxy searches and data-selection searches. Simplification searches (postdata) also are a straightforward problem in statistical inference, albeit not in the simple versions. The sixth search—postdata model construction—is either within the framework of statistical inference or not, depending on whether the models that are instigated by the data were conscious or preconscious before the analysis began. If the models were preconscious, inference may usefully proceed as if they were, in fact, conscious, and an inference problem that is necessarily outside the framework of statistical inference can be treated as if it were within.

CHAPTER 2

AN INTRODUCTION TO BAYESIAN INFERENCE

An inference is a logical conclusion drawn from a set of facts. Statistical inference is concerned with drawing conclusions about unobservables $\boldsymbol{\theta}$ from a set of facts, including observed data $\mathbf{x}$ and a conditional probability distribution $f(\mathbf{x}|\boldsymbol{\theta})$, that indicates the probability of various values of $\mathbf{x}$ given various values of $\boldsymbol{\theta}$. Bayesian inference is distinguished from classical inference by its inclusion of a "prior" probability function $f(\boldsymbol{\theta})$ in the set of facts. To a Bayesian there is no sound logical reason why the distribution $f(\mathbf{x}|\boldsymbol{\theta})$ should be regarded to be more of a "fact" than the distribution $f(\boldsymbol{\theta})$. A classicist, however, argues that the distribution $f(\mathbf{x}|\boldsymbol{\theta})$ is an objectively verifiable feature of the world, whereas any distribution $f(\boldsymbol{\theta})$ is purely a figment of someone's imagination. The foundation of the dispute between Bayesians and classicists can thus be found in their definitions of probability, discussed in Section 2.1.

A theme that is developed in this chapter and elsewhere is that data analysis involves three distinct phases. The data $\mathbf{x}$ is first *summarized*, then it is *interpreted*, and lastly *decisions* are made. Summarization and interpretation jointly constitute the process of learning or inference.[1]

[1]Of course, much of the learning activity is aimed explicitly or implicitly at some decision problem, and the sharp distinction between inference and decision is misleading.

$$\text{Data Analysis}\begin{cases}\left.\begin{array}{l}\text{1. Summarization}\\ \text{2. Interpretation}\end{array}\right\}\text{Learning}\\ \text{3. Decision}\end{cases}$$

The Bayesian method of data analysis involves each of these three phases. Learning from observations is governed by Bayes' rule

$$f(\boldsymbol{\theta}|\mathbf{x}) = \frac{f(\mathbf{x}|\boldsymbol{\theta})f(\boldsymbol{\theta})}{f(\mathbf{x})}$$

which describes how the uncertainty in $\boldsymbol{\theta}$ summarized by the probability function $f(\boldsymbol{\theta})$ is influenced by the data $\mathbf{x}$. Summarization can occur to the extent that $f(\mathbf{x}|\boldsymbol{\theta})$ depends only on some summary of the event $\mathbf{x}$. Interpretation of the evidence $\mathbf{x}$ amounts to changing the uncertainty about $\boldsymbol{\theta}$ from $f(\boldsymbol{\theta})$ to $f(\boldsymbol{\theta}|\mathbf{x})$. Lastly, decisions can be made, given the distribution $f(\boldsymbol{\theta}|\mathbf{x})$.

Classical inference lacks a formal interpretation phase. Strictly speaking, it is only a method of data summarization. Of course, practitioners are interested in learning from data and have built elaborate ad hoc methods of data interpretation. It is hardly surprising that these methods are sometimes in agreement and sometimes greatly at odds with Bayes' rule. Following a brief discussion of Bayes' rule in Section 2.2, Bayesian inferences about a proportion and about a mean are described in Sections 2.3 and 2.4, and the theme of the three phases of data analysis will be elaborated on.

One feature of Bayes' (1763) original essay that brought the greatest scorn was his choice of a prior distribution $f(\boldsymbol{\theta})$ to represent "knowing nothing" (a contradiction in terms?). Opponents of Bayesian inference focus their attacks on the problem of choosing the prior distribution, and Bayesians have responded defensively by trying to find objective subjective priors. My negative attitudes toward the likely fruitfulness of such endeavors are reported in Section 2.5.

2.1 Objective or Subjective Probability?

Classical inference, although based on a seemingly never-ending list of principles, remarkably admits only a single confusing viewpoint, and the principal statistical texts differ mostly in pedagogy and very little in substance. Paradoxically, Bayesian inference, which is based on the single principle of Bayes' rule, admits a basketful of distinctly different viewpoints. The rule straightforwardly describes a way of combining presample (prior) information packaged in a probability distribution with sample information packaged in a likelihood function. Bayesians who accept the rule as their principal commandment find time in their busy missionary schedules to argue vigorously over the Correct Interpretation.

The apparently innocuous rule is simply the conditional probability rule

$$P(A|B)=\frac{P(A\cap B)}{P(B)}=\frac{P(B|A)P(A)}{P(B)}$$

describing the probability of an uncertain event A given the uncertain event B in terms of the probability of B given A, the probability of A and the probability B. What distinguishes Bayesians from non-Bayesians is not their acceptance of the conditional probability rule but rather their willingness to apply it to events A that clearly admit no frequency interpretation. For example, A may be the hypothesis that the gravitational force between two objects decreases with the square of the distance between them. To a Bayesian $P(A)$ summarizes the weight of evidence in favor of A before B is observed, and $P(A|B)$ summarizes the weight of evidence after B is observed. Non-Bayesians argue instead that A is either true or false and that $P(A)$ is appropriately either one or zero, depending on whether A is true or not. Bayes' rule under those circumstances amounts to either $1=1$ or $0=0$.

The distinction between Bayesians and non-Bayesians should thus be understood in terms of the definitions of probability, and it is, therefore, necessary here to discuss the various definitions. The number of conflicting opinions is enormous; for a fuller treatment the reader should consult Barnett (1973, Chap. 3). The viewpoint offered in this book is that probabilists are naturally divided into objectivists, who believe that a probability is usefully regarded as an objective description of physical reality, and subjectivists, who believe that a probability ought to be defined explicitly as a subjective description of man's perception of his surroundings.

THE PROBABILITY AXIOMS

From the standpoint of mathematical theory, probability is a set function that obeys certain axioms. To use the theorems of mathematical probability, it is enough to satisfy yourself that these axioms apply. However, as is discussed subsequently the interpretation of the results of such exercises depends on your understanding of the primitive concept of probability.

Mathematically, probability is described as follows. Let U be a universal or reference set. A function P that associates to every subset $A\subset U$ a real number, $P(A)$, is said to be a *probability measure* on U provided it satisfies the following:

AXIOM 1 For every $A\subset U$, $P(A)\geqslant 0$
AXIOM 2 $P(U)=1$
AXIOM 3 If A and B are disjoint, then $P(A\cup B)=P(A)+P(B)$.

These axioms apply in many circumstances in which no one would use the word probability. For example, your arm may contain 10 percent of the weight of your body, but it is unlikely that you would report that the probability of your arm is .1. Objectivists and subjectivists have quite different ideas about the use of the word "probability," and their debate is now to be discussed.

OBJECTIVE PROBABILITY

Objectivists define probability with reference to repetitive phenomena such as dice, roulette, and cards. Although gamblers doubtlessly had some idea how to compute odds in games of chance long before the sixteenth century, it was the Italian mathematician Gerolamo Cardano (1501–1565) who is given credit for the first correct probability calculations. To Cardano, the probability of an event A such as pulling a red card from a deck is simply the ratio of the number of (equally likely) outcomes that lead to the event A divided by the total number of (equally likely) outcomes. This may have been intended only as a formula for calculating probabilities, but deMoivre in 1718 and later Laplace adopted it as a definition, and it is now called the classical definition of probability. As such, it has obvious deficiencies.

To give an example, two flips of a coin can lead to one of three events: a pair of heads, a pair of tails, or a head and a tail. The classical definition might lead us to say that the probability of two heads is one-third. Not so, you "probably" would object; these three events are not equally likely. There are, in fact, four equally likely events: two heads, two tails, a head followed by a tail, and a tail followed by a head. But how are we to know which events are equally likely? And if by equally likely we mean equally probable, have we not circularly presupposed a definition of probability when we defined probability?

Although probability was defined by early writers as a ratio of favorable cases to the total number of cases, the frequency interpretation of probability lurked informally in the background to check the appropriateness of what was meant by "equally likely." This naturally led to a definition of probability in terms of the frequency itself: Let n be the number of trials or experiments (tosses of a coin, rolls of a die) and let m be the number of occurrences of the event A (coin lands heads up, die stops ace up); then we will define the probability of A as

$$P(A)=\frac{\text{number of occurrences of } A}{\text{number of experiments}}=\frac{m}{n}.$$

The ratio m/n, by definition, changes as n changes. If we want to avoid the embarrassment of having our probability assignment to A depend on the number of hypothetical trials, then we must let n grow hypothetically without bound

$$P(A) = \lim_{n \to \infty} \frac{m}{n},$$

that is, we define probability in terms of the *limit* of a *relative frequency*. This, of course, requires that the limit exists, which is, by its definition, impossible to verify. In practice, one always checks his probability assignment by observing the (converging) behavior of m/n in a finite number of trials.

Although it was not until 1837 that Denis Poisson formally defined probability as a limit of a long-run relative frequency, surely gamblers long before the time of Poisson—and before Cardano for that matter—would quote odds based on the relative frequency of occurrence in a limited number of trials. As long as one dealt with repetitive phenomena of a standardized variety—such as in games of chance, actuarial science, genetics, and statistical mechanics—the relative frequency point of view and the classical view equating probability to the ratio of favorable to total cases were adequate. For nonstandardized, nonrepetitive phenomena, however, the frequency definition of probability simply does not usefully apply. A frequentist cannot calculate a nontrivial probability that Andrew Jackson was the eighth President of the United States, or that someone named Andrew Jackson will be the President in the year 2000. To calculate a frequency, we must define the class of relevant experiments. For Andrew Jackson there is (apparently) only one relevant experiment, and the relative frequency is necessarily either one or zero. A frequentist, therefore, would make the following trivial statement. The probability that Andrew Jackson was the eighth President of the United States is either one or zero, one if he was, zero if he was not.

In other cases it may be difficult to define exactly the class of relevant experiments. We often appeal to the vague adjectives "standardized" and "repetitive." All flips of a coin may (intuitively?) be regarded as repetitions of the same experiment. But in a trivial sense, repetition of the same experiment must lead to the same outcome. And a great many things are undeniably different each time we flip a coin. This discussion leads to the conclusion that in order to calculate a relative frequency we must subjectively define the class of events over which to count the frequency. We may all agree on the class of events and in that sense have an "objective" frequency, but that objectivity is something in us and not in nature.

A related problem is that a frequency is a property of a class of events, not a property of individual events. It is quite unclear whether a frequency probability may then be applied to individual events. For example, may we legitimately discuss the probability of getting a head on the next flip of a coin? There apparently are three positions that frequentists have taken on this issue:

f_1: Probabilities are defined for events that have not occurred. They are not defined for events that have already occurred. We may, therefore, talk about the probability of a head on the next flip of a coin until it is flipped. After that it is either a head or a tail, and no probability applies.

f_2: Probabilities are *not* defined for individual events regardless of whether they have occurred. Probabilities are objective properties of *classes* of events, and they do not apply to individual events.

f_3: Probabilities *are* defined for individual events, both before and after the event occurs under certain circumstances to be discussed subsequently.

The position f_1 makes special reference to the time of occurrence of the event, which is something we may not even know. Imagine that a coin is flipped on a star so distant that it takes 1 hour for light to arrive here. *We* will *see* the flip an hour after it occurs. Is it then the case that for the previous hour we will have the mistaken impression that the probability of a head is one-half, when in fact the probability was not defined since the event had already occurred? More generally, when is an event determined? Precisely when does the probability cease to be defined? When the coin stops vibrating? When it stops rolling but is still vibrating? When it is sailing through the air? When it is resting on the flipping thumb?

Another difficulty with position f_1 is that it can generate probabilities for individual events only in a somewhat circular manner. The frequency is said to apply to an individual event when the experiments determining the hypothetical sequence of events are "standardized and repetitive." If by that statement we have elliptically asserted that all sequences are equally likely, the frequency definition of the probability of an individual event degenerates into the classical definition, with the universe of potential outcomes being all sequences that have a certain frequency. But, as we have already pointed out, that presupposes a definition of probability. It is thus not clear how a probability can be defined for an individual event without first defining probability.

Position f_2 amounts to a negation of the concept of probability as indicating the likelihood of an uncertain event. The usefulness of probability theory is thereby greatly reduced, but there are certainly circumstances in which properties of classes of events are sufficient. It is possible to write insurance, for example, given only the knowledge that 10% of enrollees will suffer some loss. Probabilities applying to particular enrollees are unnecessary. For particular inferences, probabilities applying to particular events seem absolutely essential, however.

The third position f_3 applies the frequency to an individual member of the class when there are no *recognizable subsets* within the class. A subset is recognizable if its frequency is known to differ from the frequency of the complete class of events.

The following illustrates the notion of recognizable subsets. Of all flips of a coin 50% may land heads up, yet it is possible that 75% of those flips that rotate in a primarily north-south direction are heads and only 25% of those flips that rotate in a primarily east-west direction are heads. If this were true and if we knew in which direction a particular coin rotates, then we, clearly, would not say that the probability of a head is one-half, even though the frequency in a large number of trials is one-half. The probability *is* one-half if we do not know which way the coin rotates, that is, if there are no recognizable subsets in the class of all coin flips.

The existence of recognizable subsets is clearly personal, and the position f_3 implies a personal definition of probability. Classical inference built around frequency probability takes either f_1 or f_2. That is, probability statements are made about classes of events not individual events (f_2) or about events that have not occurred (f_1). It is useful here to recall the standard confidence interval statements that students learn to repeat but rarely understand. A 95% confidence interval comes from a class of intervals, 95% of which cover the true value. A particular interval either covers or does not cover the true value, and no probability statement can be made concerning whether it does or does not. This seems to be the position f_2. The position of f_1 is also possible: the probability that an interval covers the true value is .95 until a particular interval is generated. Then the interval either covers, or it does not.

If you are scratching your head in confusion, you have perfectly understood the problem with a frequency definition of probability. I just do not see how a frequentist can make meaningful probability statements at all. He can talk about classes of events and their respective physical, possibly objective, frequencies. But if we reserve the adjective "probability" for set functions that both obey the probability axioms and also indicate the likelihood of uncertain events, a frequentist statement that two times out of ten we will pull a red ball from the urn is no more a probability statement than the statement that 20% of the balls are red or that the red balls make up 20% of the total mass of the balls. Only under certain subjective circumstances can we allow the frequency of .2 to be translated into the statement "The probability of drawing a red ball from the urn is .2."

SUBJECTIVE PROBABILITY

An alternative to the objectivist view that probability is a physical concept such as a limiting relative frequency or a ratio of physically described possibilities is the view first enunciated by James Bernoulli in *Ars Conjectandi* (1713) that probability is a "degree of confidence"—later writers use degree of "belief"—that an individual attaches to an uncertain event and that this degree depends on his knowledge and can vary from individual to individual. Three important questions arise: (1) Why should they be or why in fact are degrees of belief also probabilities, in the sense of obeying the probability axioms? (2) What is the relationship between degrees of belief and relative frequencies? (3) Are degrees of belief measureable?

Concerning the first question, there are two competing answers. Some statisticians and philosophers assert that a degree of belief is an inherent property of a body of knowledge in the same sense that height is an inherent property of a physical body. As an axiom, they assert that given two uncertain propositions A and B and some body of evidence that relates to the "likelihood" of A and B, one of the following relationships holds: A is more likely, B is more likely, or A and B are equally likely. The word "likely" is not defined by such a statement; only the existence of an ordering is asserted. This is to be compared to the statement that given two individuals it is possible to order them by their height: A is taller than B, B is taller than A, or A and B are equally tall. Height is not thereby defined, but rather is taken as a primitive concept.

Given an individual's ordering of the "likelihood" of uncertain events, we may ask if this ordering is consistent with a probability ordering. That is, do there exist probabilities such that $P(A)<P(B)$ if and only if event A is judged to be less likely than B? Clearly, some orderings of the events rule out probabilities. For example, suppose A is judged more likely than B, B more likely than C, and C more likely than A. It is simply not possible to find numbers $P(A)$, $P(B)$, and $P(C)$ such that $P(A)>P(B)$, $P(B)>P(C)$, and $P(C)>P(A)$. Such an intransitive ordering of events must be ruled out either directly or indirectly.

In fact, several assumptions are necessary for the ordering of uncertain events to be consistent with a probability interpretation. The interested reader is referred to DeGroot (1970, Chap. 6) for a discussion. It is enough here to understand that degrees of belief are asserted to exist and to obey the probability axioms in the sense that certain, apparently compelling, simpler assumptions can be shown to imply the probability axioms. One of these assumptions is briefly discussed: for the probabilities to be unique there must be a reference standard capable of generating all possible probabilities.

Imagine a pointer perfectly balanced on a pin in the middle of a perfect circle as in Figure 2.1. The pointer is to be spun and allowed to come to rest pointing somewhere on the circle. Propositions are arcs such as A and B on the circumference of the circle. The word perfect refers to those requirements that make the degree of belief you hold in any arc dependent

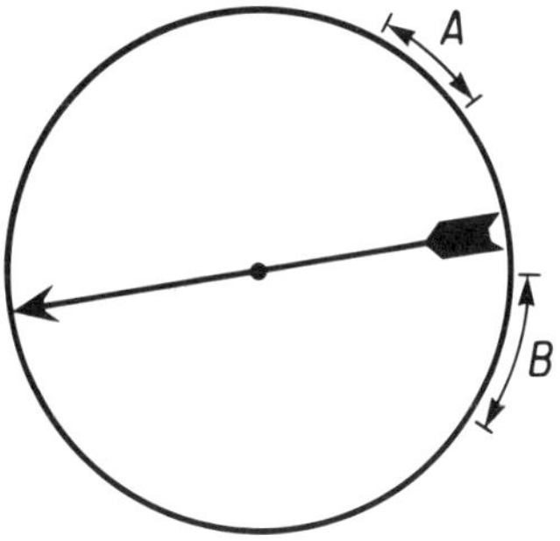

Fig. 2.1 A canonical experiment.

on the length of the arc alone. More than that, a proposition (an arc) A is more probable, equally probable, or less probable than a proposition B if the length of A is greater than, equal to, or shorter than the length of B. If c is the circumference of the circle and $d(A)$ is the length of arc A, we may arbitrarily assign to the degree of belief in A the numerical measure $P(A)$ called the probability of A

$$P(A) = \frac{d(A)}{c}.$$

We now have a reference standard, sometimes called a canonical experiment which can be used to assign particular numbers to the degrees of belief we hold in any propositions. That is, an individual i is said to have degree of belief $P_i(X) = P(A)$ in a proposition X if he regards the proposition X to be equally likely as the arc A in our canonical experiment, where $P(A)$ is the length of arc A divided by the circumference of the circle.

Subjective probability can also be defined in the context of a real or hypothetical problem of decision making under uncertainty. Frank Ramsey (1926) was the first to give a theory of action based on the dual, intertwining notions of judgmental probability and utility. To Ramsey, probability is defined operationally in terms of a person's willingness to act in some decision-making situations in which eventual rewards are uncertain. As an example of such a definition we may say that two uncertain events A and B have the same probability if you are indifferent between winning a dollar if A occurs and winning a dollar if B occurs. For some events there are obvious problems with this definition. Suppose A is the event "Brown will be President of the United States in 1990" and B is the event "the consumer price index will more than quadruple between 1974 and 1990." Although you may regard A and B to be equally likely in some primitive sense, you may prefer to bet on event A since the dollar is likely to be worth more if A is true than if B is true. To give an analogy in

MANKATO STATE UNIVERSITY
MEMORIAL LIBRARY
MANKATO, MINNESOTA

another measurement problem, would you say that individual A is "taller" than B if you would prefer A on your basketball team?

Keeping in mind this difficulty with any decision-based definition of probability, we would like to review here deFinetti's (1937) analysis of betting odds and show in particular that betting odds obey the probability axioms. Suppose you are asked to quote betting odds on a set of uncertain events $A, B, \ldots$, and accept any wagers others may desire to make about these events. That is, you must assign to each event A a "probability" $P(A)$, thereby indicating a willingness to sell lottery tickets that pay $\$S_a$ if A occurs for the price $\$P(A)S_a$ where S_a are the stakes (positive or negative) to be selected by your opponent. What properties seem desirable for these "probabilities"? Well, you certainly do not want to assign probabilities such that your opponent can select the stakes to guarantee that you will lose regardless of the eventual outcome. This simple *coherence principle* is sufficient to imply the three fundamental axioms of probability:

(a) $1 \geqslant P(A) \geqslant 0$. If your opponent bets only on A and if A occurs, his winnings are $W_1 = S(1 - P(A))$, where S may be negative. If A does not occur, he wins $W_2 = -SP(A)$. Coherence requires that $W_1 W_2 \leqslant 0$ for all S. (If $W_1 W_2$ is positive for some S, then W_1 and W_2 have the same sign. If they are both positive, you are a sure loser. If they are both negative, your opponent may change the sign of S to make them both positive.) The condition $W_1 W_2 \leqslant 0$ implies $[1 - P(A)]P(A) \geqslant 0$, which implies $0 \leqslant P(A) \leqslant 1$.

(b) $P(U) = 1$. The universal set U is certain to occur. Thus your losses on bets about U are, necessarily, $W_u = S_u[1 - P(U)]$. By coherence, there must be no S_u such that $W_u < 0$. This implies $P(U) = 1$.

(c) If $A \cap B = \phi$, then $P(A \cup B) = P(A) + P(B)$. Suppose you make bets on the events A, B, and $C = A \cup B$. The following events and winnings are possible:

Event	Winnings
$A \cap \sim B$	$W_1 = S_a[1 - P(A)] - S_b P(B) + S_c[1 - P(C)]$
$B \cap \sim A$	$W_2 = -S_a P(A) + S_b[1 - P(B)] + S_c[1 - P(C)]$
$\sim A \cap \sim B$	$W_3 = -S_a P(A) - S_b P(B) - S_c P(C)$

Coherence requires that there be no values of the stakes (S_a, S_b, S_c) such that the winnings (W_1, W_2, W_3) are all positive. If this linear system of equations expressing the winnings as a function of the stakes is invertible, it is possible to specify stakes to make the winnings take on any values whatsoever. To avoid this, the determinant must be zero. Setting the determinant to zero yields

$$0=\begin{vmatrix} 1-P(A) & -P(B) & 1-P(A\cup B) \\ -P(A) & 1-P(B) & 1-P(A\cup B) \\ -P(A) & -P(B) & -P(A\cup B) \end{vmatrix}$$

$$=-P(A\cup B)+P(A)+P(B).$$

Thus $P(A\cup B)=P(A)+P(B)$.

This treatment of subjective probability, although terribly appealing, has two flaws which have received a certain amount of attention. The first concerns the units in which the stakes S are measured; the second concerns the possibility that the other party to the wager has better information about the event in question. Both are related to the fact that you are not allowed to drop out of the game (when the stakes are too high or when the cards are marked).

Suppose that you are asked to quote odds on the flipping of a coin, with the stakes being a penny. Even odds is the natural choice, and the bet seems fair, if a bit dull. What about stakes of a thousand dollars? Or a million? The nature of the game seems to change as the stakes go up, and what is an acceptable wager for low stakes becomes unacceptable (for most of us) at high stakes. Does this mean that the probabilities change as the stakes go up? No, there is a better explanation. You are interested in happiness, not dollars. If you lose a penny you lose almost the same amount of happiness as you would gain if you were to win a penny. For larger stakes, this is not true. The stakes in terms of happiness are asymmetrical, with even odds of gaining a little and losing a lot. This is exactly the problem discussed previously when probabilities were defined in terms of indifference between lotteries. Expected dollar winnings is not the only thing that matters in choosing lotteries.

We express this graphically in Figure 2.2, which depicts a utility (or happiness) function in terms of dollars. The utility function has the characteristic that continued increments to your wealth provide ever decreasing increments to your happiness. If you do not play the game, you attain happiness level U_0. If you do play, you are equally likely to be at U_w and U_l with U_w only marginally above U_0 but with U_l considerably below U_0. When the stakes are small the discrepancy between (U_w-U_0) and (U_0-U_l) becomes imperceptible, and the choice between playing and not playing becomes ambiguous.

The point of all this is simply to demonstrate that probabilities can be independent of stakes and also that the stakes can influence the acceptability of a gamble.

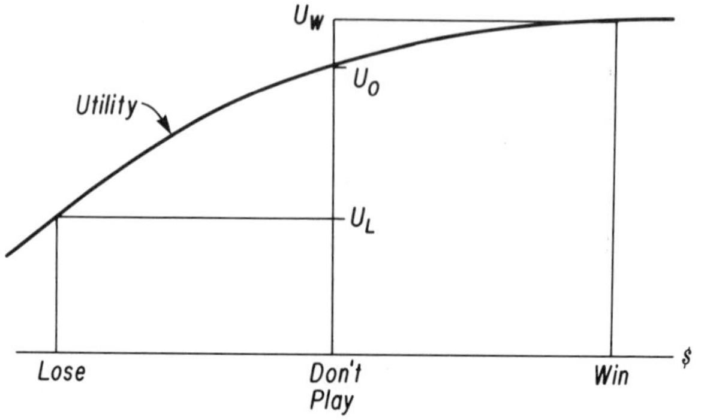

Fig. 2.2 A utility function that discourages gambling.

There is a second problem with this decision-based definition of probability. Although the true probability is naturally thought to be independent of the stakes, the announced probability may not be. The previous discussion implicitly assumes that the second individual has the same opinions as the first and that failure to quote the true probability would result in an advantage to the second player. When the second player has a different probability, the first may want to quote a probability different from the one he truly believes, either to take advantage of the second player or to avoid being taken advantage of himself. The problem is further complicated if the first individual is allowed to drop out of the game. We conclude that, at least for our purposes, it is better not to define probability in the context of some specific decision problem.

This concludes the answer to the first question: why should they be or why in fact are degrees of belief also probabilities? Among the answers are that (1) measures of uncertain knowledge are probabilities for intuitively compelling reasons; or (2) anyone who makes decisions under uncertainty in a "rational" way will act as if he had degrees of belief that obeyed the probability axioms.

Our second question concerns the relationship between degrees of belief and relative frequencies. More generally, what further constraints other than the probability axioms must degrees of belief obey? Consider, for example, the tossing of a coin. It has been generally accepted that a head and a tail are equally likely and are to be assigned an "objective" probability of one-half. The subjective description of probability casts doubt on that statement. After all the central function of a scientific inquiry is to eliminate the uncertainty and to be able to quote longer odds. It is certainly conceivable that a scientist could carefully analyze the coin, landing surface, the size and strength of the flipping thumbs, and the like, and could with confidence conclude that the head is more likely than the tail on this particular flip. A subjectivist would, therefore, assert that there is no compelling reason why the probability of a head must be one-half.

In fact, subjectivists are divided into two warring factions: the *personalists* and the *necessarists*. Personalists such as deFinetti (1937), Ramsey (1926), and Savage (1954) argue that since knowledge obviously varies from individual to individual, the quantitative measure of knowledge must vary from individual to individual, also. Any individual is advised to constrain his degrees of belief to obey the probability axioms but is otherwise free to assign them as he sees fit (or, more accurately, as is appropriately determined by his knowledge). *Necessarists* such as Jeffreys (1961) and Keynes (1921), on the other hand, argue that a probability is the degree of belief it is *rational* to hold regarding some uncertain proposition, given some other propositions. This is a subtle distinction, if it is a distinction at all. Just what do we mean by "rational"?

Two personalists who had the same joint probability function over all uncertain events would, conditional on the same propositions, have the same (posterior) degrees of belief. Do we mean, then, by rationality that everyone's primitive (preobservation) degrees of belief are identical? This is a conceivable proposition, but it does not seem to have any practical consequences, given that everyone's knowledge (i.e., set of known propositions) is distinctly different. Another possible definition of rationality is errorless computation of degrees of belief. For example, it is "irrational" to believe something to be true merely because you want it to be. But on this point, personalists agree with necessarists; probabilities are descriptions of knowledge measured hypothetically without error, indicating the weight of evidence in favor of uncertain propositions and independent of the desires (or the decision problems) of the individual. A third definition of rational belief is the one that most clearly distinguishes personalists from necessarists. There is a set of propositions that are socially known to be true in the sense that some "reliable" observer includes them in the set he regards to be true. This whole set of propositions is to be used to compute a social or public set of degrees of belief. Probability calculus applies only to these "rational" degrees of belief, because any other degrees of belief are "mistaken." For example, consider the proposition "Andrew Jackson was the eighth President of the United States." A personalist might select an individual whose knowledge about this proposition is incomplete and attempt to measure his degree of belief in the proposition. A necessarist might retort that Jackson either was or was not the eighth President, and the "rational" degree of belief is either one or zero. Any other probability except the correct one is mistaken.

My own feeling is that information transmission among individuals is so imperfect and so poorly understood that it makes little sense to consider public, "rational" degrees of belief. At best, the probability calculus could apply to personal beliefs, although a substantial part of this book implicitly makes the point that for a variety of reasons real learning is poorly described by this mechanistic mathematical model. At any rate, there are no well-defined "necessary" probabilities.

Although I take the personalist view that probabilities describe personal knowledge, I recognize that there are situations in which many or most people have essentially the same degree of belief in some uncertain proposition. This uniformity of subjective opinion should not be confused with objectivity. The personalist James Dickey has suggested that the words "public" and "private" might informatively replace the words "objective" and "subjective." Public probabilities most often are based on a given relative frequency, in which case a personalist will constrain his probability to be consistent with the publicly known frequency. For example, if it is known that the relative frequency of heads in the next two flips of a coin is exactly one-half, then the personal probabilities of the sequences *HH* and *TT* must be zero. Or suppose there is a class of n propositions. The ith proposition, for example, might be "the ith flip of a coin will land heads up." If it is known that exactly f of these propositions are true and if each proposition is equally likely, then the probability of each proposition is, necessarily, f/n. That is, the probability is equal to the relative frequency.

This constraint may be expressed more generally and more formally in the following result, stated in deFinetti (1937). Let the universe consist of m events, $U=\{e_1,e_2,\ldots,e_m\}$, and let the probability function defined on this universe be $P(\{e_k\})$. Suppose that there are n compound events A_1, $A_2,\ldots,A_n$ defined as unions of one or more simple events. Let $P_i=P(A_i)$ and $\pi_j=P$ (exactly j of these events occur). Then it can be shown that

$$\sum_{i=1}^{n}\frac{P_i}{n}=\sum_{j=0}^{n}\frac{j\pi_j}{n}, \tag{2.1}$$

the average probability is equal to the expected relative frequency.

Before this result is proved, note that if the frequency of occurrence j of the n compound events is known to be f with probability one, the right-hand side of this expression becomes just f and the formula can be written

$$\sum_{i=1}^{n}\frac{P_i}{n}=\frac{f}{n}.$$

In words, the average probability is just equal to the relative frequency. Furthermore, if each of the events A_i is equally probable, the formula becomes

$$P_i = \frac{f}{n}.$$

In words, the probability is equal to the relative frequency.

This is an important result that indicates when a personal probability is necessarily equal to a relative frequency. It is a formalization of the notion of recognizable subsets discussed previously. When a class of events is known to have relative frequency f/n and when it is impossible to divide that class into subsets that have greater or lesser probability of occurrence than the class as a whole, we are obligated to adopt the relative frequency as our personal probability.

The proof of the proposition (2.1) is straightforward. Let a_{ik} indicate whether the simple event e_k is included in A_k

$$a_{ik} = \begin{cases} 1 \text{ if } e_k \in A_i \\ 0 \text{ otherwise.} \end{cases}$$

Then

$$P_i = \sum_{k=1}^{m} a_{ik} P(e_k).$$

The number of events A_i that occur if the simple event e_k occurs is

$$j(e_k) = \sum_{i=1}^{n} a_{ik}$$

and the expected frequency can be written

$$\begin{aligned} \sum_{j=0}^{n} j\pi_j &= \sum_{k=1}^{m} j(e_k) P(e_k) \\ &= \sum_{k=1}^{m} \sum_{i=1}^{n} a_{ik} P(e_k) \\ &= \sum_{i=1}^{n} \left(\sum_{k=1}^{m} a_{ik} P(e_k) \right) \\ &= \sum_{i=1}^{n} P_i. \end{aligned}$$

This concludes our answer to the second question: what constraints must degrees of belief obey? The answer is, they must be probabilities but are otherwise free. If the relative frequency is known and if each event is judged to be equally likely, a personal probability will coincidentally equal a frequency.

The third question to be discussed is "are degrees of belief measurable?" Although we hypothesize a complete ordering of uncertain events, we have not indicated how well, if at all, this ordering can be determined. It is useful here to think, again, in terms of heights. We believe that individual A is either taller or shorter than individual B, but we may be unable to identify which one is, in fact, the taller. This is especially true if they are at opposite ends of a football field, but even if we stand them right next to each other, we may have difficulty comparing their heights. Thus the existence of an ordering need not imply the existence of measuring devices sufficiently accurate to disclose the ordering.

The absence of perfect measuring devices does not prevent the use of such concepts as "length," "weight," or "temperature." A designer of a house, for example, at times proceeds as if perfect measurement were possible—he calls this angle a right angle and this length 5 inches, even though no physical angle could be exactly 90° nor any physical length exactly 5 inches. It is essential for him also to consider the effect of likely departures from his design. If the house would fall down unless the angle were 90° to the fifth decimal place, he would alter his design to be more "robust" to departures. We can thus idealize the design of a house into those phases that proceed as if measurement were perfect and those phases that consider the consequences of imperfect measurement.

We take a similar attitude toward probabilities. Although we hypothesize the existence of degrees of belief, it is clearly impossible to measure them without error. It nonetheless makes sense in constructing a theory of inference or a theory of decision making to proceed sometimes as if degrees of belief were measured perfectly. But it is also essential to consider the consequences of measurement error. If our inferential house were to fall down with the slightest discrepancy between the measured and the true degrees of belief, we would surely want to build a different kind of house.

When measurement is very imprecise, it may be argued that we should disdain design altogether and proceed directly to the building of the house. We can expect to learn effective construction in the process, since we will be encouraged to build houses somewhat differently when they fall down. House-building would be taught under such circumstances not by textbooks and lectures but by apprenticeship. In terms of data analysis, it may be argued that the impossibility of measuring degrees of belief makes learning an art, and we should concern ourselves not with designing mathematical models of learning that presuppose perfect measurement of probabilities, but rather with trying different "styles" of learning and selecting those that turn out best according to some criteria. In principal, I agree with this position, but I think it is useful nonetheless to describe actual learning in formal terms. Mathematical models can be used as teaching devices and perhaps also as guides for improving the learning processes.

We, therefore, consider the inferential implication of certain probabilistic structures suggested by the techniques of data analysis used by economists. We consider the consequences of minor and major changes in the probabilistic structures. And we analyze some forms of measurement error, for example, memory failures.

OBJECTIVISM VERSUS SUBJECTIVISM

We are now in a position to draw the battle line clearly between the objectivist and subjectivist schools. The following list defines functions that obey the probability axioms and which might be called probabilities in the sense of indicating the likelihood of uncertain events:

1. Proportions.
2. Relative frequencies.
3. Degrees of belief.
4. Betting odds.

Representative objectivists and subjectivists were asked to comment on the following pair of statements:

A. The probability of getting a six in the roll of a die is one-sixth.
B. The probability that Andrew Jackson was the eighth president of the United States is one-sixth.

Objectivist. "Statement *A* is a perfectly good probability statement. Of the six ways that a fair die may land, there is one way favorable to the event 'six'. There is, moreover, no experience with rolls of a die that suggest that "six" occurs more or less often than that. Statement *B* may reflect someone's betting odds or even be someone's 'degree of belief,' but it is certainly not a probability statement in the same sense as statement *A*. Jackson either was or was not the eighth President of the United States, a fact we can look up in a book."

Subjectivist. "Both statements may represent someone's degrees of belief or someone's betting odds, and may, therefore, be proper probability statements. Statement *A* may, on the other hand, be a frequency statement or even just a statement about the proportions of events favorable to the event 'six.' If so, it is no more a probability statement than is the statement 'my legs make up 20% of the weight of my body.' By this I mean that I think the word probability should be restricted to describe the 'likelihood' of uncertain events. A frequency statement indicates only the proportion of a class of propositions that is true. It does not unambiguously indicate the likelihood—the probability—of particular propositions such as 'a six on the next roll of a die.' Of course, there are situations when a frequency statement should be directly translated into a probability statement in my sense. I would think rolling dice would be one of those situations and can well imagine that statement *A* describes a degree of belief (a probability) as well as a frequency of occurrence."

It is important here to notice that neither the objectivist nor the subjectivist questioned whether the four functions indicated above do, might, or should obey the probability axioms. The subjectivist, in fact, acknowledges that both proportions and frequencies satisfy these requirements. At issue is only which of these functions should be called probability functions in the special sense of indicating the "likelihood" of uncertain events. The objectivist, or more specifically, the frequentist, takes the position that there is a special class of phenomena that lend themselves to probability (frequency) descriptions. He is either uninterested or sees no meaningful mathematical content in statements such as "Andrew Jackson was probably the eighth President." The diametrically opposite position is taken by the subjectivist. He regards the frequency as uninteresting except under those special circumstances when it is also a degree of belief.

In an effort to achieve greater clarity and perhaps a resolution of this conflict we have asked our witnesses the following questions, which as yet have not been answered.

To the *subjectivist*: You have admitted the difficulty of assessing or measuring these abstractions you call degrees of belief. Would you not admit that a person with some confidence can say he holds degree of belief one-sixth in the proposition "a six will occur on the next roll of a die." What person could be so precise about the degree of belief he holds in the proposition "Andrew Jackson was the eighth President" (assuming he was uncertain about it)? In particular, can you not imagine that a person who thinks this proposition is unlikely might *only* be able to say that he holds degree of belief in it less than one-half? Is the frequency/nonfrequency distinction that the objectivist draws thus not also useful to you in thinking about the precision with which degrees of belief may be determined?

To the *objectivist*: The subjectivist has admitted the difficulty of assigning a particular number to many degrees of belief. Does your lack of interest in degrees of belief derive from the apparent difficulty of translating statements such as "Jackson was unlikely to have been the eighth President" into a precise number? Do you not also recognize that any frequency statement makes a number of implicit assumptions concerning the replication of an event? Is it not true that these assumptions are at best approximate and that a frequency statement is thus necessarily somewhat imprecise? And if so, is not your distinction between frequencies and degrees of belief based on a distinction in degree and not in kind?

2.2 Bayes' Rule

The probabilistic rule that plays the central role in Bayesian inference is the conditional probability axiom or "Bayes' rule"

$$P(A|B)=\frac{P(A\cap B)}{P(B)}$$

$$=\frac{P(B|A)P(A)}{P(B)}.$$

From the personalist point of view, it indicates that the degree of belief in proposition A given proposition B is equal to the joint probability of A and B divided by the probability of B. It is sometimes called the rule of inverse probability, since it describes how a conditional probability, B given A, can be turned into or inverted into a conditional probability, A given B.

The rule can be written in terms of odds ratios as

$$\frac{P(A|B)}{P(\sim A|B)}=\frac{P(B|A)}{P(B|\sim A)}\frac{P(A)}{P(\sim A)},$$

indicating the posterior odds ratio (given B) equal to the relative likelihood of B under the two hypotheses times the prior odds ratio. The evidential content of B can be completely summarized in terms of the relative likelihood $P(B|A)/P(B|\sim A)$.

The principle of coherence can be used to derive the conditional probability rule (deFinetti, 1937). The first individual establishes his betting odds by announcing a willingness to sell for $\$P(A)S$ lottery tickets which pay $\$S$ in the event that A occurs, where S is the stakes selected by the second individual. The conditional probability $P(A|B)$ should be taken to indicate a willingness to wager on A if B occurs; otherwise, the bet is called off. For ease of notation, we indicate $P(A|B)$, $P(A\cap B)$, and $P(B)$ by p, p', and p'', with the corresponding stakes S, S', and S'' to be chosen by the second individual.

If A is a subset of B, three distinctly different events can occur, and the winnings of the second individual are

$$\begin{array}{rl} A\cap B & W_1=S'(1-p')+S''(1-p'')+S(1-p) \\ \sim A\cap B & W_2=-S'p' \quad +S''(1-p'')-Sp \\ \sim B & W_3=-S'p' \quad -S''p''. \end{array}$$

If this linear system is invertible, the second individual will be able to find values of S, S', and S'' which will make W_1, W_2, and W_3 take on any values whatsoever; in particular, he will be able to inflict arbitrarily large losses on the first individual regardless of the eventual outcome. Thus, coherence requires that p, p', and p'' be selected so that the system is not invertible, that is, so that the determinant $p'-pp''$ is zero. But this is just the conditional probability axiom.

A complete theory of learning is implied by Bayes' rule. A "primitive" joint distribution indicating the personal probability of every event is updated by conditioning on observed events. Learning amounts to merely selecting the appropriate conditional probability. It should be obvious that it is impossible to construct the required joint distribution consciously. Although Bayes' rule may be used unconsciously, it seems unlikely that learning proceeds unconsciously strictly according to Bayes' rule. Three features of this book make overt reference to this author's lack of complete belief in the rule. In Chapter 10 we analyze some simple models of memory failure. In Chapter 9 we discuss "concept formation." In many chapters we report sensitivity analyses which are intended to identify the probabilistic assumptions that crucially determine the nature of the inferences that may be made from a given body of data. In so doing we implicitly admit that no probabilistic assumptions can be made with complete confidence.

2.3 Inference About a Proportion

Consider a population that, of its elements, has a proportion p that possess a given attribute. For example, the population might consist of United States residents of voting age, and p might be the proportion who are currently registered voters. Suppose that the proportion p is not known with certainty, although there is some more or less vague information concerning p. It seems unlikely, for example, that as many as 90% or as few as 10% of the eligible voters are, in fact, registered. Suppose, finally, that n members of the population are asked sequentially if they have the given attribute. What information does the sequence of answers to this question contain concerning the unknown proportion p? What if, for example, no one whom we asked was a registered voter?

The meaning of this sequence of answers depends first of all on how we found the members of the population that we questioned. If we had decided only to ask convicted murderers, it would not be surprising to find no one registered, and the fact that we received such a set of answers would have little impact on our opinions about the proportion p who are registered in the whole population.

Some definitions are now in order. The answers to our query given by the selected population members are called a *sample*. The process by which members of the population are selected for questioning (or sampling) is called the *sampling process*. The set of possible sequences of answers is called the *sample space*. The probability function defined on the sample space indicating the probability of each possible sample is called a *sampling distribution*. Thus the foregoing paragraph indicates that the information content of a particular sample—the extent to which it influences our opinions about p—depends on the sampling process.

The meaning of a sample also depends on prior information about p. Interpretation of a particular sample is quite different if before sampling we thought p was almost exactly .9 than if we thought p was almost exactly .1. We proceed as if our prior opinions about p could be put into a precise distribution, with density, say, $f_1(p)$, indicating that the degree of belief that we hold in the uncertain proposition $a<p<b$ is $\int_a^b f_1(p)\,dp$. We do, however, want to analyze the extent to which the interpretation of a sample depends on minor changes in the prior density function $f_1(p)$, since, of course, a prior distribution cannot be specified unambiguously.

A word about assumptions is in order. Anyone who insists that personal probabilities be assessed precisely cannot perform statistical inference for the same reason that no spaceship would ever have reached the moon if measurement of lengths had to be perfectly precise before construction could commence. To say something is 6.5 centimeters long is only to say that for the purposes at hand we may proceed *as if* it were. An assumption is not, therefore, a statement of unquestioned truth. It is a tentative statement on which initial action can usefully be based.

We make assumptions about sampling distributions and prior distributions not because they could possibly represent accurately anyone's degrees of belief but rather because they seem sufficiently representative of a class of interesting opinions that they may be used as a useful starting point for an analysis. A statistical analysis is most emphatically a two-way street, however. It involves both the mathematical process of inference given assumptions, and also the artful process of challenging and discarding inadequate assumptions. More is said of this in Chapter 9.

SAMPLING DISTRIBUTIONS

Suppose, first, that only two members of the population are to be sampled. Indicating a positive answer, that is, possession of the given attribute, by S (mnemonic for success) and the contrary event by F (mnemonic for failure) there are four possible outcomes: SS, SF, FS, FF. Both, one, or neither individual may have the attribute. The *sample space* is the set of all possible samples

$$\text{Sample space:} \qquad \{SS, SF, FS, FF\}.$$

The probability of obtaining one of the samples—one element in the sample space—depends on how members of the population are identified. The probability function defined on the sample space indicating the probability of each sample is called the *sampling distribution*.

Sampling distribution: $P(SS|p), P(SF|p), P(FS|p), P(FF|p)$.

A special sampling distribution applies if "independent random" sampling is the sampling process. This requires the dual assumption that the probability of selecting a member of the population is independent of the other selections and also that each member is equally likely to be selected at each "draw" from the population. If each member is equally likely to be selected at any given draw, we know from Section 2.1 that the probability of a success on the particular trial (draw, experiment) must be equal to the class frequency p, in this case,

$$P(S|p)=p, \quad P(F|p)=1-p.$$

Furthermore, if each draw is independent of the others, the sampling distribution becomes

$$\begin{aligned} P(SF|p)&=P(S|p)P(F|p)=p(1-p)\\ P(SS|p)&=P(S|p)P(S|p)=pp\\ P(PS|p)&=P(F|p)P(S|p)=(1-p)p\\ P(FF|p)&=P(F|p)P(F|p)=(1-p)(1-p). \end{aligned}$$

How do we know if the sampling process yields independent random samples? The answer is, we don't. Remember that these probabilities are personal degrees of belief that, necessarily, are difficult to specify precisely. We *can* identify processes that clearly do not generate independent random samples. The sequence SF may be unambiguously more or less likely to occur than an arc of our canonical experiment covering $p(1-p)$ 100% of the circumference of the circle. We will never be able to say that SF is exactly as likely as this arc, any more than we can say that two rulers are exactly the same length. We will, however, be willing to proceed in many circumstances *as if* we were observing an independent random sample. We must be aware, however, that this is a working hypothesis that ought not to be retained too tenaciously.

Some examples of dependence in the sampling scheme are worth discussing. Suppose the population in question consists of a finite number of members, say, N, with proportion $p=R/N$ possessing the attribute. Suppose we draw from this population one member who has the attribute. If we do not return him to the population, the probability of another success is $(R-1)/(N-1)$. Thus the outcome of the second draw is dependent on the outcome of the first draw. The sampling distribution would then be

$$
\begin{aligned}
P(SF|p) &= \frac{R}{N}\frac{N-R}{N-1} &&= p(1-p)(1-N^{-1})^{-1} \\
P(SS|p) &= \frac{R}{N}\frac{R-1}{N-1} &&= p(p-N^{-1})(1-N^{-1})^{-1} \\
P(FS|p) &= \frac{N-R}{N}\frac{R}{N-1} &&= (1-p)p(1-N^{-1})^{-1} \\
P(FF|p) &= \frac{N-R}{N}\frac{N-R-1}{N-1} &&= (1-p)(1-p-N^{-1})(1-N^{-1})^{-1}.
\end{aligned}
$$

Note, however, that for N large this sampling distribution is inconsequentially different from the independent random sampling distribution.

Another form of dependent sampling occurs if bunching of successes and failures is likely. Suppose we wanted to know the proportion of families in Boston that have incomes less than $5,000, and suppose we identified a "random home" as the first element in our sample. Suppose also that, for the second element of the sample, instead of a random choice, we selected the family who lived next door. The fact that poor people tend to live next door to each other almost guarantees that these neighbors will answer the question identically. The resultant sampling distribution would then be

$$
\begin{aligned}
P(SF|p) &= 0 \\
P(SS|p) &= p \\
P(FS|p) &= 0 \\
P(FF|p) &= 1-p.
\end{aligned}
$$

We henceforth assume independent random sampling. A sample of size n will consist of a sequence of Ss and Fs, n in all. The sample space consists of the 2^n different possible samples, and the sampling distribution is

$$
\begin{aligned}
P(\text{Sample: } SFFSF\cdots, S|p) &= p(1-p)(1-p)p(1-p)\cdots p \\
&= p^r(1-p)^{n-r} \qquad (2.2)
\end{aligned}
$$

where r is the number of successes and n the sample size.

PRIOR AND POSTERIOR DISTRIBUTIONS OF A PROPORTION

The sampling distributions just discussed describe degrees of belief in propositions about samples given that p is known. The proportion p is also uncertain, and degrees of belief in propositions concerning p may be fully represented in a density function $f_1(p)$, thereby indicating that for any $a \leqslant b$, $P(a \leqslant p \leqslant b) = \int_a^b f_1(p)\,dp$.

Given the sampling distribution (2.2), the prior distribution, and Bayes' rule, we may write the posterior distribution as

$$f_2(p)=f(p|\text{sample})=\frac{P(\text{sample}|p)f_1(p)}{P(\text{sample})}$$

$$=\frac{p^r(1-p)^{n-r}f_1(p)}{P(\text{sample})} \tag{2.3}$$

where P indicates a probability function, f a density function, and where we have been mathematically sloppy in not distinguishing the two more carefully.

The part of this function that deals with the outcome of the sample, namely, $p^r(1-p)^{n-r}$, is called the *likelihood function* of p given that the sample resulted in r successes in n trials. The extent to which the posterior distribution depends on the sample is completely determined by the likelihood function. Multiplying the likelihood function by a constant will not alter the posterior distribution, since it has to be normalized to integrate to one. The likelihood function is, therefore, defined only up to a multiplication constant, and we indicate it as

$$L(p|r,n)\propto p^r(1-p)^{n-r}$$

where $\propto$ indicates "proportional to." Observe that *as a function of* p we have

$$L(p|\text{sample})\propto P(\text{sample}|p).$$

The right-hand side is apparently a probability function defined on the sample space for a particular value of p. The likelihood function is a function of p, however; it is computed from the sampling distribution by identifying how the probability of the particular observed sample depends on p. The fact that the observed sample is twice as probable if $p=p_1$ than if $p=p_2$ is taken as evidence that $p=p_1$. Given this sample, we would say that p_1 is twice as likely as p_2.

The fact that both the likelihood function and the sampling distribution can be written with the same expression causes great confusion for the beginning student, and the reader should examine and understand the following example if he is at all confused by this phenomenon.

Example. A sample consisting of one observation has a sampling distribution $P(S|p)=p$, $P(F|p)=1-p$ depicted graphically in Figure 2.3. If a success is observed, the likelihood function is $L(p|S)\propto P(S|p)=p$, also depicted in Figure 2.3. That is, $p=1$ is most likely; $p=1$ is twice as likely as $p=.5$; $p=0$ is impossible. If, however, a failure is observed, the likelihood function is proportional to $(1-p)$, and $p=0$ is most favored.

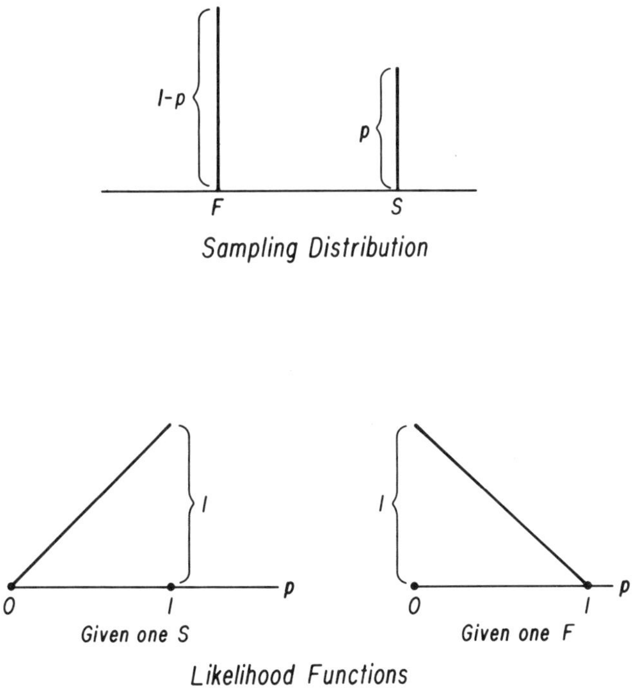

Fig. 2.3 Sampling distributions and likelihood functions.

Now consider again Equation 2.3. Suppose our prior information concerning p is extremely vague in the sense that we have no reason to believe that p is more likely to lie in one interval rather than any other interval of the same length. More specifically, suppose $f_1(p)$ is taken to be rectangular over the interval from zero to one. Then $f_1(p)=1$, and hence by (2.3)

$$f_2(p)=Cp^r(1-p)^{n-r}.$$

By Appendix 2 this p.d.f. is recognized as a beta p.d.f. with parameters $r+1$ and $n+2$, $f_2(p)=f_\beta(p|r+1,n+2)$. More generally, suppose our prior p.d.f. is itself a beta p.d.f. with parameters n_1 and r_1. Then the posterior distribution is also a beta distribution:

THEOREM 2.1. (BINOMIAL SAMPLING AND BETA PRIOR). *If the prior p.d.f. on p is beta with parameters r_1 and n_1, $f_\beta(p|r_1, n_1)$, and a random sample yields n observations, r of which are Ss, then the posterior p.d.f. is*

$$f_2(p)=\frac{1}{C_2}p^{r_2-1}(1-p)^{n_2-r_2-1}=f_\beta(p|r_2,n_2)$$

where

$$n_2 = n_1 + n,$$

and

$$r_2 = r_1 + r.$$

Proof: Substitute

$$f_1(p) = \frac{1}{C_1} p^{r_1 - 1}(1-p)^{n_1 - r_1 - 1}$$

in (2.3) and collect terms.

SOME QUALITATIVE CONCLUSIONS

From (2.3) we see that the shape of the posterior p.d.f. is obtained by multiplying the likelihood function, $L(p|\text{sample})$, by the prior and scaling so that the area under f_2 is unity. The likelihood function is shaped like a beta p.d.f. with parameters $r+1$, $n+2$. Now, by observing graphs of beta p.d.f.'s with different (r,n) combinations, a number of general observations can be made that agree with one's intuition. For example:

(a) As the sample size becomes larger, one's prior distribution and information are "washed out," in the sense that the prior has less and less influence on the posterior.

(b) If one's prior is vague or diffuse, then even with little sample information, the posterior corresponds closely with sample results. (See subsequent discussion.)

(c) If one's prior information is great or his prior distribution "tight," that is, has small variance, say, then reinforcing sample information serves to make the posterior even "tighter."

(d) If one's prior is "tight" and the sample evidence is contradictory, then increasing amounts of such sample information (increasing sample sizes) tends at first to leave the prior relatively unchanged, then may cause the posterior to become more diffuse (representing increasing uncertainty), and finally increasingly large amounts of contradictory information cause a gradual shift to a "tight" posterior position conforming to this sample evidence.

SUMMARIZATION AND INTERPRETATION

We should like for emphasis to point out again that learning can be usefully separated into a summarization phase and an interpretation phase. In this chapter we are discussing inference from a sequence of observable events that can be entirely summarized in terms of two numbers: the number of successes and the number of trials. More formally, let us define two concepts.

Our data consists of an n-dimensional unknown vector of ones and zeroes where a one indicates success and a zero failure, $(X_1, X_2, \ldots, X_n, X_i = 0, 1)$ distributed as $f(x_1, x_2, \ldots, x_n | p)$ where p is an unknown parameter.

Definition. A *statistic* t is a multivalued real function of the data $X_1, \ldots, X_n$.

Example. In this chapter the data are a sequence of successes and failures. The number of failures is a statistic. Another statistic is the whole sample, that is, an n-dimensional vector of ones and zeroes indicating the trials on which the successes and failures occurred.

Definition. A statistic t is a *sufficient statistic* if the density function of the data can be written as

$$f(x_1, x_2, \ldots, x_n | p) = k\big(t(x_1, x_2, \ldots, x_n); p\big)u(\mathbf{x}),$$

where u is a function independent of p and k depends only on $t(\mathbf{x})$, that is, if the likelihood function depends only on t (up to a factor of proportionality).

Example. We have already seen that r and n are sufficient statistics for a binomial p.

A posterior distribution is the product of a likelihood function times a prior distribution. The data can affect the posterior only as they affect the likelihood function, and the likelihood function depends only on a sufficient statistic. Thus a sufficient statistic offers a complete summary of the data evidence. To the extent that we can agree on the process that generates the data, we can agree on the sufficient statistics and thus can agree on the appropriate *summary* of the evidence.

Interpretation of the summarized evidence means changing one's mind, that is, discarding a prior distribution and adopting a posterior distribution. This interpretation obviously depends on the prior distribution in the same way that the summarization depends on the sampling distribution. To illustrate this fact, consider the meaning of 2 successes in 10 trials if p represents one of the following five phenomena:

1. The proportion of times a head occurs in flips of a coin.
2. The proportion of trees in Cambridge that have green leaves on a randomly selected day in 1978.
3. The proportion of Harvard students who have IQs over 150.
4. The proportion of Harvard students who have IQs under 150.
5. The proportion of Martians who weigh more than 150 Marspounds.

Table 2.1

Interpretation of 2 Successes in 10 Trials

Phenomenon	r_1	n_1	95% Prior Interval	P (success on first trial)
Coin lands heads up	50	100	[.419 .581]	1/2
Green leaves	.25	.5	[0, .1] ∪ [.9, 1.0]	1/2
IQ over 150	1	22	[0, .13]	1/22
IQ under 150	21	22	[.867, 1.0]	21/22
Martian weights > 150	0	0	?	?
Phenomenon	r_2	n_2	**95% Posterior Interval**	P **(another success)**
Coin lands heads up	52	110	[.417, .579]	52/110
Green leaves	2.25	10.5	[.009, .43]	2.25/10.5
IQ over 150	3	32	[.12, .20]	3/32
IQ under 150	23	32	[.54, .87]	23/32
Martian weights > 150	2	10	[.009, .433]	.2

We have summarized our opinions about these five phenomena in terms of beta distributions with parameters r_1 and n_1 given in Table 2.1. That is:

1. We are almost certain that the proportion of heads is nearly one-half, and our 95% prior interval runs from .419 to .581.
2. Depending on whether the randomly selected day is in winter or in summer, almost all or almost none of the trees will have green leaves. Our 95% interval is, therefore, the union of the intervals [0, .1] and [.9, 1.0].
3. Smart as they are, we do not think many Harvard students have IQs over 150. Our 95% interval runs from 0 to .13.
4. The mirror image of (3).
5. We have very little information about Martians, and have adopted the diffuse prior that is nonintegrable and has no well-defined 95% interval.

The probability of a success on the first trial given p is just p. This is a conditional probability. The marginal probability of a success is just

$$P(\text{success}) = \int P(\text{success}|p) f_\beta(p|r_1, n_1)\, dp$$

$$= \int p f_\beta(p|r_1, n_1)\, dp$$

$$= Ep = \frac{r_1}{n_1}.$$

This may be found in the fourth column of Table 2.1.

The posterior parameters, the posterior 95% interval, and the probability of another success (r_2/n_2) may all be found in the same table. The interpretation of 2 successes in 10 trials is seen to depend on the process that generated the outcomes in the following way.

1. If we are talking about flips of a coin, 2 in 10 is hardly evidence at all. Both the 95% interval and the probability of a success are hardly changed.
2. Two trees in ten with green leaves suggests that it is winter, but there are more green trees than we would have guessed for winter. We thus essentially eliminate the summer branch of our 95% interval [.9, 1.0], but lengthen the winter branch to [.009, .43]. The probability of another success is greatly reduced from .5 to approximately .2.
3. Two in ten students with IQs over 150 is largely corroborative evidence, although it is somewhat more than we expected. We accordingly shorten but adjust rightward our 95% interval from [0, .13] to [.12, .20].
4. Two in ten students with IQs under 150 is startling evidence. [Note this is a different sample from (3), since success is defined differently.] We react to this with confusion by increasing greatly our interval from [.86, 1] to [.54, .86].
5. Where before we were "ignorant" about Martians, we now know a great deal. Notice that Martians and green leaves end up with essentially the same posterior, and arguments over fine adjustments to our definition of ignorance are likely to imply only minor adjustments to the posterior distribution.

EXCHANGEABLE EVENTS AND PREDICTIONS

In formulating Equation (2.2) we have used the independence assumption that if p is known, observation of one outcome does not alter our opinions about the probability of other outcomes. In his paper on probability cited earlier, deFinetti argues that independence is not a phenomenon about which we have very useful intuition. In place of independence, he substitutes the notion of *exchangeability*. A sequence of random events is said to be an *exchangeable sequence* if the probability assigned to particular sequences does not depend on the order of successes and failures, for example, if the sequence *SF* has the same probability as the sequence *FS*.

The following remarkable theorem is due to deFinetti (1937):

THEOREM 2.2 (INFINITE EXCHANGEABLE SEQUENCE). *Any coherent probability assignment to an infinite exchangeable sequence of binomial events is equivalent to the (marginal) assignment derived from the following joint distribution*

(*a*) *Conditional on p, the events are drawn independently*

$$P(SSF\cdots FS|p)=p^r(1-p)^{n-r}$$

(*b*) *p has a (unique) (prior) distribution* $f_1(p)$.

The marginal is, of course,

$$P(SSF\cdots FS)=\int_0^1 p^r(1-p)^{n-r}f_1(p)\,dp.$$

Example. A proper proof of this theorem is beyond the scope of this book, but an example can illustrate the ideas. Suppose coins are flipped in such a way that the probability of a head on any flip is one-half. Restricting ourselves to the first two flips and assuming that the sequence is exchangeable, $P(HT)=P(TH)$, the class of proper probability assignments is

Event	Probability	Probability if independent
HH	a	p^2
HT	$\frac{1}{2}-a$	$p(1-p)$
TH	$\frac{1}{2}-a$	$p(1-p)$
TT	a	$(1-p)^2$

for $\frac{1}{2}\geqslant a\geqslant 0$. Theorem 2.2 asserts that there exists a distribution for p such that the expected value of the third column is equal to the second. You may verify that this is true for any distribution such that

$$Ep=\tfrac{1}{2},\ Ep^2=a$$

with $Vp=Ep^2-(Ep)^2=a-\frac{1}{4}$.

Note that this cannot be satisfied for $a<\frac{1}{4}$. But for $a<\frac{1}{4}$, we do not have an infinite exchangeable sequence. Suppose $a=0$, for example, then the events HHT, HHH, TTH, and TTT have zero probability, and by the exchangeability assumption, so must HTH, THH, HTT, THT, leaving no event with positive probability. The value $a=0$ is, therefore, not allowable. For two events, two moments of the distribution are determined. For n events, n are determined; hence, the distribution in an infinite sequence is unique.

Suppose now that we have a proper probability assignment on an infinite sequence of exchangeable events. Given that we have observed r

successes in the first n trials, the probability of another success is

$$P(S_{n+1}|S_1S_2F_3\cdots F_n)=\frac{P(S_1S_2F_3\cdots F_nS_{n+1})}{P(S_1S_2F_3\cdots F_n)}$$

$$=\frac{\int_0^1 p^{r+1}(1-p)^{n+1-r-1}f_1(p)\,dp}{\int_0^1 p^r(1-p)^{n-r}f_1(p)\,dp}$$

$$=\int_0^1 pf_2(p)\,dp$$

where

$$f_2(p)=\frac{p^r(1-p)^{n-r}f_1(p)}{\int_0^1 p^r(1-p)^{n-r}f_1(p)\,dp}.$$

That is, our prediction of the next outcome is *as if* there were a true proportion p and a prior distribution on p, $f_1(p)$, which we updated according to Bayes' rule to $f_2(p)$ with independent sampling. In fact, both the proportion p and the notion of independence are mathematical fictions (or at least unnecessary accoutrements), since our subjective probability distribution, in fact, applies (need only apply?) to the observable exchangeable sequence of successes and failures. Of course, once we decide that we are observing an infinite exchangeable sequence, it is terribly convenient in formulating our probability assignment to make use of the fictional p and to concentrate our efforts on formulating a prior distribution for p. This sometimes turns back on itself, since in attempting to formulate this prior distribution we properly may ask ourselves questions about the implied distribution on the observable events.

2.4 Inference About a Mean

We have considered in the previous section how inferences may be made about p, the proportion of a given population that possess a given attribute. For attributes that are numerically defined another interesting number is μ, the population mean. For example, μ may be average income, average height, or average IQ. In this section we discuss how inferences may be made about a population mean.

The essential principles of subjectivist inference have been completely established in the previous section. We first select a *sampling* distribution

$P(\text{sample}|\,\mu)$ that describes our predictions about the sample if we knew the population mean μ. For a particular sample this distribution is treated as a function of μ and is called a *likelihood function*. The *posterior distribution* that describes opinions about μ after the sample is observed is computed by multiplying the likelihood function times a *prior distribution* that is selected to summarize our knowledge of μ before we see the sample.

There is one new wrinkle in this section. Interpretation of the evidence about the population mean μ depends on the population variance σ^2. When σ^2 is not known with certainty, we make inferences jointly about both μ and σ^2. That is, we have a two-parameter problem. Although the principles of inference already described do not change, the two-parameter problem is subtly different from the one-parameter problem.

SAMPLING DISTRIBUTIONS

Our first item of business is the construction of a sampling distribution. This is a two-step problem involving, first, the selection of the sample space and, second, the assignment of a probability distribution over that space. In general, we may think of the attribute under study (height, income, etc.) as necessarily lying between minus infinity and plus infinity. If n members of the population are to be sampled, the set of all possible samples is an n-dimensional space

$$\text{sample space} = \{(x_1, x_2, \ldots, x_n) | -\infty < x_i < \infty\}.$$

The very special distribution defined on this space that will take up most of our time is

$$f(x_1, x_2, \ldots, x_n | \,\mu, \sigma^2) = f_N(x_1 | \,\mu, \sigma^2) f_N(x_2 | \,\mu, \sigma^2) \cdots f_N(x_n | \,\mu, \sigma^2)$$

where $f_N(\cdot\,|\,\mu, \sigma^2)$ indicates a normal distribution with mean μ and variance σ^2. This distribution follows from two assumptions:

(1) The population is normal with mean μ and variance σ^2.
(2) The sampling process yields independent random samples.

Both of these assumptions are worth discussing. Let us consider the first. When can we know that the distribution of attributes in a population is normal? Never. Nonetheless, the assumption is a useful starting point. A wide variety of phenomena do have symmetric unimodal distributions that are well approximated by bell-shaped normal curves. In other cases a transformation of the attribute may be approximately normally distributed, for example, the logarithm of income.

Like the assumption of normality, the assumption of independent random sampling can never be asserted to be certainly valid. Various forms of dependence and nonrandomness are likely to be inherent in any sampling process. In some cases departures from independent random sampling wil be obvious, but when they are not, assumption (2) makes a useful starting point for the analysis. Again, we emphasize that it is a working hypothesis, not a true proposition.

The sampling distribution can be written as

$$\begin{aligned} f(x_1, x_2, \ldots, x_n | \mu, \sigma^2) &= \prod_{i=1}^{n} f_N(x_i | \mu, \sigma^2) \\ &= \prod_{i=1}^{n} (\sqrt{2\pi}\,\sigma)^{-1} \exp - \frac{(x_i - \mu)^2}{2\sigma^2} \qquad (2.4) \\ &= (\sqrt{2\pi}\,\sigma)^{-n} \exp \frac{-\sum_i (x_i - \mu)^2}{2\sigma^2}. \end{aligned}$$

Letting the sample mean be $m = \Sigma x_i / n$, the exponent in (2.4) can be written as

$$\begin{aligned} \sum_{i=1}^{n} (x_i - \mu)^2 &= \sum_{i=1}^{n} [(x_i - m) + (m - \mu)]^2 \\ &= \sum_{i=1}^{n} (x_i - m)^2 + 2(m - \mu) \sum_{i=1}^{n} (x_i - m) + n(m - \mu)^2 \qquad (2.5) \\ &= D^2 + n(m - \mu)^2, \end{aligned}$$

where $D^2 \equiv \Sigma_{i=1}^{n} (x_i - m)^2$, and where we made use of the result that $\Sigma(x_i - m) = 0$. Substituting (2.5) into (2.4) we obtain

$$\begin{aligned} f(\mathbf{x} | \mu, \sigma^2) &= \left\{ (2\pi\sigma^2)^{-(n-1)/2} n^{-1/2} e^{-D^2/2\sigma^2} \right\} \\ &\quad \left\{ (2\pi\sigma^2/n)^{-1/2} e^{-(\mu - m)^2 n / 2\sigma^2} \right\} \qquad (2.6) \\ &= C f_N(\mu | m, \sigma^2/n) \end{aligned}$$

where C stands for the first expression in braces, and where the second expression in braces if $f_N(\mu | m, \sigma^2/n)$. Note that: (1) the expression for C does not depend on μ, and (2) the second expression could equally well have been written as $f_N(m | \mu, \sigma^2/n)$ where μ and m are formally interchanged. We prefer to use the form in (2.6) because of what comes next.

PRIOR TO POSTERIOR ANALYSIS: KNOWN VARIANCE

We have now specified a conditional probability distribution for the sample given the population parameter μ with σ^2 assumed known. A joint distribution over all the uncertain events f (sample, μ) is the product of the distribution above and a marginal on μ denoted by $f(\mu)$. Symbolically,

$$f(\text{sample}, \mu) = f(\text{sample}|\mu)f(\mu)$$

where $f(\mu)$ simply describes our opinions about the population mean before any observations are made. From this joint distribution we may straightforwardly calculate the conditional distribution

$$f(\mu|\text{sample}) = \frac{f(\text{sample}, \mu)}{f(\text{sample})}$$

$$= \frac{f(\text{sample}|\mu)f(\mu)}{f(\text{sample})}$$

indicating opinions about μ given any particular sample.

In this context, for obvious reasons, $f(\mu)$ is called a *prior* distribution and $f(\mu|\text{sample})$ is called a *posterior* distribution. Although $f(\text{sample}|\mu)$ is a conditional probability distribution, once a particular sample is observed it is treated as a function of μ and called a *likelihood function* denoted by

$$L(\mu;\text{sample}) \propto f(\text{sample}|\mu).$$

Note that the *posterior distribution is proportional to the product of the likelihood function and the prior distribution*. The normalizing constant $P(\text{sample})$ is often called a *predictive p.d.f.*, because it summarizes our opinions about the sample before the sample is observed; that is, it predicts the sample.

Making use of (2.6) and Bayes' rule, we may write

$$f_2(\mu) = kf_N(\mu|m, \sigma^2/n)f_1(\mu) \tag{2.7}$$

where k is a proportionality constant and f_2 and f_1 are posterior and prior p.d.f.'s.

Now consider the following two special cases:

CASE 1. Suppose our information concerning μ is extremely vague in the sense that we have no reason to believe that μ lies in any interval rather than any other interval of the same length. Then $f_1(\mu)$ can be taken as a constant in (2.7), and we get

$$f_2(\mu) = f_N(\mu|m, \sigma^2/n) \tag{2.8}$$

since the constant in (2.7) must be selected so that $\int_{-\infty}^{\infty} f_2(\mu)\,d\mu = 1$ and since $\int_{-\infty}^{\infty} f_N(\mu|m, \sigma^2/n)\,d\mu = 1$.

Strictly speaking, a flat prior distribution is not a valid p.d.f. since it does not integrate to one. It therefore violates the principles used to police our opinions about uncertain events. Nonetheless, there are, as we shall see, valid prior p.d.f.'s so nearly flat that for all practical purposes (2.8) is the corresponding posterior distribution.

CASE 2. Suppose our prior is normal with mean m^* and variance v^*. Then by (2.7),

$$f_2(\mu) = k f_N(\mu|m, \sigma^2/n) f_N(\mu|m^*, v^*) \tag{2.9}$$

and substituting the analytical forms for f_N we get, after a lot of elementary algebra, the following result

THEOREM 2.3 (NORMAL SAMPLES AND NORMAL PRIORS). *If the conditional distribution of the vector* $\mathbf{x}$ *given the parameters* μ *and* σ^2 *is normal with mean vector* $E(\mathbf{x}) = \mathbf{1}_n \mu$, *where* $\mathbf{1}_n$ *is a vector of n ones, and variance matrix* $V(\mathbf{x}) = \sigma^2 \mathbf{I}_n$, *and if* μ *is normal with mean* m^* *and variance* v^*, *then the distribution of* μ *given* $\mathbf{x}$ *and* σ^2 *is normal with moments*

$$E(\mu|\mathbf{x}, \sigma^2) = \left(v^{*-1} + (\sigma^2/n)^{-1}\right)^{-1}\left(m^* v^{*-1} + m(\sigma^2/n)^{-1}\right) = m^{**} \tag{2.10}$$

$$V(\mu|\mathbf{x}, \sigma^2) = \left(v^{*-1} + (\sigma^2/n)^{-1}\right)^{-1} = v^{**} \tag{2.11}$$

where m is the sample mean, $m = \mathbf{x}'\mathbf{1}_n / n$.

These prior and posterior distributions deserve further comment. (1) Note first that the posterior distribution is in the same family as the prior distribution; that is, both are normal. This is terribly convenient, since the sample evidence can be straightforwardly interpreted in terms of the change in the (two) parameters. If the distribution were to be altered by the sample outcome, there would be rather serious difficulties in discussing the nature of the sample evidence. (2) The posterior mean m^{**} is a weighted average of prior mean and sample mean with weights proportional to v^{*-1} and (n/σ^2), which may be called the precisions of the information sources. Note that for very diffuse priors, that is, for v^* very large, the posterior mean and variance are effectively m and σ^2/n, the sample summaries. This is the situation described in case 1. (3) Suppose that our sample came in two bundles: the first n^* observations yielded a mean of m^*, and the remaining n observations yielded a mean of m. If we started with a uniform prior, after we had seen only the first bundle, we would have opinions about μ that are normal with mean m^* and variance σ^2/n^*. Thus we may think of our prior information with mean m^* and variance v^* as being equivalent to a certain preliminary experiment with sample size $n^* = \sigma^2/v^*$ and with mean m^*. Note that if formulas (2.10) and (2.11) are defined in terms of $n^* \equiv \sigma^2/v^*$ instead of v^* they become

$$v^{**} = \frac{\sigma^2}{(n^* + n)}$$

$$m^{**} = (n^* + n)^{-1}(n^* m^* + nm),$$

that is, the posterior mean is a weighted average of prior mean m^* and sample mean m with weights proportional to effective prior sample size n^* and sample size n. (4) The posterior variance v^{**} does not depend on the sample mean m. A seemingly desirable property of a posterior distribution is that when the sample and prior distribution are in conflict with neither information source dominating the other, then the posterior distribution should be fairly diffuse. We might also want it to be bimodal. This is not a property of normal sampling with known variance and normal priors, since the posterior distribution is normal with a variance that depends on sample size alone. We may view this as a shortcoming of the normality assumption. (5) The difference between v^* and σ^2 should be clearly understood. The former indicates the uncertainty about μ, and the latter indicates the extent to which a particular observation can wander from μ. There is no reason why these numbers should be the same, or even related.

Sufficiency of the Sample Mean. Given the prior distribution of μ, the posterior distribution of μ depends on the sample outcome $(x_1, \ldots, x_n)$ only through the sample mean

$$m = \sum_{i=1}^{n} x_i / n.$$

We, therefore, say that for prior-to-posterior analysis it is *sufficient* to know *only* the sample mean. This strongly depends on the fact that the sample is known to have been drawn from a normal population with known variance of σ^2 and unknown mean μ.

SUMMARIZATION AND INTERPRETATION

Again, it is useful to emphasize the difference between summarizing and interpreting data evidence. We have assumed a sampling process—the process that generates the data—that admits the sufficient statistic $m = \Sigma x_i/n$. That is, the evidence about μ (when σ^2 and n are known) is completely summarized in the single number m, and, for example, we need not remember the particular value of the first observation x_1 if we remember m. The way that we interpret this summarized evidence depends on our prior information about μ.

To illustrate this, let us form prior distributions for μ when μ is

a. Average age in years of Harvard undergraduates
b. Average age in years of Harvard faculty members
c. Average height in feet of American males
d. Average number of cars per day in thousands in Cambridge during 1978.

We have chosen normal prior distributions for these four averages with means and variances indicated in Table 2.2. The content of this table is

a. We think Harvard undergraduates are most likely to have an average age of 20 years. We are willing to bet at roughly 20:1 odds that the average age is between 19 and 21.
b. Faculty members tend to be older, say, around 50, but most likely have an average age between 40 and 60.
c. The average height of American males is likely to be between 5 and 7 feet.
d. We have very little information about the average number of cars in Cambridge but would guess that it is between 0 and 100,000.

Suppose now we observe a sample with mean $m=20$ and variance $\sigma^2/n=5$. The posterior parameters and posterior 95% intervals for each of the four phenomena are also indicated in Table 2.2. The content of these

Table 2.2

Interpreting Evidence From A Normal Sample, $m=20$, $\sigma^2/n=5$

	Prior			Posterior		
Phenomenon	m^*	v^*	95% Interval	m^{**}	v^{**}	95% Interval[a]
(a) Age of Harvard undergraduates	20	.25	[19, 21]	20	.24	[19.1, 20.9]
(b) Age of Harvard faculty members	50	25	[40, 60]	25	4.2	[21, 29]
(c) Height of American males	6	.25	[5, 7]	6.7	.24	[5.8, 7.6]
(d) Number of cars in Cambridge	50	625	[0,100]	~20	~5.	[15.5, 24.5]

[a]Approximately.

results is

a. Our opinion about the average age of Harvard undergraduates is supported, though weakly, by the evidence and we narrow our 95% interval to [19.1, 20.9].
b. Observing a sample of Harvard faculty members with average age 20 is astounding. Naive application of the formulas leads to a 95% posterior interval for average age of [21, 29]. It may be more meaningful to question the nature of the process that generated the data.
c. Observing males with average height of 20 feet is even more astounding. Again, we might want to examine the rulers that did the measuring. If we are satisfied with both the sampling process and our prior, we will have a 95% posterior internal of [5.8, 7.6], slightly taller than our prior interval.
d. We began with essentially no information about the number of cars in Cambridge. We now have much better information with our 95% interval reduced from [0,100] to [15.5, 24.5].

We have seen, therefore, that although in all cases the summary of the data is identical, the interpretation of the evidence depends on prior information about the process. We have also seen that naive application of Bayes' rule can lead to some silly results, especially when we have incorrectly described the sampling process. Note also the fact that although the sample supports the prior in Case (a) and contradicts it in Case (c), the posterior variance is the same for both. This is the property of normal sampling with known σ^2 and normal priors that makes us question the wisdom of the normal priors. An apparently desirable property of a learning model is that contradictory evidence induces confusion.

NORMAL SAMPLING WITH UNKNOWN MEAN AND UNKNOWN VARIANCE

When σ^2 is unknown the problem of inference is subtly altered. Beginning with a diffuse prior for μ with σ^2 known, a posterior 95% "credible" interval is given approximately by $m-2\sigma/\sqrt{n} \leqslant \mu \leqslant m+2\sigma/\sqrt{n}$, where m is the sample mean and n is the sample size. The center of this interval m does *not* depend on σ^2, but the width of it does. The smaller is σ^2, the shorter is our credible interval and consequently the more precise is our opinion about μ. In this sense we may say that m represents the information we have about μ, and that σ^2/n represents the *quality* of that information.

With σ^2 unknown, the *quality* of the information afforded by the sample is necessarily uncertain. As we shall see, the sample contains information about σ^2 we want to use to help "estimate" the quality of the sample information. We also want to use any available prior information. The interesting feature of this two-parameter problem is that the interpretation of the sample evidence about μ depends on uncertain prior information about σ^2. To the extent that the prior for σ^2 is difficult to specify, the interpretation of the evidence about μ is ambiguous.

For convenience alone, we now would like to replace the process variance σ^2 with a new parameter $h=1/\sigma^2$, called the process precision. We express the prior, likelihood, and posterior in terms of h, keeping in mind that distributions in terms of σ^2 may be easily calculated by a suitable transformation of variables. The likelihood then becomes (from 2.6)

$$L(\mu, h; \text{sample}) = kh^{n/2}e^{-\frac{1}{2}hD^2}e^{-\frac{1}{2}hn(m-\mu)^2}. \tag{2.12}$$

The prior distribution which is most convenient for this likelihood function is a normal-gamma distribution

$$f_1(\mu,h)=f_N\left(\mu|m^*, (hn^*)^{-1}\right)f_\gamma(h|s^{*2},\nu^*)$$

indicating that conditional on the process precision h, μ has a normal distribution with mean m^* and precision hn^*

$$f_N\left(\mu|m^*,(hn^*)^{-1}\right)=\frac{1}{\sqrt{2\Pi}}(hn^*)^{\frac{1}{2}}e^{-\frac{1}{2}hn^*(\mu-m^*)^2}$$

and h has a gamma distribution

$$f_\gamma(h|s^{*2},\nu^*)=\frac{\left(\frac{1}{2}\nu^*s^{*2}\right)^{\frac{1}{2}\nu^*}}{\left(\frac{1}{2}\nu^*-1\right)!}e^{-\frac{1}{2}\nu^*s^{*2}h}h^{\frac{1}{2}\nu^*-1}.$$

Relevant properties of this gamma distribution are discussed in Appendix 2. Here it may be observed that the marginal on μ is a Student density

$$\begin{aligned} f(\mu) &= \int_0^\infty f_N\left(\mu|m^*,(hn^*)^{-1}\right)f_\gamma(h|s^{*2},\nu^*)\,dh \\ &= f_S\left(\mu|m^*,\frac{s^{*2}}{n^*},\nu^*\right) \end{aligned}$$

with mean and variance

$$E(\mu)=m^* \text{ and } V(\mu)=\frac{s^{*2}}{n^*}\frac{\nu^*}{\nu^*-2}.$$

We may rewrite the likelihood function as

$$L(\mu,h;\text{sample})=ke^{-\frac{1}{2}h\nu s^2}h^{\frac{1}{2}\nu}e^{-\frac{1}{2}hn(m-\mu)^2}h^{\frac{1}{2}}$$

where

$$\nu = n - 1$$
$$\nu s^2 = D^2 = \sum (x_i - m)^2.$$

The posterior distribution, as usual, is formed by normalizing the product of the likelihood function and the prior distribution

$$f_2(\mu, h) = kL(\mu, h; \text{sample}) f_1(\mu, h)$$
$$\propto e^{-\frac{1}{2}h\nu s^2} h^{\frac{1}{2}\nu} e^{-\frac{1}{2}hn(m-\mu)^2} h^{\frac{1}{2}}$$
$$h^{\frac{1}{2}} e^{-\frac{1}{2}hn^*(\mu - m^*)^2} e^{-\frac{1}{2}\nu^* s^{*2} h} h^{\frac{1}{2}\nu^* - 1}$$

where irrelevant constants have been subsumed into k. This may be rewritten as

$$f_2(\mu, h) \propto (hn)^{\frac{1}{2}} \exp\left[-\tfrac{1}{2}hn^{**}(\mu - m^{**})^2\right] \exp\left[-\tfrac{1}{2}\nu^{**}s^{**2}h\right] h^{\frac{1}{2}\nu^{**}-1} \tag{2.12}$$

where

$$n^{**} = n^* + n$$
$$m^{**} = (n^* + n)^{-1}(n^* m^* + nm) \tag{2.13}$$
$$\nu^{**} = \nu^* + n$$
$$\nu^{**}s^{**2} = \nu^* s^{*2} + n^* m^{*2} + \nu s^2 + nm^2 - n^{**}m^{**2}.$$

By inspection, (2.12) is a normal-gamma distribution. This establishes the following result.

THEOREM 2.4 (NORMAL SAMPLES AND NORMAL-GAMMA PRIORS). *If the conditional distribution of the vector* $\mathbf{x}$ *given* μ *and* σ^2 *is normal with moments* $E(\mathbf{x}|\mu, \sigma^2) = \mathbf{1}_n\mu$ *and* $V(\mathbf{x}|\mu, \sigma^2) = \sigma^2\mathbf{I}$, *and if* (μ, σ^2) *has a normal-gamma distribution with parameters* $(m^*, n^*, s^{*2}, \nu^*)$, *then* (μ, σ^2) *given* $\mathbf{x}$ *has a normal-gamma distribution with parameters* (2.13).

The first two comments made about sampling with known variance apply here as well. The prior and posterior distributions are in the same four-parameter family of distributions. The posterior mean is a weighted average of prior and sample means. However, the posterior variance now depends on the sample and prior means, since

$$\nu^{**}s^{**2} = \nu^* s^{*2} + \nu s^2 + n^* m^{*2} + nm^2 - n^{**}m^{**2}$$
$$= \nu^* s^{*2} + \nu s^2 + \frac{(m - m^*)^2 n^* n}{(n^* + n)},$$

and the greater the discrepancy between m and m^*, the larger is s^{**2}, and correspondingly the flatter is the distribution on μ. Thus conflict between the prior and the sample may be so great that the posterior variance of μ exceeds the prior variance. Note also that the variance of μ depends on the prior about σ^2.

2.5 Noninformative Priors

Can a Bayesian compute a posterior distribution for a parameter if he has no prior distribution? Decidedly not; Bayes' rule requires a prior distribution. Well, then, is there a prior distribution that represents ignorance, and can you use Bayesian inference if you are ignorant? Like the issue of original sin, this is a question that remains unresolved, that attracts a fair amount of theological interest, but that seems rather remote from the concerns of the man on the street.

The argument begins chronologically with Bayes' *An Essay Towards Solving a Problem in the Doctrine of Chances*, which suggests by the principle of insufficient reason that ignorance is represented by a probability function that assigns equal probability to all events. Of course, there can be no such probability function, since if mutually exclusive events A and B are assigned equal probability, the event A union B is implicitly assigned twice the probability. Or if a continuous random variable θ is assigned a uniform distribution, the variable $\gamma=\theta^{-1}$ has density function proportional to γ^{-2}. In a situation of "real ignorance" there is insufficient reason to select one event space rather than another, or one parameterization rather than another, and the principle of insufficient reason is apparently insufficient to determine probabilities.[2]

Jeffreys (1961) is the modern-day proponent of the Bayes–LaPlace principle of insufficient reason. He suggests that for a parameter μ defined from $-\infty$ to ∞, one should use the (improper) uniform prior distribution

$$f(\mu)\,d\mu \propto d\mu, \quad -\infty<\mu<\infty.$$

This seems to me to be implying that μ almost certainly is either enormously large positive or enormously large negative, since the ratio of the mass outside any finite interval to the mass inside any finite interval is one. A researcher usually can get away with making such a ridiculous statement in practice, since the data will imply even more strongly that the parameter μ is not enormous.

[2]Are individuals A and B equally tall, merely because there is insufficient reason to regard one to be taller than the other?

This last observation suggests a philosophically sound method of computing a diffuse prior distribution. Assuming that a particular experiment is about to be observed, we can find a prior distribution that will have little impact on the posterior distribution. It is important to understand that this is not a prior intended to represent ignorance. It is defined only in the context of a particular experiment. It is a prior that represents information that will be dominated by the sample information. For example, referring to formula (2.10) that describes the parameters of the posterior distribution for μ, observe that as the sample size increases and the sample information dominates the prior information, the parameters converge to n and σ^2; but if we set v^* to zero and use a uniform prior for μ, the posterior and prior will bare this same relationship to each other regardless of the sample size. Thus a uniform distribution is a prior that is dominated by any sample from a normal population.[3]

In practice, the sample may dominate the prior information and the posterior distribution may be inconsequentially different from a posterior distribution corresponding to an improper noninformative prior. A prior density that is relatively uniform where the likelihood function attains its maximum is likely to imply such a posterior. This is discussed by Savage (1962) under the title, "stable estimation." But for any proper prior, there are samples that do not generate information that dominates the prior, and it is impossible to know in advance of observation whether a diffuse prior is an adequate approximation to the proper prior that is truly representative of your opinions.

Of course, there are difficulties in precisely defining a noninformative prior. But much more important is the fact that the use of improper, noninformative priors can sometimes lead to very undesirable outcomes, many of which are discussed by Lindley (1971). In the hypothesis-testing problem of Chapter 4, noninformative priors imply that all hypotheses with more than the minimal number of parameters should certainly be rejected. A noninformative prior can lead to inadmissible decision rules, as discussed in Chapter 5. Quite pragmatically, however, since we have gone to all this trouble to develop a method that formally incorporates subjec-

[3]Another way of defining priors in relation to anticipated sample information is due to Jeffreys. He suggests that if a researcher is ignorant about a parameter θ, then his opinions about θ given some evidence $\mathbf{x}$ should be the same whether he regards θ to be the parameter or some one-to-one differentiable transformation of it, $g(\theta)$. A prior that has this invariance property is

$$f(\theta) \propto |-E(\partial^2 \log f(x|\theta)/\partial\theta^2)|^{1/2}$$

where the expectation is with respect to the density $f(\mathbf{x}|\boldsymbol{\theta})$. In words, the prior is proportional to the square root of Fisher's information measure. For a proof and discussion see Zellner (1971) where minimal information priors are also discussed.

tive prior information, it seems a shame in the end to use priors that are intended to represent ignorance. Is the Bayesian logic to be used only to rename the likelihood function?

One reason statisticians have spent so much time looking for priors to represent ignorance is that in practice it is extremely difficult actually to select a prior. The word "ignorance" is suggestive of a prior that is not only relatively uninformed but also difficult to determine with adequate accuracy. It may make sense in that case to let the data help in measuring the prior by determining a class of dominated priors, all of which imply essentially the same posterior distribution as does the diffuse improper prior. One need only ask himself if this particular sample outcome seems "peculiar." If the answer is no, one might as well use the posterior distribution implied by the diffuse prior, since careful measurement of the prior can lead to only minor adjustment of the posterior.

CHAPTER 3

THE LINEAR-REGRESSION MODEL

This chapter reviews theories of inference with the linear-regression model. The first section summarizes the standard theorems of classical inference. The second section deals with pooling information from two samples. The third section describes Bayesian pooling of information from one sample with prior information. And the fourth section describes Bayesian inference about the mean vector and variance-covariance matrix of a multivariate normal distribution.

For more complete introductory material the reader is referred to classical treatments by Johnston (1973), Theil (1971), and Rao (1965), and to Bayesian treatments by Lindley (1965), Zellner (1971), and Box and Tiao (1973).

3.1 Classical Inference with the Linear-Regression Model: A Review

Theoretical descriptions often amount to nothing more than the statement that some hypothetical variable η might depend on some vector of hypothetical variables χ. Empirical workers daringly translate this into a statement about observable variables such as

$$Y_t = \beta_1 X_{1t} + \beta_2 X_{2t} + \cdots + \beta_k X_{kt} + u_t$$

where Y_t is the observable counterpart of η, $\mathbf{X}_t' = (X_{1t}, X_{2t}, \ldots, X_{kt})$ is the observable counterpart of χ, t indexes a set of T observations on each variable $t=1$,

$\ldots, T$, and where the unobservables $\beta_1, \beta_2, \ldots, \beta_k$ and u_t represent features of the precise relationship between η and χ that are not disclosed by the theory. The functional dependence of Y_t on χ_t is thereby assumed to be linear with an unknown parameter vector $\boldsymbol{\beta}' = (\beta_1, \beta_2, \ldots, \beta_k)$, and the theory is asserted to be incomplete, $u_t \neq 0$, or at least the linear approximation to it is incomplete.

The model may be written in matrix form as

$$\mathbf{Y} = \mathbf{X}\boldsymbol{\beta} + \mathbf{u} \tag{3.1}$$

where $\mathbf{Y}$ is a $T \times 1$ vector of observables, $\mathbf{X}$ is a $T \times k$ matrix of observables, $\boldsymbol{\beta}$ is a $k \times 1$ vector of unobservables, and $\mathbf{u}$ is a $T \times 1$ vector of unobservables. Traditionally, $\boldsymbol{\beta}$ is called a parameter vector and $\mathbf{u}$ a disturbance or error vector. By definition, $\mathbf{u}$ is all of those things that determine $\mathbf{Y}$, excluding $\mathbf{X}\boldsymbol{\beta}$, and (3.1) is merely a tautological definition of $\mathbf{u}$, $\mathbf{u} \equiv \mathbf{Y} - \mathbf{X}\boldsymbol{\beta}$. To be more explicit, write $\mathbf{Y}$ as

$$\mathbf{Y} = \mathbf{X}\boldsymbol{\beta} + \mathbf{Z}\boldsymbol{\gamma}$$

where $\mathbf{Z}\boldsymbol{\gamma}$ is the part of $\mathbf{Y}$ left out of Equation (3.1), with $\mathbf{Z}$ a $(T \times m)$ matrix of variables and $\boldsymbol{\gamma}$ an $(m \times 1)$ vector of fixed effects, and where m need not be finite.

Substantive content is introduced into the analysis by assigning to $\mathbf{u} \equiv \mathbf{Z}\boldsymbol{\gamma}$ a frequency distribution, say, multivariate normal with mean zero and variance matrix $\boldsymbol{\Sigma}$. It is thereby asserted that if the matrix $\mathbf{X}$ were held fixed and the matrix $\mathbf{Z}$ allowed to vary within the confines of some more-or-less well-defined experimental conditions, the vector $\mathbf{Z}\boldsymbol{\gamma}$ would appear to have been drawn from a particular normal distribution. More importantly, it is assumed that there is no *tendency for any of the T residual effects to exceed or fall short of zero on the average*, $E(\mathbf{Z}\boldsymbol{\gamma}|\mathbf{X}) = \mathbf{0}$. Or in the more familiar parlance, the left-out effects are assumed to be uncorrelated with the included effects.

The requirement $E(\mathbf{Z}\boldsymbol{\gamma}|\mathbf{X}) = \mathbf{0}$ is a crucial and quite unlikely assumption. To give a trivial example, suppose in a sample of individuals the dependent variable Y measures sunburn susceptibility and the explanatory variable $\mathbf{X}$ measures hair color. The correlation between these two variables may lead us erroneously to conclude that red hair causes sunburn, when in fact, genetic inheritance (Z) determines both sunburn susceptibility and hair color.

The most foolish errors of inference derive from this kind of model misspecification; yet books about statistical inference hardly mention it. Of course, as discussed in Chapter 1, the choice of specification in this sense is not within the purview of the theory of statistical inference, which almost by definition takes the model as well defined. This enormously restricts the usefulness of statistical theory in nonexperimental research. It is sensible sometimes during a data analysis to take the model as well specified; but it is senseless always to do so. Any inference will be thoroughly discredited by the identification of particular left-out variables. In devoting a whole chapter to post-data model construction, this book places relatively great emphasis on this topic. Even here, the heavy allocation of space to problems that presume the definition of the underlying causal model reflects what *can* be said about the various topics, not what *should* be said. More is said of this in Chapter 9, but until then the specification assumption $E(\mathbf{u}|\mathbf{X})=\mathbf{0}$ should be taken as given.

FINDING A COMPLETE THEORY

An initial question that may be asked of the data is whether the theory is, in fact, complete, $\mathbf{u}=\mathbf{0}$. In particular, is there a value of $\boldsymbol{\beta}$, say, $\mathbf{b}$, that makes the theory appear to be complete, $\mathbf{Y}-\mathbf{Xb}=\mathbf{0}$? Whenever the number of observations T exceeds the number of parameters k, such a value of $\mathbf{b}$ is quite unlikely to exist. We can, however, find a value that makes the theory appear as complete as possible by making the residual vector $\mathbf{e}=\mathbf{Y}-\mathbf{Xb}$ as small as possible. One (arbitrary) measure of smallness is the square of the length of $\mathbf{e}, \mathbf{e}'\mathbf{e}$. Minimizing $\mathbf{e}'\mathbf{e}$ by setting its derivatives to zero implies the equations

$$\mathbf{0}=\frac{\partial}{\partial\boldsymbol{\beta}}(\mathbf{Y}-\mathbf{X}\boldsymbol{\beta})'(\mathbf{Y}-\mathbf{X}\boldsymbol{\beta})=-2\mathbf{X}'\mathbf{Y}+2\mathbf{X}'\mathbf{X}\boldsymbol{\beta}.$$

Solving this with the assumption that $\mathbf{X}'\mathbf{X}$ is invertible yields the least squares value of $\boldsymbol{\beta}$

$$\mathbf{b}=(\mathbf{X}'\mathbf{X})^{-1}\mathbf{X}'\mathbf{Y}. \tag{3.2}$$

THE MULTIPLE CORRELATION COEFFICIENT

Although $\mathbf{e}'\mathbf{e}$ is one measure of the completeness of the theory, it is more common to use an R^2 defined by

$$R^2=\left[\sum_t\left(Y_t-\bar{Y}\right)^2-\mathbf{e}'\mathbf{e}\right]\Big/\sum_t\left(Y_t-\bar{Y}\right)^2, \tag{3.3}$$

where $\bar{Y}=\Sigma Y_t/T$. The R^2 can be shown to lie between zero and one. A model with only a constant term $Y_t=\beta_1+u_t$ can be written in vector notation as $\mathbf{Y}=\mathbf{1}_T\beta_1+\mathbf{u}$, where $\mathbf{1}_T$ is a $T\times 1$ vector of ones. The least-squares estimate of β_1 is $\hat{\beta}_1=(\mathbf{1}_T'\mathbf{1}_T)^{-1}\mathbf{1}_T'\mathbf{Y}=\bar{Y}$, where $\bar{Y}$ is the mean of the observations Y_t, $t=1,\ldots,T$. The corresponding residual sum of squares is $(\mathbf{Y}-\mathbf{1}_T\hat{\beta}_1)'(\mathbf{Y}-\mathbf{1}_T\hat{\beta}_1)=\Sigma(Y_t-\bar{Y})^2$. *Assuming that the first column of* $\mathbf{X}$ *is a vector of ones,* it is always possible to make $\mathbf{e}'\mathbf{e}$ equal to $\Sigma(Y_t-\bar{Y})^2$ by setting $\boldsymbol{\beta}'=(\bar{Y},0,0,\ldots,0)$. Since we are choosing $\mathbf{b}$ to make $\mathbf{e}'\mathbf{e}$ as small as possible, it must be true that $0\leqslant\mathbf{e}'\mathbf{e}\leqslant\Sigma(Y_t-\bar{Y})^2$ and thus $0\leqslant R^2\leqslant 1$. An

R^2 of one means that the model is possibly complete, $\mathbf{e}'\mathbf{e}=0$, whereas an R^2 of zero means that the model is totally incapable of explaining the variability in the observed data.

ESTIMATING $\boldsymbol{\beta}$

The least squares vector **b** has properties other than making the theory appear as complete as possible. A frequency distribution for **u** with $E\mathbf{u}=\mathbf{0}_T$, and $V\mathbf{u}=\sigma^2\mathbf{I}_T$ implies a frequency distribution for **b**. ($\mathbf{I}_T$ is the $T\times T$ identity matrix. $\mathbf{0}_T$ is a $T\times 1$ vector of zeros.) Since **b** is a linear function of **u**,

$$\mathbf{b}=(\mathbf{X}'\mathbf{X})^{-1}\mathbf{X}'(\mathbf{X}\boldsymbol{\beta}+\mathbf{u})=\boldsymbol{\beta}+(\mathbf{X}'\mathbf{X})^{-1}\mathbf{X}'\mathbf{u},$$

the moments of **b** are straightforwardly calculated as

$$E\mathbf{b}=\boldsymbol{\beta}+(\mathbf{X}'\mathbf{X})^{-1}\mathbf{X}'E(\mathbf{u})=\boldsymbol{\beta},$$

$$V\mathbf{b}=(\mathbf{X}'\mathbf{X})^{-1}\mathbf{X}'V(\mathbf{u})\mathbf{X}(\mathbf{X}'\mathbf{X})^{-1}=\sigma^2(\mathbf{X}'\mathbf{X})^{-1}.$$

It is important to understand that these moments describe the behavior of **b** in repeated samples. This is, if the "experiment" is repeated over and over with fixed **X** matrix, the average value of **b** would be $\boldsymbol{\beta}$. In that sense only, a particular value of **b** is taken as indicating where $\boldsymbol{\beta}$ lies. These results are summarized as follows:

THEOREM 3.1 (MOMENTS OF LEAST-SQUARES ESTIMATOR). *Assuming* $\mathbf{X}'\mathbf{X}$ *is invertible,* $E\mathbf{u}=\mathbf{0}_T$ *and* $V(\mathbf{u})=\sigma^2\mathbf{I}_T$, *the least-squares estimator of* $\boldsymbol{\beta}$ *is* $\mathbf{b}=(\mathbf{X}'\mathbf{X})^{-1}\mathbf{X}'\mathbf{Y}$, *with mean* $\boldsymbol{\beta}$ *and variance* $\sigma^2(\mathbf{X}'\mathbf{X})^{-1}$.

A desirable feature of the estimator **b** from the standpoint of calculation is that it is a linear function of **Y**. It is also an unbiased estimator of $\boldsymbol{\beta}$, $E(\mathbf{b})=\boldsymbol{\beta}$. It is of interest to deterrmine if some other unbiased linear estimator might be better. Suppose we wished to estimate the scalar parameter $\theta=\boldsymbol{\psi}'\boldsymbol{\beta}$ where $\boldsymbol{\psi}$ is a k-dimensional vector of constants. The least-squares estimator of θ is taken to be $\boldsymbol{\psi}'\mathbf{b}$. An alternative linear estimator is

$$\hat{\theta}=\mathbf{A}'\mathbf{Y}+a$$

where **A** is a vector of constants, and a is a scalar constant. The estimator is unbiased only if

$$\boldsymbol{\psi}'\boldsymbol{\beta}=E\hat{\boldsymbol{\theta}}, \qquad \text{for all} \quad \boldsymbol{\beta},$$

that is, only if

$$\boldsymbol{\psi}'\boldsymbol{\beta}=A'E\mathbf{Y}+a=\mathbf{A}'\mathbf{X}\boldsymbol{\beta}+a.$$

For this to be true for all $\boldsymbol{\beta}$ we must have

$$a=0, \qquad \boldsymbol{\psi}'=\mathbf{A}'\mathbf{X}.$$

The variance of the estimator $\hat{\boldsymbol{\theta}}$ is

$$V\hat{\boldsymbol{\theta}}=\mathbf{A}'V(\mathbf{Y})\mathbf{A}=\sigma^2\mathbf{A}'\mathbf{A}.$$

Minimization of the variance of $\hat{\boldsymbol{\theta}}$ subject to the unbiasedness restriction is a simple constrained maximization problem. With $\boldsymbol{\lambda}$ as the vector of k Lagrange multipliers, this requires the derivatives of the function $\sigma^2\mathbf{A}'\mathbf{A}+2\boldsymbol{\lambda}'(\boldsymbol{\psi}-\mathbf{X}'\mathbf{A})$ to be set to zero; that is,

$$\begin{aligned}\mathbf{0}&=\frac{\partial f}{\partial \mathbf{A}}=2\sigma^2\mathbf{A}'-2\boldsymbol{\lambda}'\mathbf{X}'\\ \mathbf{0}&=\frac{\partial f}{\partial \boldsymbol{\lambda}}=\boldsymbol{\psi}'-\mathbf{A}'\mathbf{X}\end{aligned} \tag{3.4}$$

Postmultiplying the first equation by $\mathbf{X}$ we obtain

$$\sigma^2\mathbf{A}'\mathbf{X}-\boldsymbol{\lambda}'\mathbf{X}'\mathbf{X}=\mathbf{0}$$

which implies

$$\boldsymbol{\lambda}'=\sigma^2\mathbf{A}'\mathbf{X}(\mathbf{X}'\mathbf{X})^{-1}=\sigma^2\boldsymbol{\psi}'(\mathbf{X}'\mathbf{X})^{-1}.$$

Substituting this last expression into (3.4) yields

$$\mathbf{0}=2\sigma^2\mathbf{A}'-2\sigma^2\boldsymbol{\psi}'(\mathbf{X}'\mathbf{X})^{-1}\mathbf{X}'$$

or

$$\mathbf{A}'=\boldsymbol{\psi}'(\mathbf{X}'\mathbf{X})^{-1}\mathbf{X}'.$$

Thus the minimum variance linear unbiased estimator of $\boldsymbol{\psi}'\boldsymbol{\beta}$ is

$$\hat{\boldsymbol{\theta}}=\mathbf{A}'\mathbf{Y}=\boldsymbol{\psi}'(\mathbf{X}'\mathbf{X})^{-1}\mathbf{X}'\mathbf{Y},$$

which is, of course, just the least-squares estimator, and we have established the following result.

THEOREM 3.2 (GAUSS–MARKOV). *Assuming* $\mathbf{X}'\mathbf{X}$ *is invertible,* $E(\mathbf{u})=\mathbf{0}$ *and* $V(\mathbf{u})=\sigma^2\mathbf{I}$ *with* σ^2 *finite, the least-squares estimator* $\boldsymbol{\psi}'\mathbf{b}$ *of the linear combination of coefficients* $\boldsymbol{\psi}'\boldsymbol{\beta}$ *has minimum variance among the class of unbiased linear estimators.*

If $\mathbf{u}$ has a normal distribution, the least-squares estimator $\mathbf{b}$ which is a linear function of $\mathbf{u}$ is also normally distributed. Furthermore, $\mathbf{b}$ is the maximum likelihood estimator of $\boldsymbol{\beta}$. The distribution of $\mathbf{Y}$ is

$$f(\mathbf{Y}|\mathbf{X},\boldsymbol{\beta},\sigma^2)=(2\pi\sigma^2)^{-T/2}\exp\left[-\frac{1}{2\sigma^2}(\mathbf{Y}-\mathbf{X}\boldsymbol{\beta})'(\mathbf{Y}-\mathbf{X}\boldsymbol{\beta})\right]$$

Using the fact that

$$(\mathbf{Y}-\mathbf{X}\boldsymbol{\beta})'(\mathbf{Y}-\mathbf{X}\boldsymbol{\beta})=(\mathbf{Y}-\mathbf{X}\mathbf{b}+\mathbf{X}\mathbf{b}-\mathbf{X}\boldsymbol{\beta})'(\mathbf{Y}-\mathbf{X}\mathbf{b}+\mathbf{X}\mathbf{b}-\mathbf{X}\boldsymbol{\beta})$$
$$=(\mathbf{Y}-\mathbf{X}\mathbf{b})'(\mathbf{Y}-\mathbf{X}\mathbf{b})+(\boldsymbol{\beta}-\mathbf{b})'\mathbf{X}'\mathbf{X}(\boldsymbol{\beta}-\mathbf{b}) \quad (3.5)$$

since $\mathbf{X}'(\mathbf{Y}-\mathbf{X}\mathbf{b})=\mathbf{0}$, we may write the likelihood function as

$$L(\boldsymbol{\beta},\sigma^2;\mathbf{Y},\mathbf{X})$$
$$\propto(\sigma^2)^{-T/2}\exp\left[-\frac{1}{2\sigma^2}(\mathbf{Y}-\mathbf{X}\mathbf{b})'(\mathbf{Y}-\mathbf{X}\mathbf{b})\right]\exp\left[-\frac{1}{2\sigma^2}(\boldsymbol{\beta}-\mathbf{b})'\mathbf{X}'\mathbf{X}(\boldsymbol{\beta}-\mathbf{b})\right] \quad (3.6)$$

which attains its maximum at

$$\boldsymbol{\beta}=\mathbf{b}, \qquad \sigma^2=\frac{(\mathbf{Y}-\mathbf{X}\mathbf{b})'(\mathbf{Y}-\mathbf{X}\mathbf{b})}{T}=\hat{\sigma}^2.$$

Summarizing these results in a theorem:

THEOREM 3.3 (MAXIMUM LIKELIHOOD ESTIMATOR). *Assuming that* $\mathbf{X}'\mathbf{X}$ *is invertible and that* $\mathbf{u}$ *is normally distributed with mean vector* $\mathbf{0}_T$ *and variance matrix* $\sigma^2\mathbf{I}_T$, *the least-squares estimator* $\mathbf{b}$ *is the maximum likelihood estimator and is normally distributed with mean* $\boldsymbol{\beta}$ *and variance* $\sigma^2(\mathbf{X}'\mathbf{X})^{-1}$.

ESTIMATING σ^2

The maximum likelihood estimator of σ^2 is $\hat{\sigma}^2=(\mathbf{Y}-\mathbf{X}\mathbf{b})'(\mathbf{Y}-\mathbf{X}\mathbf{b})/T$. With $\mathbf{M}_x=\mathbf{I}-\mathbf{X}(\mathbf{X}'\mathbf{X})^{-1}\mathbf{X}'$ this estimator can be written as $\hat{\sigma}^2=\mathbf{Y}'\mathbf{M}_x\mathbf{Y}/T=\mathbf{u}\mathbf{M}_x\mathbf{u}/T$, since $\mathbf{M}_x\mathbf{X}=\mathbf{0}$. The expected value of $\hat{\sigma}^2$ is $E(\hat{\sigma}^2)=E(\mathbf{u}'\mathbf{M}_x\mathbf{u}/T)$ $=E(tr[\mathbf{u}'\mathbf{M}_x\mathbf{u}]/T)=E(tr[\mathbf{M}_x\mathbf{u}\mathbf{u}']/T)=\sigma^2 tr\mathbf{M}_x/T=\sigma^2(T-tr[\mathbf{X}(\mathbf{X}'\mathbf{X})^{-1}\mathbf{X}'])$ $/T=\sigma^2(T-tr[(\mathbf{X}'\mathbf{X})^{-1}\mathbf{X}'\mathbf{X}])/T=\sigma^2(T-k)/T$. Therefore $\hat{\sigma}^2$ is biased estimator, $E(\hat{\sigma}^2)\neq\sigma^2$. The unbiased estimator is

$$s^2\equiv\frac{(\mathbf{Y}-\mathbf{X}\mathbf{b})'(\mathbf{Y}-\mathbf{X}\mathbf{b})}{T-k}.$$

CONFIDENCE REGIONS AND HYPOTHESIS TESTS

The construction of confidence intervals and hypothesis tests makes use of the following result.

LEMMA. *If the* $p\times1$ *vector* $\mathbf{z}$ *is normally distributed with mean* $E(\mathbf{z})$ *and variance* $V(\mathbf{z})$, *then the quantity* $[\mathbf{z}-E\mathbf{z}]'[V(\mathbf{z})]^{-1}[\mathbf{z}-E\mathbf{z}]$ *is the sum of squares of* p *independent, standard, normal, random variables, which by definition has a chi-square distribution on* p *degrees of freedom.*

The proof of this lemma follows straightforwardly from the observation that there exists a matrix $\mathbf{C}$ such that $\mathbf{C}V(\mathbf{z})\mathbf{C}'$ is the identity matrix, and therefore, $\mathbf{C}\mathbf{z}-\mathbf{C}E(\mathbf{z})$ has mean vector zero and an identity variance matrix. Thus $[\mathbf{C}\mathbf{z}-\mathbf{C}E(\mathbf{z})]'[\mathbf{C}\mathbf{z}-\mathbf{C}E(\mathbf{z})]=[\mathbf{z}-E(\mathbf{z})]'\mathbf{C}'\mathbf{C}[\mathbf{z}-E(\mathbf{z})]=[\mathbf{z}-E(\mathbf{z})][V^{-1}(\mathbf{z})][\mathbf{z}-E(\mathbf{z})]$ is the sum of squares of p independent standard normal random variables.

Partitioning the matrices conformably,

$$\mathbf{b}'=(\mathbf{b}_I',\mathbf{b}_J')$$

$$\boldsymbol{\beta}'=(\boldsymbol{\beta}_I',\boldsymbol{\beta}_J')$$

$$(\mathbf{X}'\mathbf{X})^{-1}=\begin{bmatrix}\left[(\mathbf{X}'\mathbf{X})^{-1}\right]_{II} & \left[(\mathbf{X}'\mathbf{X})^{-1}\right]_{IJ}\\ \left[(\mathbf{X}'\mathbf{X})^{-1}\right]_{JI} & \left[(\mathbf{X}'\mathbf{X})^{-1}\right]_{JJ}\end{bmatrix},$$

and making use of the fact that marginal distributions of multivariate normal random variables are themselves normal, we have the following consequence of this lemma.

THEOREM 3.4 (THE CHI-SQUARE TEST STATISTIC). *The scalar random variable*

$$\left[\mathbf{b}_I-\boldsymbol{\beta}_I\right]'\left[\sigma^2(\mathbf{X}'\mathbf{X})^{-1}\right]_{II}^{-1}\left[\mathbf{b}_I-\boldsymbol{\beta}_I\right]=\chi^2 \tag{3.6}$$

has a chi-square distribution with p degrees of freedom, where p is the dimension of $\mathbf{b}_I$.

Thus if $\chi^2_\alpha(p)$ is the upper α percentage point of the chi-square distribution with p degrees of freedom, then the ellipsoid [1]

$$\left[\mathbf{b}_I-\boldsymbol{\beta}_I\right]'\left[\sigma^2(\mathbf{X}'\mathbf{X})^{-1}\right]_{II}^{-1}\left[\mathbf{b}_I-\boldsymbol{\beta}_I\right]\leqslant\chi^2_\alpha(p) \tag{3.7}$$

is a $1-\alpha\%$ confidence region for $\boldsymbol{\beta}$. Similarly, an α-level test of the point null hypothesis $\boldsymbol{\beta}_I=\boldsymbol{\beta}_I^*$ versus the alternative $\boldsymbol{\beta}_I\neq\boldsymbol{\beta}_I^*$, would reject the null hypothesis if $\boldsymbol{\beta}_I=\boldsymbol{\beta}_I^*$ were outside this region.

If σ^2 is unknown, these statements, although still true, lose their usefulness, since the regions described are functions of σ^2. It seems natural to use some estimate in place of the unknown σ^2. Remarkably enough, for the normal linear regression model this is approximately correct. To show that, we need the following result.

[1] Actually, this is an ellipsoidal cylinder with $\boldsymbol{\beta}_J$ unconstrained.

THEOREM 3.5. *The quantity* $(\mathbf{Y}-\mathbf{Xb})'(\mathbf{Y}-\mathbf{Xb})/\sigma^2=(T-k)s^2/\sigma^2$ *has a chi-square distribution with* $T-k$ *degrees of freedom and is independent of* $\boldsymbol{\beta}-\mathbf{b}$.

The residual sum of squares $(\mathbf{Y}-\mathbf{Xb})'(\mathbf{Y}-\mathbf{Xb})$ can be written as $\mathbf{Y}'\mathbf{M}_x\mathbf{Y}$ where $\mathbf{M}_x=\mathbf{I}_T-\mathbf{X}(\mathbf{X}'\mathbf{X})^{-1}\mathbf{X}'$. Making use of $\mathbf{M}_x\mathbf{X}=\mathbf{0}$, the residual sum of squares can also be written

$$(\mathbf{Y}-\mathbf{Xb})'(\mathbf{Y}-\mathbf{Xb})=\mathbf{u}'\mathbf{M}_x\mathbf{u}.$$

The idempotent matrix $\mathbf{M}_x$ is shown in Appendix 1 to have k characteristic values equal to zero and the remaining $T-k$ equal to one. Thus $\mathbf{u}'\mathbf{M}_x\mathbf{u}$ can be written as $\mathbf{u}'\mathbf{C}'\boldsymbol{\Lambda}\mathbf{Cu}=\mathbf{u}^{*\prime}\boldsymbol{\Lambda}\mathbf{u}^*$ where $\mathbf{u}^*$ is normally distributed with mean zero and variance matrix $\sigma^2\mathbf{I}_T$, and where $\boldsymbol{\Lambda}$ is a diagonal matrix with k zero diagonal elements and $T-k$ elements equal to one. We have thus written $(\mathbf{Y}-\mathbf{Xb})'(\mathbf{Y}-\mathbf{Xb})/\sigma^2$ as the sum of squares of $T-k$ independent, standard, normal, random variables, which by definition is distributed chi-square with $T-k$ degrees of freedom.

The independence of the two random variables in the theorem follows from the independence of $\mathbf{Y}-\mathbf{Xb}=\mathbf{M}_x\mathbf{u}$ and $(\boldsymbol{\beta}-\mathbf{b})=(\mathbf{X}'\mathbf{X})^{-1}\mathbf{X}'\mathbf{u}$. Given the normality assumption, these quantities may be shown to be independent by computing their covariance $E[\mathbf{M}_x\mathbf{u}][(\mathbf{X}'\mathbf{X})^{-1}\mathbf{X}'\mathbf{u}]'=\sigma^2\mathbf{M}_x\mathbf{X}(\mathbf{X}'\mathbf{X})^{-1}=\mathbf{0}$.

Definition. If the random variables χ_1^2 and χ_2^2 are independent and have chi-square distributions with p_1 and p_2 degrees of freedom, respectively, then the ratio

$$F=\frac{(\chi_1^2/p_1)}{(\chi_2^2/p_2)}$$

has an F-distribution with degrees of freedom parameters p_1 and p_2.

THEOREM 3.6 (THE F-TEST STATISTIC). *Assuming that* $\mathbf{X}'\mathbf{X}$ *is invertible and that* $\mathbf{u}$ *is normally distributed with mean vector* $\mathbf{0}_T$ *and variance matrix* $\sigma^2\mathbf{I}_T$, *the quantity*

$$\frac{[\mathbf{b}_I-\boldsymbol{\beta}_I]'\left[(\mathbf{X}'\mathbf{X})^{-1}\right]_{II}^{-1}[\mathbf{b}_I-\boldsymbol{\beta}_I]}{ps^2}=F \tag{3.8}$$

has an F distribution with p and $T-k$ *degrees of freedom, where p is the dimension of* $\mathbf{b}_I$.

This is a direct consequence of Theorems 3.4 and 3.5 and should be compared with Theorem 3.4, which applies when σ^2 is known. The statement implies $1-\alpha\%$ confidence ellipsoids of the form

$$[\mathbf{b}_I-\boldsymbol{\beta}_I]'\left[(\mathbf{X}'\mathbf{X})^{-1}\right]_{II}^{-1}[\mathbf{b}_I-\boldsymbol{\beta}_I] \leqslant ps^2F_\alpha(p,T-k)$$

where p is the dimension of $\mathbf{b}_I$; $p \leqslant k$, and $F_\alpha(p,T-k)$ is the upper α percentage point of an F distribution with p and $T-k$ degrees of freedom. In the special case when $p=1$, $F^{1/2}$ is said to have a Student's t distribution, and this ellipsoid becomes a confidence interval for a particular coefficient

$$|b_i-\beta_i| \leqslant s\left[(\mathbf{X}'\mathbf{X})^{-1}\right]_{ii}^{-1} t_\alpha(T-k).$$

Furthermore, an α-level test of the hypothesis $\boldsymbol{\beta}_I=\boldsymbol{\beta}_I^*$ versus the alternative $\boldsymbol{\beta}_I \neq \boldsymbol{\beta}_I^*$ would reject the hypothesis if $\boldsymbol{\beta}_I^*$ were not within the confidence region.

The quadratic form (3.3) can be written in another informative way. We may partition $\mathbf{X}'\mathbf{X}$ as above,

$$\mathbf{X}'\mathbf{X}=\begin{bmatrix} \mathbf{X}_I'\mathbf{X}_I & \mathbf{X}_I'\mathbf{X}_J \\ \mathbf{X}_J'\mathbf{X}_I & \mathbf{X}_J'\mathbf{X}_J \end{bmatrix}$$

and, by the partitioned inverse rule, we have

$$\left[(\mathbf{X}'\mathbf{X})^{-1}\right]_{II}^{-1}=\mathbf{X}_I'\mathbf{X}_I-\mathbf{X}_I'\mathbf{X}_J(\mathbf{X}_J'\mathbf{X}_J)^{-1}\mathbf{X}_J'\mathbf{X}_I.$$

Let the minimum error sum of squares be

$$ESS=(\mathbf{Y}-\mathbf{X}\mathbf{b})'(\mathbf{Y}-\mathbf{X}\mathbf{b}),$$

and the constrained minimum error sum of squares with constraint $\boldsymbol{\beta}_I=\boldsymbol{\beta}_I^*$ be

$$ESS_0=(\mathbf{Y}-\mathbf{X}_I\boldsymbol{\beta}_I^*-\mathbf{X}_J\mathbf{b}_J^*)'(\mathbf{Y}-\mathbf{X}_I\boldsymbol{\beta}_I^*-\mathbf{X}_J\mathbf{b}_J^*)$$

with $\mathbf{b}_J^*=(\mathbf{X}_J'\mathbf{X}_J)^{-1}\mathbf{X}_J'(\mathbf{Y}-\mathbf{X}_I\boldsymbol{\beta}_I^*)$. By some tedious manipulations we may show that

$$(\boldsymbol{\beta}_I^*-\mathbf{b}_I)'\left[(\mathbf{X}'\mathbf{X})^{-1}\right]_{II}^{-1}(\boldsymbol{\beta}_I^*-\mathbf{b}_I)=ESS_0-ESS \tag{3.9}$$

and therefore the numerator of the F statistic is just the increase in the error sum of squares implied by the restriction $\boldsymbol{\beta}_I=\boldsymbol{\beta}_I^*$

$$F=\frac{ESS_0-ESS}{ps^2} \tag{3.10}$$

Furthermore, using the definition of R^2 given by Equation (3.3) and letting R^2 correspond to the unconstrained regression and R_0^2 correspond to the constrained regression, the F statistic may also be written as

$$F=\left(\frac{R^2-R_0^2}{1-R^2}\right)\left(\frac{T-k}{p}\right) \tag{3.11}$$

CONSTRAINED ESTIMATION

The consequences of constraints other than $\boldsymbol{\beta}_I=\boldsymbol{\beta}_I^*$ are easy to compute. Minimization of the sum of squares $(\mathbf{Y}-\mathbf{X}\boldsymbol{\beta})'(\mathbf{Y}-\mathbf{X}\boldsymbol{\beta})$ subject to restriction $\mathbf{R}\boldsymbol{\beta}=\mathbf{r}$ is a Lagrangian problem requiring the derivatives of the following expression to be set to zero.

$$(\mathbf{Y}-\mathbf{X}\boldsymbol{\beta})'(\mathbf{Y}-\mathbf{X}\boldsymbol{\beta})-2\boldsymbol{\lambda}'(\mathbf{R}\boldsymbol{\beta}-\mathbf{r}).$$

That is

$$\mathbf{R}\boldsymbol{\beta}-\mathbf{r}=\mathbf{0} \tag{3.12}$$

$$\mathbf{X}'(\mathbf{Y}-\mathbf{X}\boldsymbol{\beta})-\mathbf{R}'\boldsymbol{\lambda}=\mathbf{0}. \tag{3.13}$$

Premultiplying (3.13) by $\mathbf{R}(\mathbf{X}'\mathbf{X})^{-1}$ yields

$$\mathbf{R}(\mathbf{X}'\mathbf{X})^{-1}\mathbf{X}'\mathbf{Y}-\mathbf{R}(\mathbf{X}'\mathbf{X})^{-1}(\mathbf{X}'\mathbf{X})\boldsymbol{\beta}=\mathbf{R}(\mathbf{X}'\mathbf{X})^{-1}\mathbf{R}'\boldsymbol{\lambda}$$

and using (3.12)

$$\boldsymbol{\lambda}=\left(\mathbf{R}(\mathbf{X}'\mathbf{X})^{-1}\mathbf{R}'\right)^{-1}(\mathbf{Rb}-\mathbf{r}),$$

which can be reinserted into (3.13) to yield, finally, the constrained least-squares value

$$\begin{aligned}\hat{\boldsymbol{\beta}}(\mathbf{R},\mathbf{r})&=(\mathbf{X}'\mathbf{X})^{-1}\mathbf{X}'\mathbf{Y}-(\mathbf{X}'\mathbf{X})^{-1}\mathbf{R}'\left(\mathbf{R}(\mathbf{X}'\mathbf{X})^{-1}\mathbf{R}'\right)^{-1}(\mathbf{Rb}-\mathbf{r})\\&=\mathbf{b}-(\mathbf{X}'\mathbf{X})^{-1}\mathbf{R}'\left(\mathbf{R}(\mathbf{X}'\mathbf{X})^{-1}\mathbf{R}'\right)^{-1}(\mathbf{Rb}-\mathbf{r}).\end{aligned} \tag{3.14}$$

Using this value of $\hat{\boldsymbol{\beta}}$ in the sum of squares expression we have

$$(\mathbf{Y}-\mathbf{X}\hat{\boldsymbol{\beta}})'(\mathbf{Y}-\mathbf{X}\hat{\boldsymbol{\beta}})=(\mathbf{Y}-\mathbf{Xb})'(\mathbf{Y}-\mathbf{Xb})+(\mathbf{Rb}-\mathbf{r})'\left(\mathbf{R}(\mathbf{X}'\mathbf{X})^{-1}\mathbf{R}'\right)^{-1}(\mathbf{Rb}-\mathbf{r}) \tag{3.15}$$

where the last term in the last line is, therefore, the increase in the error sum of squares implied by the restriction $\mathbf{R}\boldsymbol{\beta}=\mathbf{r}$. By inspection, this establishes the validity of Equation (3.9) with $\mathbf{r}=\boldsymbol{\beta}_I^*$ and $\mathbf{R}=(\mathbf{I}_p,\mathbf{0})$. Computation of the mean and variance of $\hat{\boldsymbol{\beta}}$ is left as an exercise.

TREATMENT OF THE CONSTANT TERM

A regression function ordinarily has a constant as one of the "variables." Equivalently, one of the columns of the $\mathbf{X}$ matrix is a vector of ones. Although everything that has been said still applies, it is informative to treat the constant somewhat differently.

Let us remove from the $\mathbf{X}$ matrix the vector of ones and write the regression as

$$\mathbf{Y}=\mathbf{X}\boldsymbol{\beta}+\mathbf{1}_T\alpha+\mathbf{u}$$

where $\mathbf{1}_T$ is a vector of ones and α is the scalar constant. The error sum of squares may be minimized with respect to α as

$$0=\frac{\partial}{\partial\alpha}(\mathbf{Y}-\mathbf{X}\boldsymbol{\beta}-\mathbf{1}_T\alpha)'(\mathbf{Y}-\mathbf{X}\boldsymbol{\beta}-\mathbf{1}_T\alpha)$$
$$=-2\mathbf{1}_T'(\mathbf{Y}-\mathbf{X}\boldsymbol{\beta})-2\mathbf{1}_T'\mathbf{1}_T\alpha$$

Thus as a function of $\boldsymbol{\beta}$, the least-squares estimate of α is

$$a(\boldsymbol{\beta})=\mathbf{1}_T'(\mathbf{Y}-\mathbf{X}\boldsymbol{\beta})(\mathbf{1}_T'\mathbf{1}_T)^{-1}.$$

Inserting this value into the error sum of squares yields

$$\left(\mathbf{Y}-\mathbf{X}\boldsymbol{\beta}-\mathbf{1}_T\mathbf{1}_T'[\mathbf{Y}-\mathbf{X}\boldsymbol{\beta}]T^{-1}\right)'\left(\mathbf{Y}-\mathbf{X}\boldsymbol{\beta}-\mathbf{1}_T\mathbf{1}_T'[\mathbf{Y}-\mathbf{X}\boldsymbol{\beta}]T^{-1}\right)$$
$$=(\mathbf{MY}-\mathbf{MX}\boldsymbol{\beta})'(\mathbf{MY}-\mathbf{MX}\boldsymbol{\beta})$$

where $\mathbf{M}=\mathbf{I}_T-\mathbf{1}_T(\mathbf{1}_T'\mathbf{1}_T)^{-1}\mathbf{1}_T'$. The least-squares estimate of $\boldsymbol{\beta}$ can, therefore, be computed by transforming the observations $(\mathbf{Y},\mathbf{X})$ by the matrix $\mathbf{M}$ and minimizing the transformed error sum of squares. It is straightforward to show that this yields

$$\mathbf{b}=(\mathbf{X}'\mathbf{M}'\mathbf{MX})^{-1}(\mathbf{X}'\mathbf{M}'\mathbf{MY}).$$

But now notice the effect of the matrix $\mathbf{M}$ on a vector $\mathbf{Y}$:

$$\mathbf{MY}=\mathbf{Y}-\mathbf{1}_T(\mathbf{1}_T'\mathbf{Y})T^{-1}=\mathbf{Y}-\mathbf{1}_T\bar{Y}$$

where $\bar{Y}$ is the average value of the observed variables. In words, the matrix $\mathbf{M}$ subtracts out the mean of a variable, and the least-squares estimate of the slope coefficients can be computed by first subtracting the means from all the variables and then computing the least-squares value in the usual way.

Parenthetically, if the regression process is written $\mathbf{Y}=\mathbf{X}\boldsymbol{\beta}+\mathbf{Z}\boldsymbol{\gamma}+\mathbf{u}$ where $\mathbf{X}$ and $\mathbf{Z}$ are observable matrices, the least-squares estimate of $\boldsymbol{\beta}$ is $\mathbf{b}=(\mathbf{X}'\mathbf{M}_z'\mathbf{M}_z\mathbf{X})^{-1}\mathbf{X}'\mathbf{M}_z\mathbf{M}_z\mathbf{Y}=(\mathbf{X}'\mathbf{M}_z\mathbf{X})^{-1}\mathbf{X}'\mathbf{M}_z\mathbf{Y}$ where $\mathbf{M}_z=\mathbf{I}_T-\mathbf{Z}(\mathbf{Z}'\mathbf{Z})^{-1}\mathbf{Z}'$. The residual vector formed from the regression of $\mathbf{Y}$ on $\mathbf{Z}$ is $\mathbf{Y}-\mathbf{Z}(\mathbf{Z}'\mathbf{Z})^{-1}\mathbf{Z}'\mathbf{Y}=\mathbf{M}_z\mathbf{Y}$ and the residual matrix formed from the regression of each of the columns of $\mathbf{X}$ on $\mathbf{Z}$ is $\mathbf{X}-\mathbf{Z}(\mathbf{Z}'\mathbf{Z})^{-1}\mathbf{Z}'\mathbf{X}=\mathbf{M}_z\mathbf{X}$. The least-squares estimate $\mathbf{b}$ is then just the regression of the residual vector $\mathbf{M}_z\mathbf{Y}$ on the residual matrix $\mathbf{M}_z\mathbf{X}$.

THE ADJUSTED R^2 AND REGRESSION SELECTION STRATEGIES

The R^2 given by (3.3) has been suggested as a measure of the "completeness" of a theory. For that function it has the defect that if there are as many independent variables as there are observations, the R^2 takes on the value one. Moreover, as variables are added to the equation, the R^2 necessarily increases. This creates an unwarranted preference for models with many explanatory variables, and some adjustment for the number of explanatory variables seems desirable. An analogous problem occurs when estimating σ^2. The maximum likelihood estimator of σ^2 is biased upward, and the unbiased estimator, $s^2 = ESS/(T-k)$, is usually preferred.[2] If we think of $1-R^2$ as estimating σ^2/σ_y^2 it is by the same logic natural to define the adjusted R^2 by the equation

$$1-\bar{R}^2 \equiv \frac{s^2}{s_y^2} = \frac{ESS/(T-k)}{\Sigma(y_t-\bar{y})^2/T-1}. \tag{3.16}$$

The relationship between R^2 and $\bar{R}^2$ is then

$$1-\bar{R}^2 = \frac{(T-1)}{(T-k)}(1-R^2). \tag{3.17}$$

No book on specification searches would be complete without the following two results. The first, due to Theil (1971, p. 543), has been used to justify search strategies that maximize the $\bar{R}^2$. The second defines a simple algorithm for increasing the $\bar{R}^2$.

Theorem 3.7 (Expected $\bar{R}^2$). *Given two normal regression models, one of which is assumed to be true, the expected value of s^2 for the true model is less than or equal to the expected value of s^2 for the other model.*

To prove this, suppose that $\mathbf{Y} = \mathbf{X}\boldsymbol{\beta} + \mathbf{u}$ is the true model and $\mathbf{Y} = \mathbf{Z}\gamma + \mathbf{u}$ is the alternative, where $\mathbf{X}$ and $\mathbf{Z}$ are $T \times k_x$ and $T \times k_z$ matrices. Then

$$s_x^2 = (T-k_x)^{-1}\mathbf{Y}'\mathbf{M}_x\mathbf{Y}, \qquad s_z^2 = (T-k_z)^{-1}\mathbf{Y}'\mathbf{M}_z\mathbf{Y}$$

where

$$\mathbf{M}_x = \mathbf{I} - \mathbf{X}(\mathbf{X}'\mathbf{X})^{-1}\mathbf{X}' \quad \text{and} \quad \mathbf{M}_z = \mathbf{I} - \mathbf{Z}(\mathbf{Z}'\mathbf{Z})^{-1}\mathbf{Z}'.$$

The expected error sum of squares for the false model is

$$\begin{aligned} E(\mathbf{Y}'\mathbf{M}_z\mathbf{Y}|\boldsymbol{\beta}) &= E((\mathbf{X}\boldsymbol{\beta}+\mathbf{u})'\mathbf{M}_z(\mathbf{X}\boldsymbol{\beta}+\mathbf{u})|\boldsymbol{\beta}) \\ &= \boldsymbol{\beta}'\mathbf{X}'\mathbf{M}_z\mathbf{X}\boldsymbol{\beta} + E(\mathbf{u}'\mathbf{M}_z\mathbf{u}|\boldsymbol{\beta}) = \boldsymbol{\beta}'\mathbf{X}'\mathbf{M}_z\mathbf{X}\boldsymbol{\beta} + (T-k_z)\sigma^2 \\ &\geqslant (T-k_z)\sigma^2. \end{aligned}$$

Thus $E(\mathbf{Y}'\mathbf{M}_z\mathbf{Y}|\boldsymbol{\beta})/(T-k_z) \geqslant \sigma^2 = E(\mathbf{Y}'\mathbf{M}_x\mathbf{Y}|\boldsymbol{\beta})/(T-k_x)$.

Proof of the next result is left to the reader.

Theorem 3.8 (Increasing the $\bar{R}^2$). *Omitting a variable from an equation will increase the $\bar{R}^2$ if and only if the square of the t-statistic (the F) for that coefficient is less than one.*

[2]It is also preferred by this Bayesian, but not of course because of the bias. See section 7.1.

3.2 Pooling Two Samples

Bayesian inference is different from classical inference in that it makes use of information that is not contained in the sample under study. Bayesian theory is concerned with the optimal pooling of sample information with nonsample information. Classical theory can also be used to pool information from more than one source, provided the information is generated by experiments that admit a frequency interpretation.

Suppose, in particular, that a given data set was arbitrarily split into two parts. Then the regression equation could be written as

$$\begin{bmatrix} \mathbf{Y}^* \\ \mathbf{Y} \end{bmatrix} = \begin{bmatrix} \mathbf{X}^* \\ \mathbf{X} \end{bmatrix} \boldsymbol{\beta} + \begin{bmatrix} \mathbf{u}^* \\ \mathbf{u} \end{bmatrix}$$

where the $*$ indicates the first subset of the data. The usual least-squares estimate of $\boldsymbol{\beta}$ would be

$$\mathbf{b}^{**} = \left(\begin{bmatrix} \mathbf{X}^* \\ \mathbf{X} \end{bmatrix}' \begin{bmatrix} \mathbf{X}^* \\ \mathbf{X} \end{bmatrix} \right)^{-1} \begin{bmatrix} \mathbf{X}^* \\ \mathbf{X} \end{bmatrix}' \begin{bmatrix} \mathbf{Y}^* \\ \mathbf{Y} \end{bmatrix} = (\mathbf{X}^{*\prime}\mathbf{X}^* + \mathbf{X}'\mathbf{X})^{-1}(\mathbf{X}^{*\prime}\mathbf{Y}^* + \mathbf{X}'\mathbf{Y})$$

$$= (\mathbf{X}^{*\prime}\mathbf{X}^* + \mathbf{X}'\mathbf{X})^{-1}(\mathbf{X}^{*\prime}\mathbf{X}^*\mathbf{b}^* + \mathbf{X}'\mathbf{X}\mathbf{b}) \tag{3.18}$$

where $\mathbf{b}^* = (\mathbf{X}^{*\prime}\mathbf{X}^*)^{-1}\mathbf{X}^*\mathbf{Y}^*$ and $\mathbf{b} = (\mathbf{X}'\mathbf{X})^{-1}\mathbf{X}'\mathbf{Y}$. In words, the least-squares estimate of $\boldsymbol{\beta}$ is a matrix-weighted average of the pair of estimates computed from each of two subsets of the data, $\mathbf{b}^*$ and $\mathbf{b}$.

Another possibility is that the variance of the residuals in the first part of the data set is different from the variance in the second. Indicating these variances by σ^{*2} and σ^2, respectively, the model may be transformed to make the residual covariance matrix equal the identity matrix by dividing $\mathbf{Y}$ and $\mathbf{X}$ by σ and $\mathbf{Y}^*$ and $\mathbf{X}^*$ by σ^*. The resulting estimate of $\boldsymbol{\beta}$ is

$$\mathbf{b}^{**} = (\sigma^{*-2}\mathbf{X}^{*\prime}\mathbf{X}^* + \sigma^{-2}\mathbf{X}'\mathbf{X})^{-1}(\sigma^{*-2}\mathbf{X}^{*\prime}\mathbf{X}^*\mathbf{b}^* + \sigma^{-2}\mathbf{X}'\mathbf{X}\mathbf{b}). \tag{3.19}$$

The notation just used may seem confusing, but it is chosen to anticipate the Bayesian analysis of the next section, in which personal prior opinions end up being equivalent to the preliminary observations $\mathbf{Y}^*, \mathbf{X}^*$. The prior information is pooled with the current information exactly in accordance with formula (3.19).[3]

The variances σ^2 and σ^{*2} may also be estimated. The likelihood function implied by the model is

$$L(\boldsymbol{\beta}, \sigma^2, \sigma^{*2}; \mathbf{Y}, \mathbf{Y}^*, \mathbf{X}, \mathbf{X}^*) \propto (\sigma^2)^{-T/2} \exp\left[-\frac{1}{2\sigma^2}(\mathbf{Y} - \mathbf{X}\boldsymbol{\beta})'(\mathbf{Y} - \mathbf{X}\boldsymbol{\beta}) \right]$$

$$(\sigma^{*2})^{-T/2} \exp\left[-\frac{1}{2\sigma^2}(\mathbf{Y}^* - \mathbf{X}^*\boldsymbol{\beta})'(\mathbf{Y} - \mathbf{X}\boldsymbol{\beta}) \right].$$

[3] A schizophrenic attitude toward probabilities allows Theil and Goldberger (1961) to introduce prior information into classical inference in this way. An earlier paper of the same genre is Durbin (1953).

The reader may verify that maximizing this likelihood function implies choosing $\boldsymbol{\beta}$ in accord with Equation (3.19) and choosing the variances as

$$\hat{\sigma}^2 = \frac{(\mathbf{Y}-\mathbf{X}\mathbf{b}^{**})'(\mathbf{Y}-\mathbf{X}\mathbf{b}^{**})}{T} \tag{3.20}$$

$$\hat{\sigma}^{*2} = \frac{(\mathbf{Y}^*-\mathbf{X}^*\mathbf{b}^{**})'(\mathbf{Y}^*-\mathbf{X}^*\mathbf{b}^{**})}{T^*}. \tag{3.21}$$

3.3 Bayesian Inference with the Linear-Regression Model

Bayesian inference with the linear-regression model is now introduced and shown to be formally equivalent to the pooling problem just discussed. Bayes' rule applied to the uncertain parameters $(\boldsymbol{\beta},\sigma^2)$ is

$$f(\boldsymbol{\beta},\sigma^2|\mathbf{Y},\mathbf{X}) = \frac{f(\mathbf{Y},\mathbf{X}|\boldsymbol{\beta},\sigma^2)f(\boldsymbol{\beta},\sigma^2)}{f(\mathbf{Y},\mathbf{X})},$$

where $\mathbf{Y}$ and $\mathbf{X}$ are observable data. The linear-regression model $\mathbf{Y}=\mathbf{X}\boldsymbol{\beta}+\mathbf{u}$, where $\mathbf{u}$ is normally distributed with mean $\mathbf{0}$ and variance $\sigma^2\mathbf{I}$, implies a conditional distribution for $\mathbf{Y}$: $f_N(\mathbf{Y}|\mathbf{X}\boldsymbol{\beta},\sigma^2\mathbf{I})$, indicating that $\mathbf{Y}$ is normally distributed with mean $\mathbf{X}\boldsymbol{\beta}$ and variance $\sigma^2\mathbf{I}$. If, furthermore, $\mathbf{X}$ is distributed independently of $\boldsymbol{\beta}$ and σ^2, the posterior distribution can be written as

$$f(\boldsymbol{\beta},\sigma^2|\mathbf{Y},\mathbf{X}) = \frac{f_N(\mathbf{Y}|\mathbf{X}\boldsymbol{\beta},\sigma^2\mathbf{I})f(\mathbf{X})f(\boldsymbol{\beta},\sigma^2)}{f(\mathbf{Y}|\mathbf{X})f(\mathbf{X})}$$

$$= \frac{f_N(\mathbf{Y}|\mathbf{X}\boldsymbol{\beta},\sigma^2\mathbf{I})f(\boldsymbol{\beta},\sigma^2)}{f(\mathbf{Y}|\mathbf{X})}.$$

Before discussing the choice of prior, $f(\boldsymbol{\beta},\sigma^2)$, a few things may be said about the immediately preceding formula. It implies that randomness in the matrix $\mathbf{X}$ is irrelevant for inference if $\mathbf{X}$ is distributed independently of $\boldsymbol{\beta}$ and σ^2. Classical inference, in contrast, cannot easily ignore the randomness in $\mathbf{X}$, since sampling properties are necessarily affected. Another important point is that both the vectors of unobservables, $\boldsymbol{\beta}$ and $\mathbf{u}$, are assigned personal probability distributions. Classically, the residual vector $\mathbf{u}$ is thought to have a frequency distribution and the parameter vector $\boldsymbol{\beta}$ is thought to be a fixed vector of constants. But from the Bayesian point of view, your personal opinion about both $\boldsymbol{\beta}$ and $\mathbf{u}$ is described in probabilistic terms, and the extreme distinction between $\boldsymbol{\beta}$ and $\mathbf{u}$ is regarded as unwarranted.

Having assumed that the residual vector is normally distributed, only the choice of a prior distribution for $(\boldsymbol{\beta},\sigma^2)$ remains. Two alternatives are suggested here. The resulting posterior distributions are described in the following pair of theorems, and further discussion follows. Both results make use of the assumption that $\mathbf{Y}$ given $\mathbf{X}$, $\boldsymbol{\beta}$ and σ^2 is normal with mean $\mathbf{X}\boldsymbol{\beta}$ and variance $\sigma^2\mathbf{I}$.

CONJUGATE PRIOR

THEOREM 3.9. *Given the normal linear-regression model and a normal-gamma prior distribution for* $(\boldsymbol{\beta},\sigma^{-2})$:

$$f(\boldsymbol{\beta},\sigma^{-2})=f_N\left(\boldsymbol{\beta}|\mathbf{b}^*,\sigma^2(\mathbf{N}^*)^{-1}\right)f_\gamma(\sigma^{-2}|s^{*2},\nu^*),$$

which indicates that $\boldsymbol{\beta}$ *given* σ^2 *is distributed normally with mean* $\mathbf{b}^*$ *and variance* $\sigma^2(\mathbf{N}^*)^{-1}$ *and that* σ^{-2} *is distributed gamma with parameters* s^{*2} *and* ν^*, *then the posterior distribution of* $(\boldsymbol{\beta},\sigma^{-2})$ *is in the normal-gamma family with parameters*

$$\mathbf{b}^{**}=(\mathbf{N}^*+\mathbf{X}'\mathbf{X})^{-1}(\mathbf{N}^*\mathbf{b}^*+\mathbf{X}'\mathbf{X}\mathbf{b}) \tag{3.22}$$

$$\mathbf{N}^{**}=\mathbf{N}^*+\mathbf{X}'\mathbf{X} \tag{3.23}$$

$$\nu^{**}=\nu^*+T \tag{3.24}$$

$$(s^{**})^2=\left[\nu^{**}\right]^{-1}\left[\nu^*s^{*2}+ESS+(\mathbf{b}-\mathbf{b}^*)'\mathbf{N}^*(\mathbf{X}'\mathbf{X}+\mathbf{N}^*)^{-1}\mathbf{X}'\mathbf{X}(\mathbf{b}-\mathbf{b}^*)\right]. \tag{3.25}$$

The marginal posterior distribution of $\boldsymbol{\beta}$ *is multivariate Student with parameters* $\mathbf{b}^{**}$, $(s^{**})^2(\mathbf{N}^{**})^{-1}$, *and* ν^{**}.

Proof: Let $h=\sigma^{-2}$ and use the likelihood function (3.6) together with the prior to obtain

$$f(\boldsymbol{\beta},h|\mathbf{Y},\mathbf{X})\propto h^{T/2}\exp\left[-\tfrac{1}{2}hESS\right]\exp\left[-\tfrac{1}{2}h(\boldsymbol{\beta}-\mathbf{b})'\mathbf{X}'\mathbf{X}(\boldsymbol{\beta}-\mathbf{b})\right]$$

$$h^{k/2}\exp\left[-\tfrac{1}{2}h(\boldsymbol{\beta}-\mathbf{b}^*)\mathbf{N}^*(\boldsymbol{\beta}-\mathbf{b}^*)\right]h^{\frac{\nu^*}{2}-1}\exp\left[-\tfrac{1}{2}\nu^*s^{*2}h\right].$$

Using result (T10) in Appendix 1, this can be rewritten as

$$f(\boldsymbol{\beta},h|\mathbf{Y},\mathbf{X})\propto h^{k/2}\exp\left[-\tfrac{1}{2}h(\boldsymbol{\beta}-\mathbf{b}^{**})'\mathbf{N}^{**}(\boldsymbol{\beta}-\mathbf{b}^{**})\right]$$

$$h^{\frac{\nu^{**}}{2}-1}\exp\left[-\tfrac{1}{2}\nu^{**}s^{**2}h\right],$$

which by inspection is the normal-gamma distribution described in the theorem above.

The normal-gamma prior is Raiffa and Schlaifer's (1961) "natural conjugate" prior. The member of this class of distributions that is sometimes used to represent prior ignorance is

$$f(\boldsymbol{\beta},\sigma^{-2})\propto(\sigma^{-2})^{k/2}$$

with $\mathbf{N}^*=\mathbf{0}$ and $\nu^*=0$. Given these prior parameters, the posterior parameters become **b**, **X′X**, T, and ESS/T. The marginal posterior distribution of $\boldsymbol{\beta}$, therefore, reproduces the classical result, except that the degrees of freedom parameter is T instead of $T-k$, a point further discussed below.

Observe also that the posterior mean (3.22) is formally the same as Equation (3.18), which describes the pooling of two samples of the same regression process. Thus the normal gamma prior is equivalent to a previous sample of the same process with estimate equal to $\mathbf{b}^*$, "**X′X**" matrix equal to $\mathbf{N}^*$, degrees of freedom equal to ν^*, and error sum of squares equal to $\nu^* s^{*2}$.

The posterior distribution for $\boldsymbol{\beta}$ associated with this normal-gamma prior has the dual features that it is unimodal and located at a fixed, weighted average of the sample location **b** and the prior location $\mathbf{b}^*$. In that sense it never distinguishes sample information from prior information, no matter how strong their apparent conflict. This is so because a conjugate prior treats prior information as if it were a previous sample of the same process. It may be argued that most prior information is distinctly different from sample information, and when they are apparently in conflict, the posterior distribution ought to be multimodal with modes at both the sample location and the prior location. A distribution that has this feature is described in the following theorem, which uses a prior suggested by Dickey (1975).

STUDENT PRIOR

THEOREM 3.10 *If $\boldsymbol{\beta}$ has a multivariate Student prior distribution independent of σ^2, and if σ^{-2} has a gamma prior distribution*

$$f(\boldsymbol{\beta},\sigma^{-2}) = f_S^k(\boldsymbol{\beta}|\mathbf{b}^*,\mathbf{H}^{*-1},\nu_\beta^*)\, f_\gamma(\sigma^{-2}|s^{*2},\nu_\sigma^*),$$

then the marginal posterior distribution for $\boldsymbol{\beta}$ is proportional to the product of two Student functions[4]

$$f(\boldsymbol{\beta}|\mathbf{Y}) \propto f_S^k(\boldsymbol{\beta}|\mathbf{b},\mathbf{H}^{-1},\nu_\sigma^{**})\, f_S^k(\boldsymbol{\beta}|\mathbf{b},\mathbf{H}^{*-1},\nu_\beta^*)$$

where **b** *is the least-squares estimate, and*

$$\mathbf{H} = (s^{**})^{-2}\mathbf{X}'\mathbf{X}$$
$$\nu_\sigma^{**} = \nu_\sigma^* + T - k$$
$$(s^{**})^2 = (\nu_\sigma^{**})^{-1}(\nu_\sigma^* s^{*2} + ESS).$$

[4]The product of two Student distributions can be written as a one-dimensional mixture of Student distributions (Dickey, 1975).

Proof: Integrate σ^2 from the product of the likelihood function and prior:

$$f(\boldsymbol{\beta}|\mathbf{Y}) \propto \int_{\sigma^{-2}} f(\mathbf{Y}|\boldsymbol{\beta},\sigma^2) f(\boldsymbol{\beta},\sigma^{-2})\,d\sigma^{-2}$$

$$= \int_{\sigma^{-2}} f_N(\mathbf{Y}|\mathbf{X}\boldsymbol{\beta},\sigma^2 I) f_\gamma(\sigma^{-2}|s^{*2},\nu_\sigma^*) f_S(\boldsymbol{\beta}|\mathbf{b}^*,\mathbf{H}^{*-1},\nu_\beta^*)\,d\sigma^{-2}$$

$$= f_S^T(\mathbf{Y}|\mathbf{X}\boldsymbol{\beta}, s^{*2}\mathbf{I},\nu_\sigma^*) f_S(\boldsymbol{\beta}|\mathbf{b}^*,\mathbf{H}^{*-1},\nu_\beta^*).$$

The Student distribution in $\mathbf{Y}$ can be written as a Student function in $\boldsymbol{\beta}$ as

$$f_S^T(\mathbf{Y}|\mathbf{X}\boldsymbol{\beta},s^{*2}\mathbf{I},\nu_\sigma^*) \propto \left[\nu_\sigma^* s^{*2} + (\mathbf{Y}-\mathbf{X}\boldsymbol{\beta})'(\mathbf{Y}-\mathbf{X}\boldsymbol{\beta})\right]^{-(\nu_\sigma^*+T)/2}$$

$$\propto \left[\nu_\sigma^* s^{*2} + ESS + (\boldsymbol{\beta}-\mathbf{b})'\mathbf{X}'\mathbf{X}(\boldsymbol{\beta}-\mathbf{b})\right]^{-(\nu^{**}+k)/2}$$

$$\propto f_S^k\left(\boldsymbol{\beta}|\mathbf{b}, s^{**2}(\mathbf{X}'\mathbf{X})^{-1},\nu_\sigma^{**}\right)$$

with ν^{**} and s^{**}defined as before.

The difference between the prior of Theorem (3.9) and the prior of Theorem (3.10) is in the latter case $\boldsymbol{\beta}$ and σ^2 are independent. The former conjugate prior implies that if information is obtained about σ^2, opinions about $\boldsymbol{\beta}$ change. Either $\boldsymbol{\beta}$ becomes more uncertain or more certain. Of course, your priors are your own business, but I can say that I tend to prefer to have $\boldsymbol{\beta}$ and σ^2 independent. A counterargument is that if you discover that the process is noisy (σ^2 large), you may come to doubt the validity of your prior information.[5]

As pointed out earlier, the conjugate prior cannot reproduce classical least squares because it does not allow loss of degrees of freedom. The Student-gamma prior with $\mathbf{H}^*=0$ and $\nu_\sigma^*=0$ does exactly reproduce classical least squares with

$$t = \frac{\beta_i - b_i}{s\left[(\mathbf{X}'\mathbf{X})^{-1}\right]_{ii}}$$

given $\mathbf{Y}$, having a t distribution with $T-k$ degrees of freedom.

The natural conjugate prior was shown to be equivalent to a previous sample of the same process. The Student prior for $\boldsymbol{\beta}$, together with the diffuse prior for $\sigma^2(\nu_\sigma^*=0)$, is equivalent to a previous sample of a regression process with a different variance. Bayesian pooling of two such samples without any other information would lead to the "double-Student" posterior of Theorem 3.10. Also, the modes of the posterior distribution

[5]To anticipate Chapter 9, it may make sense to write the regression process as $\mathbf{Y}=\mathbf{X}(\boldsymbol{\beta}+\boldsymbol{\beta}^c)+\mathbf{u}$ where $\boldsymbol{\beta}^c$ reflects the specification error. Although your prior for the true parameter $\boldsymbol{\beta}$ may be independent of σ^2, your prior for $\boldsymbol{\beta}^c$ may not be.

may be found by setting its logarithmic derivatives to zero

$$\mathbf{0} = \partial \ln f(\boldsymbol{\beta}|\mathbf{Y})/\partial\boldsymbol{\beta} = \lambda \mathbf{X}'\mathbf{X}(\boldsymbol{\beta}-\mathbf{b}) + \lambda^* \mathbf{H}^*(\boldsymbol{\beta}-\mathbf{b}^*)$$

where

$$\lambda = \frac{\nu_\sigma^* + T}{\nu_\sigma^* s^{*2} + ESS + (\boldsymbol{\beta}-\mathbf{b})'\mathbf{X}'\mathbf{X}(\boldsymbol{\beta}-\mathbf{b})} \tag{3.26}$$

$$\lambda^* = \frac{\nu_\beta^* + k}{\nu_\beta^* + (\boldsymbol{\beta}-\mathbf{b}^*)'\mathbf{H}^*(\boldsymbol{\beta}-\mathbf{b}^*)}. \tag{3.27}$$

With a suitable choice of ν_σ^*, ν_β^*, and $\mathbf{H}^*$, these equations are equivalent to (3.19), (3.20), and (3.21) which describe pooling information from two different processes.

A GRAPHICAL PRESENTATION

A graphical analysis of inference with the linear regression model is instructive. The graphs now to be discussed apply to the results just reported, but they also apply to a wider class of distributions. We need only assume that $\mathbf{A}\boldsymbol{\beta}$ and $\mathbf{u}$ have independent, spherically symmetric distributions.[6]

A random variable $\mathbf{z}$ is said to have a spherically symmetric distribution if its density function depends only on the length of $\mathbf{z}$: $f(\mathbf{z}) = cg(\mathbf{z}'\mathbf{z})$. Thus the assumption of spherical symmetry implies that the densities of $\mathbf{Y}$ and $\boldsymbol{\beta}$ may be written as

$$f(\mathbf{Y}|\boldsymbol{\beta}) \propto g_u[(\mathbf{Y}-\mathbf{X}\boldsymbol{\beta})'(\mathbf{Y}-\mathbf{X}\boldsymbol{\beta})]$$

$$f(\boldsymbol{\beta}) \propto g_\beta[(\boldsymbol{\beta}-\mathbf{b}^*)'\mathbf{A}'\mathbf{A}(\boldsymbol{\beta}-\mathbf{b}^*)].$$

Using (3.5) and setting $\mathbf{A}'\mathbf{A} = \mathbf{N}^*$, the posterior distribution can be written as

$$f(\boldsymbol{\beta}|\mathbf{Y}) \propto f(\mathbf{Y}|\boldsymbol{\beta}) f(\boldsymbol{\beta})$$

$$\propto g_u[ESS + (\boldsymbol{\beta}-\mathbf{b})'\mathbf{X}'\mathbf{X}(\boldsymbol{\beta}-\mathbf{b})]\, g_\beta[(\boldsymbol{\beta}-\mathbf{b}^*)'\mathbf{N}^*(\boldsymbol{\beta}-\mathbf{b}^*)].$$

The full posterior distribution, of course, depends on the functions g_u and g_β. The mode, however, necessarily lies on a curve. To find the equation for this curve, we may differentiate the logarithm of the posterior

$$\frac{\partial \ln f(\boldsymbol{\beta}|\mathbf{Y})}{\partial\boldsymbol{\beta}} = g_u^{-1} g_u' 2\mathbf{N}(\boldsymbol{\beta}-\mathbf{b}) + g_\beta^{-1} g_\beta' (2\mathbf{N}^*(\boldsymbol{\beta}-\mathbf{b}^*)). \quad (\mathbf{N} = \mathbf{X}'\mathbf{X})$$

[6]Hill (1969).

Setting this equal to zero and solving for $\boldsymbol{\beta}$ yields

$$\boldsymbol{\beta} = (\mathbf{N} + \lambda\mathbf{N}^*)^{-1}(\mathbf{N}\mathbf{b} + \lambda\mathbf{N}^*\mathbf{b}^*) \tag{3.28}$$

where

$$\lambda = \frac{g_\beta^{-1} g_\beta'}{g_u^{-1} g_u'}. \tag{3.29}$$

As λ varies from zero to infinity, Equation (3.28) sweeps out a curve in k-space affectionately called by Dickey (1975) the *curve décolletage*. It is anchored by the least-squares point $\mathbf{b}$ at one end and by the "prior point" $\mathbf{b}^*$ at the other. Choice of a point along the curve depends on the functions g_u and g_β, but spherical symmetry is sufficient to imply the curve.

Geometrically, the curve décolletage is the locus of tangencies between the sample family of ellipsoids, $(\boldsymbol{\beta}-\mathbf{b})'\mathbf{N}(\boldsymbol{\beta}-\mathbf{b}) = c$, and the prior family $(\boldsymbol{\beta}-\mathbf{b}^*)'\mathbf{N}^*(\boldsymbol{\beta}-\mathbf{b}^*) = c^*$. To find this locus of tangencies, set the derivatives of $(\boldsymbol{\beta}-\mathbf{b})'\mathbf{N}(\boldsymbol{\beta}-\mathbf{b}) + \lambda(\boldsymbol{\beta}-\mathbf{b}^*)'\mathbf{N}^*(\boldsymbol{\beta}-\mathbf{b}^*)$ to zero: $2\mathbf{N}(\boldsymbol{\beta}-\mathbf{b}) + 2\lambda\mathbf{N}^*(\boldsymbol{\beta}-\mathbf{b}^*) = \mathbf{0}$, which is just Equation (3.28).

A curve décolletage is illustrated in Figure 3.1. Ellipses around the least-squares point $\mathbf{b}$ are isolikelihood ellipses. The data prefer the point $\mathbf{b}$, and the data are indifferent between any points on a given likelihood ellipse. The most preferred point from the standpoint of the prior is $\mathbf{b}^*$, and an ellipse around the prior point is a prior isodensity ellipse. The curve décolletage contains all points jointly preferred by data and prior. Given any point not on the curve, there is a better point on the curve, in the sense that neither the likelihood value nor the prior density is less, and at least one is greater.[7]

The location of modes on the curve décolletage depends on the functions g_u and g_β. One possibility is the exponential family indexed by the precision parameter h:

$$g(z^2|h) = e^{-\frac{1}{2}hz^2}$$

with

$$g'g^{-1} = -\frac{h}{2}.$$

Thus letting the data and prior densities have different precision parameters, $g_u(z^2) = g(z^2|h)$ and $g_\beta(z^2) = g(z^2|h^*)$, we would have, from (3.29), $\lambda = h^*/h$; the posterior mode, since λ is independent of $\boldsymbol{\beta}$, is just (3.28), using this value of λ.

[7]Economists are reminded of the analogous Edgeworth–Bowley diagram. This analogy is pursued in Chapter 5, and the curve décolletage is there referred to as the information contract curve.

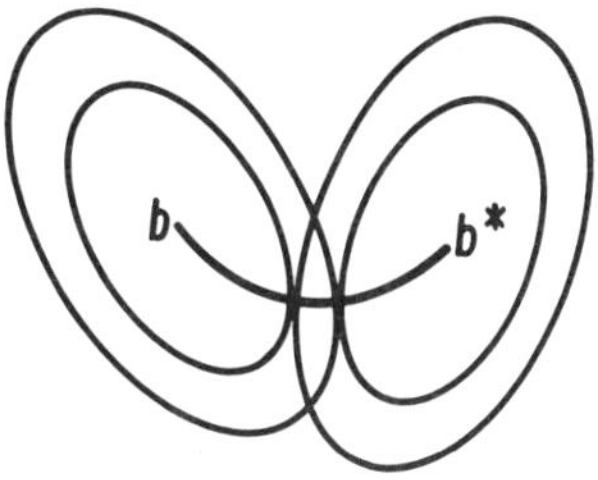

Fig. 3.1 Curve décolletage.

Describing this more traditionally, we have just assumed that the data and prior densities are normal $\mathbf{u} \sim N(0, h^{-1}(\mathbf{B}'\mathbf{B})^{-1})$, $\boldsymbol{\beta} \sim N(\mathbf{b}^*, h^{*-1}\mathbf{N}^{*-1})$, with h/h^* known. This same result has been described in Theorem 3.9.

A somewhat less restrictive class of functions for labeling the information indifference curves is the Student family

$$g_S(z^2|a,\nu) = (a+z^2)^{-\nu/2} \qquad a,\nu > 0$$

with

$$g_S' g_S^{-1} = -\frac{\nu}{2(a+z^2)}.$$

Letting $g_u(z^2) = g_S(z^2|a,\nu)$ and $g_\beta(z^2) = g_S(z^2|a^*,\nu^*)$ we have from (3.29)

$$\lambda = \frac{\nu^*/\left[a^* + (\boldsymbol{\beta}-\mathbf{b}^*)'\mathbf{N}(\boldsymbol{\beta}-\mathbf{b}^*)\right]}{\nu/\left[a + ESS + (\boldsymbol{\beta}-\mathbf{b})'\mathbf{N}(\boldsymbol{\beta}-\mathbf{b})\right]}.$$

This family of labeling distributions has the property that the logarithmic derivative $g'g^{-1}$ decreases with z^2. It is thus relatively steep around the origin and relatively flat elsewhere as indicated in Figure 3.2. This means

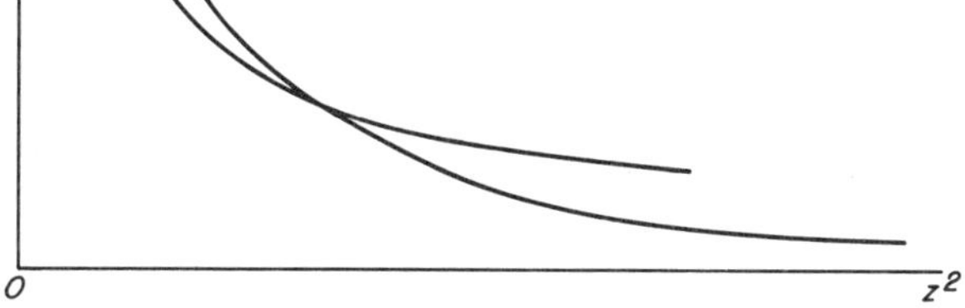

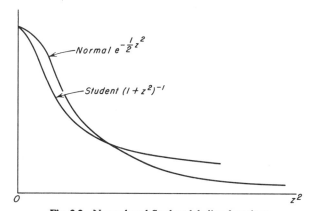

Fig. 3.2 Normal and Student labeling functions.

that an information source will be relatively resistant to moving the estimate of $\boldsymbol{\beta}$ away from its most preferred point in the neighborhood of that point and relatively indifferent to adjustments that occur farther from that point. Note also that the logarithmic derivative of the likelihood function depends not only on a and ν but on ESS and $\boldsymbol{\beta}$ as well. The location of modes on the curve décolletage is, therefore, data dependent.

These assumptions are described more traditionally in Theorem 3.10.

SWEEPING OUT THE MEANS

Ordinarily, there is a constant in a regression function about which we know very little relative to the prior information on the slopes. We show here that the effect of this structure of information is merely to sweep out the means of the observed variables.

Writing the constant explicitly, the regression process becomes

$$\mathbf{Y}=\mathbf{X}\boldsymbol{\beta}+\mathbf{1}_T\alpha+\mathbf{u}$$

with definitions as before, but with $\mathbf{1}_T$ a vector of ones and α the scalar "process level." Let $\mathbf{u}$ be normal with mean zero and covariance matrix $\sigma^2\mathbf{I}_T$, and let α be normal with mean zero and variance v. We may then write

$$\mathbf{Y}=\mathbf{X}\boldsymbol{\beta}+\boldsymbol{\varepsilon}$$

where $\boldsymbol{\varepsilon}$ is normal with mean zero and covariance matrix $\sigma^2\mathbf{I}_T+\mathbf{1}_T v\mathbf{1}'_T$. The precision matrix for $\boldsymbol{\varepsilon}$ is

$$\left(\sigma^2\mathbf{I}_T+\mathbf{1}_T v\mathbf{1}'_T\right)^{-1}=\sigma^{-2}\left(\mathbf{I}_T-\mathbf{1}_T\left[\mathbf{1}'_T\mathbf{1}_T+\frac{\sigma^2}{v}\right]^{-1}\mathbf{1}'_T\right).$$

In the limit as v increases the conditional distribution of $\mathbf{Y}$ given $\boldsymbol{\beta}$ becomes improper with singular precision matrix

$$\sigma^{-2}\mathbf{M}=\sigma^{-2}\left(\mathbf{I}_T-\mathbf{1}_T(\mathbf{1}'_T\mathbf{1}_T)^{-1}\mathbf{1}'_T\right).$$

The determinant of the variance matrix of $\mathbf{Y}$, using (T18) in Appendix 1, is $(\sigma^2)^T(1+Tv\sigma^{-2})$, which for large v behaves like $(\sigma^2)^{T-1}v$. Thus the limiting likelihood function is

$$f(\mathbf{Y}|\boldsymbol{\beta})\propto(\sigma^2)^{-(T-1)/2}\exp\left[-\frac{1}{2\sigma^2}(\mathbf{Y}-\mathbf{X}\boldsymbol{\beta})'\mathbf{M}(\mathbf{Y}-\mathbf{X}\boldsymbol{\beta})\right].$$

But $\mathbf{M}$ is the idempotent matrix that was shown in Section 3.1 to sweep out the means of the variables. That is, given that the constant α is diffuse and independent of the other random variables, inferences about the slope $\boldsymbol{\beta}$ may proceed by first subtracting out the means of the observables $\mathbf{Y}$ and $\mathbf{X}$ and then proceeding as above.

3.4 Multivariate Normal Sampling

In this section, Bayesian inference about the parameters of a multivariate normal distribution is discussed. (Several results here are referred to occasionally, but it is not necessary for the reader at this point to master this material.) The $(k\times 1)$ vector $\mathbf{z}_t$ is to be normally distributed with mean vector $\boldsymbol{\mu}$ and variance matrix $\boldsymbol{\Sigma}$. A sample of T independently distributed vectors $\mathbf{z}_t$ implies the likelihood function

$$L(\boldsymbol{\mu},\boldsymbol{\Sigma};\mathbf{z}_1,\mathbf{z}_2,\ldots,\mathbf{z}_T)\propto|\boldsymbol{\Sigma}|^{-T/2}\exp\left[-\frac{1}{2}\sum_t(\mathbf{z}_t-\boldsymbol{\mu})'\boldsymbol{\Sigma}^{-1}(\mathbf{z}_t-\boldsymbol{\mu})\right].$$

Letting $\bar{\mathbf{z}}=\Sigma_t\mathbf{z}_t/T$

$$\mathbf{S}=\sum_t(\mathbf{z}_t-\bar{\mathbf{z}})(\mathbf{z}_t-\bar{\mathbf{z}})',$$

the exponent in the likelihood function can be written as

$$\sum_t(\mathbf{z}_t-\boldsymbol{\mu})'\boldsymbol{\Sigma}^{-1}(\mathbf{z}_t-\boldsymbol{\mu})=\sum_t\left[(\mathbf{z}_t-\bar{\mathbf{z}})'\boldsymbol{\Sigma}^{-1}(\mathbf{z}_t-\bar{\mathbf{z}})+(\bar{\mathbf{z}}-\boldsymbol{\mu})'\boldsymbol{\Sigma}^{-1}(\bar{\mathbf{z}}-\boldsymbol{\mu})\right]$$

$$=\sum_t\operatorname{tr}\left[\boldsymbol{\Sigma}^{-1}(\mathbf{z}_t-\bar{\mathbf{z}})(\mathbf{z}_t-\bar{\mathbf{z}})'\right]+T(\bar{\mathbf{z}}-\boldsymbol{\mu})'\boldsymbol{\Sigma}^{-1}(\bar{\mathbf{z}}-\boldsymbol{\mu})$$

$$=\operatorname{tr}(\boldsymbol{\Sigma}^{-1}\mathbf{S})+T(\bar{\mathbf{z}}-\boldsymbol{\mu})\boldsymbol{\Sigma}^{-1}(\bar{\mathbf{z}}-\boldsymbol{\mu}).$$

A normal-Wishart distribution for $(\boldsymbol{\mu},\boldsymbol{\Sigma}^{-1})$ is a convenient prior:

$$f_{NW}(\boldsymbol{\mu},\boldsymbol{\Sigma}^{-1}|\bar{\mathbf{z}}^*,\mathbf{S}^*,T^*,\nu^*)\propto|\boldsymbol{\Sigma}^{-1}|^{1/2}\exp\left[-\frac{T^*}{2}(\boldsymbol{\mu}-\bar{\mathbf{z}}^*)'\boldsymbol{\Sigma}^{-1}(\boldsymbol{\mu}-\bar{\mathbf{z}}^*)\right].$$

$$|\boldsymbol{\Sigma}^{-1}|^{(\nu^*-k-1)/2}\exp\left[-\frac{1}{2}\operatorname{tr}\boldsymbol{\Sigma}^{-1}\mathbf{S}^*\right].$$

Combining this prior with the likelihood function and using (T10) in Appendix 1 the posterior distribution is seen to be in the normal-Wishart family with parameters

$$\bar{\mathbf{z}}^{**}=(T+T^*)^{-1}(T\bar{\mathbf{z}}+T^*\bar{\mathbf{z}}^*)$$

$$\mathbf{S}^{**}=\mathbf{S}+\mathbf{S}^*+(\bar{\mathbf{z}}-\bar{\mathbf{z}}^*)'\boldsymbol{\Sigma}^{-1}(\bar{\mathbf{z}}-\bar{\mathbf{z}}^*)TT^*(T+T^*)^{-1}$$

$$T^{**}=T+T^*$$

$$\nu^{**}=T+\nu^*.$$

The predictive distribution of the next sample vector, $\mathbf{z}_F$, is used in Chapter 6. Conditional on $\boldsymbol{\mu}$ and $\boldsymbol{\Sigma}$, $\mathbf{z}_F$ is normal with mean vector $\boldsymbol{\mu}$ and covariance matrix $\boldsymbol{\Sigma}$. Conditional on $\boldsymbol{\Sigma}$ and the data $(\bar{\mathbf{z}},\mathbf{S},\mathrm{T})$, $\boldsymbol{\mu}$ is normal with mean $\bar{\mathbf{z}}^{**}$ and covariance matrix $\boldsymbol{\Sigma}/T^{**}$. Thus integrating with respect to $\boldsymbol{\mu}$, the vector $\mathbf{z}_F$ is normal with mean $\bar{\mathbf{z}}^{**}$ and covariance matrix $\boldsymbol{\Sigma}(1+(T^{**})^{-1})$. Integrating with respect to $\boldsymbol{\Sigma}^{-1}$ as in Appendix 2, the distribution for $\mathbf{z}_F$ is Student

$$f(\mathbf{z}_F|\bar{\mathbf{z}},\mathbf{S},T)=f_S^k\left(\mathbf{z}_F|\bar{\mathbf{z}}^{**},\mathbf{S}^{**}\left(1+(T^{**})^{-1}\right)/\nu^{**},\nu^{**}-k+1\right).$$

The results needed for Chapter 6 are the conditional moments of $\mathbf{z}_F^I$ given $\mathbf{z}_F^J$ where the partition of $\mathbf{z}_F$ is $\mathbf{z}_F'=(\mathbf{z}_F^I,\mathbf{z}_F^J)$. Using the diffuse prior assumption that $\mathbf{S}^*=0$, $T^*=0$, $\nu^*=0$, the conditional moments are

$$E\left(\mathbf{z}_F^I|\mathbf{z}_F^J\right)=\bar{\mathbf{z}}^I+\mathbf{S}_{IJ}\mathbf{S}_{JJ}^{-1}\left(\mathbf{z}_F^J-\bar{\mathbf{z}}^J\right)$$

$$V\left(\mathbf{z}_F^I|\mathbf{z}_F^J\right)=\left(\mathbf{S}_{II}-\mathbf{S}_{IJ}\mathbf{S}_{JJ}^{-1}\mathbf{S}_{JI}\right)(1+T^{-1})(T-k+1+k_J)^{-1}$$

where k_J is the number of elements in $\mathbf{z}_F^J$.

CHAPTER 4

HYPOTHESES-TESTING SEARCHES

The first variety of specification search that we discuss corresponds to the familiar hypothesis-testing problem. We assume the existence of a set of M "models" or hypotheses of the form

$$H_i : \mathbf{Y} = \mathbf{X}_i \boldsymbol{\beta}_i + \mathbf{u}_i, \qquad i = 1, \ldots, M, \tag{4.1}$$

where $\mathbf{Y}$ is a $(T \times 1)$ vector of observable variables, $\mathbf{X}_i$ is a $(T \times k_i)$ matrix of observable explanatory variables, $\boldsymbol{\beta}_i$ is a $(k_i \times 1)$ vector of parameters, and $\mathbf{u}_i$ is a $(T \times 1)$ vector of unobservable disturbances assumed to be normally distributed with mean zero and variance-covariance matrix $\sigma_i^2 \mathbf{I}_T$. The statistical problem is to determine which of these M models did, in fact, generate the data and at the same time to make inferences about the coefficient vectors $\boldsymbol{\beta}_i$.

The formal classical theory of hypothesis testing describes the decision problem of selecting an action from among a set of feasible actions. An action is either right or wrong, depending on the "true state of nature," and the statistician is interested in being wrong as infrequently as possible. The problem being considered in this chapter involves a set of M actions of the form "act as if hypothesis H_i were true" and a set of M states of nature of the form "hypothesis H_i is true." An error occurs when action i is chosen but hypothesis j $(i \neq j)$ is true.

If an action entailed the potential of some specific loss, hypothesis testing should be considered to be solving a decision problem. When the losses are not stated, it is difficult to interpret hypothesis testing as a solution to a decision problem. In fact, most researchers use statements such as "the hypothesis H_i is rejected at the .05 level of significance" to mean that the data cast doubt on H_i. The statement is *not* meant to suggest that because of the data evidence, it is undesirable to act as if H_i were true, regardless of the decision problem. Thus the *language* of hypothesis testing is used to summarize the data evidence. This chapter, therefore, largely ignores the decision-theory problem, but argues that classical hypothesis testing has led to greatly distorted data summaries.

In Chapter 3 we briefly reviewed the classical approach to hypothesis testing with a pair of nested hypotheses of the form

$$H_0: \mathbf{Y}=\mathbf{X}_J\boldsymbol{\beta}_J+\mathbf{u}$$
$$H_1: \mathbf{Y}=\mathbf{X}_I\boldsymbol{\beta}_I+\mathbf{X}_J\boldsymbol{\beta}_J+\mathbf{u}$$
$$=\mathbf{X}\boldsymbol{\beta}+\mathbf{u}$$

where $\mathbf{X}_I$ is a $T\times p$ observable matrix, $\mathbf{X}_J$ is a $T\times(k-p)$ observable matrix, $\mathbf{u}$ is a $T\times 1$ unobservable vector, distributed normally with mean zero and covariance $\sigma^2\mathbf{I}_T$, and $\mathbf{X}=[\mathbf{X}_I,\mathbf{X}_J]$, $\boldsymbol{\beta}'=[\boldsymbol{\beta}_I',\boldsymbol{\beta}_J']$. With $\mathbf{b}=(\mathbf{X}'\mathbf{X})^{-1}\mathbf{X}'\mathbf{Y}$ and $s^2=(\mathbf{Y}-\mathbf{Xb})'(\mathbf{Y}-\mathbf{Xb})/(T-k)$, the statistic

$$F=\frac{\mathbf{b}_I'\left(\mathbf{X}_I'\mathbf{X}_I-\mathbf{X}_I'\mathbf{X}_J(\mathbf{X}_J'\mathbf{X}_J)^{-1}\mathbf{X}_J'\mathbf{X}_I\right)\mathbf{b}_I}{ps^2},$$

conditional on the null hypothesis that $\boldsymbol{\beta}_I=\mathbf{0}$, has an F distribution with p and $T-k$ degrees of freedom. A large F is taken as evidence against the null hypothesis.

The F statistic has been written in Equation (3.10) in terms of error sums of squares as

$$F=\frac{(ESS_0-ESS_1)/p}{ESS_1/(T-k)} \tag{4.2}$$

where ESS_i is the error sum of squares associated with the ith hypothesis. In terms of R^2, it can be written as

$$F=\left(\frac{R_1^2-R_0^2}{1-R_1^2}\right)\left(\frac{T-k}{p}\right). \tag{4.3}$$

Formal testing of the hypothesis $\boldsymbol{\beta}_I=\mathbf{0}$ involves, first, selecting a significance level for the F test, say, α, then finding from a table the α point of the F distribution, and finally recording that the hypothesis is or is not rejected at the α level, depending on whether F exceeds or falls short of this cutoff point. For example, the .05 point of the F distribution with 10 or more degrees of freedom ($T-k \geqslant 10$) varies from approximately five to one, depending on the actual number of degrees of freedom as well as on the number of restrictions p. Thus for moderate degrees of freedom, F values in excess of 5 would be regarded as evidence against the null hypothesis.

By referring to the foregoing formulas, we can see that although the R^2s of two equations may differ only in the tenth decimal place, an F may attain any arbitrarily large value if the degrees of freedom $T-k$ is large enough. In very large samples such as would be generated by surveys of individuals or firms, it thus turns out that almost any hypothesis of this form is rejected. To paraphrase a quotation of Berkson (1938),[1] since a large sample is presumably more informative than a small one, and since it is apparently the case that we will reject the null hypothesis in a sufficiently large sample, we might as well begin by rejecting the hypothesis and not sample at all.

This brings us to the first question of this chapter:

Problem 1. Is classical hypothesis testing at fixed level of significance a "good" way to summarize the evidence in favor of or against hypotheses of the form described above?

Our answer is decidedly negative—meaningful hypothesis testing requires the significance level to be a decreasing function of sample size. Incidentally, the argument that leads to this conclusion is not the same as Berkson's. He might have pointed out that it is practically certain that any series of real observations does not actually come from a regression process with $\boldsymbol{\beta}_I = \mathbf{0}$. If so, we do, indeed, want to reject the null hypothesis; and it is neither surprising nor unwarranted that a large informative sample leads to the rejection of the hypothesis. One, in those circumstances, should trouble himself not with the results of classical hypothesis testing but rather with the question of why he bothered to test an obviously false hypothesis in the first place. As it turns out, hypothesis testing does

[1] "I believe that an observant statistician who has had any considerable experience with applying the chi-square test repeatedly will agree with my statement that, as a matter of observation, when the numbers in the data are quite large, the P's tend to come out small. Having observed this, and on reflection, I make the following dogmatic statement, referring for illustration to the normal curve: 'If the normal curve is fitted to a body of data representing any real observations whatever of quantities in the physical world, then if the number of observations is extremely large—for instance, on the order of 200,000—the chi-square P will be small beyond any usual limit of signficance'."

have some validity, even when a restriction is practically certain to be false. Certainly false hypotheses are the subject of the next five chapters. In this chapter we concern ourselves with hypotheses that are not so trivially rejected; we are thus concerned with situations in which classical hypothesis testing has an unambiguously legitimate function. This distinction can be made quite clear from a Bayesian point of view: We are here considering hypotheses that are assigned positive probability; in later chapters we consider hypothesis testing in contexts when some of the hypotheses receive zero probability. The conclusion that the significance level should be a decreasing function of sample size is due largely (but not exclusively) to the fact that Bayesians who assign positive probabilities to hypotheses of this form summarize the evidence in ways that implicitly make the significance level a decreasing function of sample size.

This is a good place to indicate that I doubt that there are many instances when a regression hypothesis that involves a restricted parameter space would, in fact, be assigned positive probability. The things that we call models usually originate in other ways, and I consequently doubt the practical relevance of this chapter. This is, nonetheless, a useful topic to begin with, because it corresponds closely to the situation in which classical hypothesis testing is strictly relevant, and because many people think of specification searches exclusively in those terms.

The hypotheses we have just considered have an exceedingly simple form: the null hypothesis is a restricted version of the alternative. Practical specifications rarely have such a simple structure, and dealing with complicated sturctures of hypotheses is our second problem:

Problem 2. How should multiple hypotheses with a nonnested structure be treated?

I personally find the classical answer to this question hopelessly confusing and would prefer not to get too deeply involved in discussing it. A simple example illustrates some of the problems. Let the hypotheses be

$$H_0: \mathbf{Y} = \mathbf{x}\beta + \mathbf{u} \quad H_1: \mathbf{Y} = \mathbf{z}\gamma + \mathbf{u},$$

where $\mathbf{x}$ and $\mathbf{z}$ are vectors and β and γ are scalars. In this case it is possible to generate a .05-level test of H_0 by testing in the usual way the null hypothesis H_0 against the alternative $H_1^*: \mathbf{Y} = \mathbf{x}\beta + \mathbf{z}\gamma + \mathbf{u}$. Although this is a perfectly well-defined test, it completely ignores the fact that H_1^* is not the alternative hypothesis. It, furthermore, treats asymmetrically two hypotheses, H_0 and H_1, that are apparently symmetric.

Suppose, instead, that the hypotheses are treated symmetrically; in particular, suppose the equation with the higher R^2 is accepted. Then the conditional probability of error is $P,\ _0^2 < R_1^2|H_0) = P(\mathbf{Y}'\mathbf{x}(\mathbf{x}'\mathbf{x})^{-1}\mathbf{x}'\mathbf{Y} <$

$\mathbf{Y}'\mathbf{z}(\mathbf{z}'\mathbf{z})^{-1}\mathbf{z}'\mathbf{Y}|H_0) = P((\mathbf{x}\beta+\mathbf{u})'(\mathbf{x}(\mathbf{x}'\mathbf{x})^{-1}\mathbf{x}' - \mathbf{z}(\mathbf{z}'\mathbf{z})^{-1}\mathbf{z})(\mathbf{x}\beta+\mathbf{u}) < 0|\beta)$, which depends in a complicated way on β, $\mathbf{x}$, and $\mathbf{z}$. (It can be shown that this probability varies from one-half to zero as β^2 varies from zero to infinity.) Traditionally, the test is set up so that this conditional probability of error is a number, not a function, and it proves difficult to interpret this test in terms of its error probabilities. This should not cause great consternation, since the relationship between the error probabilities of a test and the persuasiveness of the evidence in favor of the various hypotheses is indirect and poorly understood. Of course, this barely scratches the surface of a complex problem. The Bayesian approach yields such a straightforward answer with clear intuitive appeal that it hardly seems worthwhile to pursue further the classical approach. For more on the classical approach see the summaries by Dhrymes et al. (1972) and by Gaver and Geisel (1974).

Anyone who has read any papers in applied econometrics has read statements of the form: "model A has performed the best; it has a high R^2, and all of its coefficients are the right sign and are statistically significant." Whatever is the meaning of the reference to the coefficients? Books on classical statistics do not suggest that the validity of an F test depends on the signs and statistical significance of the coefficients. Perhaps the author of this statement is thinking that there is no restriction that could be placed on this model that would not be rejected, but why the reference to "right" signs? This is a pretty obvious Bayesian problem, in that there is a priori information at least about the signs of the coefficients. This intuitively ought to have an impact on the hypothesis testing. Thus our third problem is:

Problem 3. How does the existence of a priori information about parameters influence the interpretation of evidence about models?

Parallel to the testing of hypotheses, we are interested in making inferences about parameters. There is a presumption that the ambiguity over the model should dilute the information about regression coefficients, since part of the evidence is spent to specify the model:

Problem 4. What estimates of the parameters and what measures of uncertainty should apply in a situation of uncertainty about the model?

Classical inference has little to say about this, although we review in the next chapter the pretesting literature that deals with estimation while testing. Again, the Bayesian solution is entirely straightforward. Among its conclusions are the fact that the ambiguity over the model is irrelevant for inference about a coefficient if and only if the estimated coefficients and standard errors are the same for all specifications.

The last problem is similar:

Problem 5. What measure of overall confidence analogous to an R^2 should apply to a research effort which reports many different equations with different R^2s? We subsequently propose a special kind of average R^2. Again, there is no classical counterpart.

In the first two sections of this chapter, the problem of identifying the class of admissible tests is distinguished from the problem of selecting a particular test. Classical hypothesis testing concerns itself almost exclusively with the first problem, but it has nothing meaningful to say about the second. The rule of thumb quite popular now, that is, setting the significance level arbitrarily to .05, is shown to be deficient in the sense that from every reasonable viewpoint the significance level should be a decreasing function of sample size.

A few words may now be said in anticipation of the sections to follow, which describe in detail the Bayesian approach to hypothesis testing. By Bayes' rule, the relative posterior probabilities of two hypotheses can be written as

$$\frac{P(H_i|\mathbf{Y})}{P(H_j|\mathbf{Y})}=\left[\frac{P(\mathbf{Y}|H_i)}{P(\mathbf{Y}|H_j)}\right]\left[\frac{P(H_i)}{P(H_j)}\right]. \tag{4.4}$$

The second factor in brackets is the prior odds ratio in favor of H_i. The data-dependent term in the first set of brackets is the "Bayes factor."

The data are said to favor H_i relative to H_j if the Bayes factor exceeds one, that is, if the observed data $\mathbf{Y}$ is more likely under hypothesis H_i than it is under hypothesis H_j. The densities of $\mathbf{Y}$ implied by the hypotheses (4.1) are conditional on the parameters, $\boldsymbol{\beta}_i$ and σ_i^2, but may be straightforwardly "mixed" into a marginal density as

$$f(\mathbf{Y}|H_i)=\int_{\boldsymbol{\beta}_i}\int_{\sigma_i^2} f(\mathbf{Y}|H_i,\boldsymbol{\beta}_i,\sigma_i^2)\,f(\boldsymbol{\beta}_i,\sigma_i^2)\,d\sigma_i^2\,d\boldsymbol{\beta}_i \tag{4.5}$$

where $f(\boldsymbol{\beta}_i,\sigma_i^2)$ is the prior density. The conditional p.d.f. $f(\mathbf{Y}|H_i,\boldsymbol{\beta}_i,\sigma_i^2)$ for a particular value of $\mathbf{Y}$ is a likelihood function of $(\boldsymbol{\beta}_i,\sigma_i^2)$, and (4.5) defines $f(\mathbf{Y}|H_i)$ as a weighted or marginal likelihood.

The Bayes factor is to be contrasted with the likelihood ratio, which is used classically to summarize the data evidence. The likelihood ratio is

$$L(H_i,H_j)=\frac{\max\limits_{\beta_i,\sigma_i^2} f(\mathbf{Y}|\beta_i,\sigma_i^2,H_i)}{\max\limits_{\beta_j,\sigma_j^2} f(\mathbf{Y}|\beta_j,\sigma_j^2,H_j)}.$$

The Bayes factor averages the likelihood function over all values of $(\boldsymbol{\beta}_i, \sigma_i^2)$. The likelihood ratio evaluates the likelihood function at its maximum.

Any attempt to summarize the data evidence in favor of the hypotheses (4.1) leads to an irreconcilable index number problem of the following form. If $\boldsymbol{\beta}_i$ assumed one value, the data evidence could be said unambiguously to favor the ith hypothesis, but if $\boldsymbol{\beta}_i$ assumed another value, the data unambiguously cast doubt on H_i. Since H_i allows $\boldsymbol{\beta}_i$ to assume any value, the data evidence is necessarily ambiguous.

The classical solution to this dilemma seems most appropriate for testing a point null hypothesis against a composite alternative. The null hypothesis is regarded as the favorite; it is the one that is being "tested." If there is *any* way for the alternative to look as good as the null hypothesis, we should be worried about retaining the null as the favorite. Consequently, we identify the evidence *against* the null hypothesis in terms of the evidence in favor of the alternative at the value of $\boldsymbol{\beta}$ that makes the alternative appear best. The appropriate statement however is not that the alternative is favored. All that is said is that the alternative is *conceivably* favored.

There is a great tendency in practice to forget the all-important word "conceivably" in this sentence, and as a consequence, classical tests distort the data evidence. In the more common case when the null hypothesis is a composite hypothesis, classical tests usually also evaluate the data evidence at the parameter point that makes the null hypothesis appear best. The resulting statement about the evidence is: "If each hypothesis is allowed to 'put its best foot forward,' hypothesis H_j is favored." In practice, the qualifying phrase "if...forward," is often forgotten, and the data evidence may consequently be significantly distorted.

A Bayesian approach, in contrast, presupposes a prior distribution that can be used to weight the evidence at different values of the parameters. Thus instead of letting an hypothesis "put its best foot forward," the performance at all values of the parameters is considered. The apparent problem that then arises is the construction of a nonarbitrary weight function. Here and elsewhere, we take the position that a researcher is obligated to report as fully as possible the mapping of priors into posteriors. He should describe the data evidence as favoring hypothesis H_1 if the prior takes one form, and favoring H_2 if it takes another. He thereby avoids having to make a choice that rightly belongs to his readers: the choice of prior distribution.

4.1. Hypothesis Testing: A Judicial Analogy

The subject of hypothesis testing may be usefully introduced by an analogy. Based on the evidence presented, a judge and jury in a legal proceeding decide whether a defendant should be set free or sent to jail. If they decide that the evidence favors the hypothesis of guilt, they accordingly send the defendant to jail. Otherwise he is set free. The assumption of innocence until proven guilty beyond a reasonable doubt explicitly favors the hypothesis of innocence. We refer to this favored hypothesis as the *null* hypothesis or H_0 and the hypothesis of guilt as the *alternative* hypothesis or H_1. The evaluation of the evidence and the decision either to free or jail the defendant is called a "test" of the null hypothesis *against* the alternative, and the decision is described as acceptance versus rejection of the hypothesis of innocence.

The more critical error—sending an innocent man to jail—is called an *error of the first kind* or *type I error*. Acceptance of the null hypothesis when it is in fact false—freeing a guilty man—is called an *error of the second kind* or *type II error*. Schematically we have

		Actions	
Hypotheses (States)		Set Free (accept H_0)	Send to Jail (reject H_0)
H_0:	Innocent		Type I error
H_1:	Guilty	Type II error	

If a man is innocent, we want to have a low probability of sending him to jail. Let this probability be α

$$\alpha = P(\text{jail}|\text{innocent defendant}).$$

Analogously, let

$$\beta = P(\text{set free}|\text{guilty defendant}).$$

Both α and β are defined before the judicial process commences. In effect, they predict the quality of the evidence and the ability of the court to process the evidence effectively. For example, a low value of α amounts to the prediction that if the defendant is innocent, the evidence will be so unambiguous and the process by which a verdict is rendered will be so perfect that with near certainty he will be justly found innocent.

The theory of hypotheses testing deals with defining procedures such that α and β are small. A typical choice set for α and β is depicted in Figure 4.1. Flipping a coin to decide whether to free or jail the suspect implies $\alpha = \beta = .5$. The line running from the point $(\alpha = 1, \beta = 0)$ to $(\alpha = 0, \beta = 1)$ represents the set of all such randomized decisions. The value $(\alpha = 0, \beta = 0)$ represents perfect evidence and a perfect procedure which is excluded

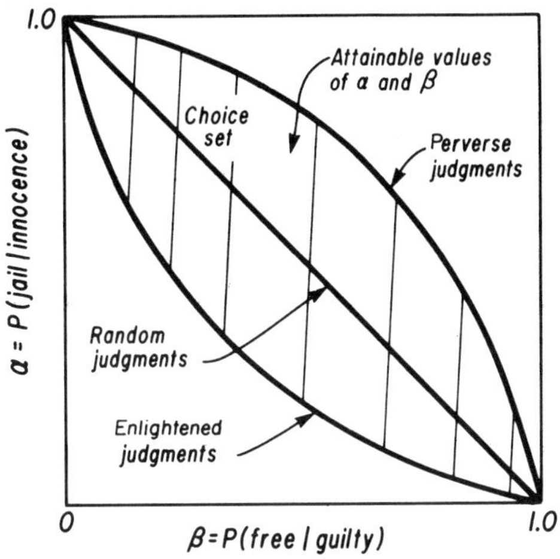

Fig. 4.1 Probabilities of error.

from the choice set in Figure 4.1. If ($\alpha=0,\beta=0$) is not available, the perfect error point ($\alpha=1,\beta=1$) is similarly not available, since if we could be sure of making an error, by doing the exact opposite we could be sure of *not* making an error. The curved line labeled "enlightened judgments" represents the best possible court procedures based on the available evidence. The curve labeled "perverse judgments" is just the mirror image of the "enlightened judgment" curve, involving the exact opposite action.

The choice of a courtroom procedure is usefully thought to involve two steps. The first step is to identify the set of procedures that involve enlightened use of the evidence, that is, those that make α and β as small as possible. The second step is to choose a particular procedure from among this set of admissible procedures. The former is a logical mathematical problem that admits a clear-cut uncontroversial solution; the latter is not. Let us consider the latter problem.

The essential problem the court faces once the line of enlightened judgments is computed is that stricter interpretation of the given body of evidence and a greater tendency to send men to jail which could reduce the probability β of freeing guilty men necessarily increases the probability α of jailing innocent men. By assumption, it is desirable to have both α and β small. The choice dilemma is that reduction of one necessarily leads to an increase in the other. Actual choice can be said to depend on a preference function $U(\alpha,\beta)$ indicating numerically the level of satisfaction attained if the courtroom procedure yields probabilities α and β. Several "contour" lines of a typical preference function are indicated in Figure 4.2.

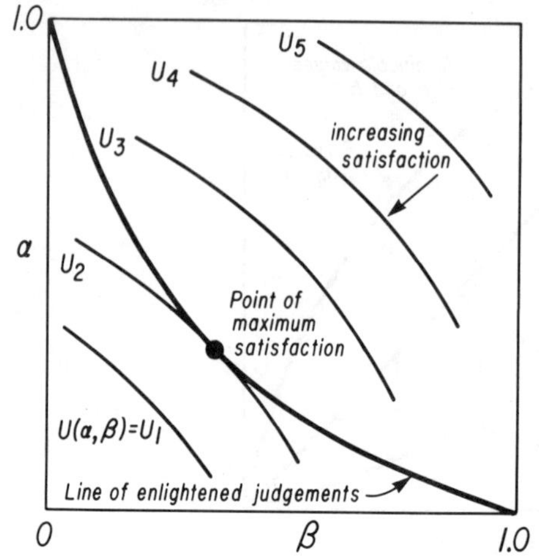

Fig. 4.2 Indifference curves of a preference function.

The line $U(\alpha,\beta)=U_1$ indicates all values of (α,β) that yield a level of satisfaction U_1. Increasing preference (lower α and β) is in the direction indicated, and maximum satisfaction on the enlightened judgment line occurs at the point indicated.

One way to alleviate the choice dilemma is to gather more evidence, thereby making *both* α and β smaller. Without any information, the line of enlightened judgments is just a straight line from corner to corner. As more and more evidence is accumulated, this line shifts in toward the origin. As this occurs, the point of maximum satisfaction also travels in toward the origin tracing out an *information expansion path* depicted in Figure 4.3. This represents the values of α and β that the court would actually choose depending on the amount of evidence that is amassed.

Three possible ways of selecting α and β have been suggested, and their preference functions and information expansion paths are indicated in Figure 4.4.

(1) Set $\alpha=.05$. The most commonly practiced procedure is to set $\alpha=.05$ and minimize β. The type I error is considered more important, and by setting $\alpha=.05$ it necessarily assumes a small value. Note the peculiar information expansion path that allows β to be infinitesimally small with α still at .05.

(2) Minimize the maximum of $l_1\alpha$ and $l_2\beta$, where l_1 is the loss associated with a type I error and l_2 is the loss associated with a type II error. Note that the expansion path moves continuously toward the origin and that an increase in the evidence is used to reduce *both* α and β. The relative probabilities are $\alpha/\beta=l_2/l_1$, and if type I loss l_1 is relatively great, the type I probability α is correspondingly relatively small.

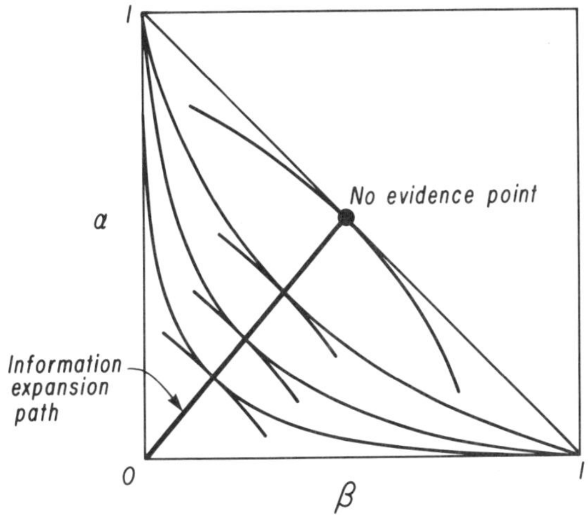

Fig. 4.3 Information expansion path.

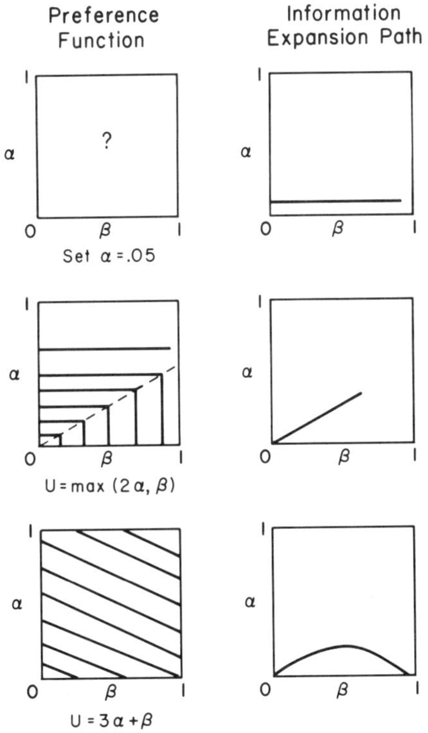

Fig. 4.4 Preference functions and information expansion paths.

(3) Minimize the expected loss. Let p be the probability that the defendant is innocent and $(1-p)$ the probability of guilt. The expected loss is $l_1\alpha p + l_2\beta(1-p)$. The slope of a typical contour or indifference curve depicted in Figure 4.4 is $-l_2(1-p)/l_1 p$. If either l_1 or p is relatively large, these lines are relatively flat, and α is set to a relatively small value. Thus type I error is avoided in the sense of setting α to some small number, if either the defendant is quite likely to be innocent or if the cost of sending an innocent man to jail is high.

It should now be clear that the choice of α and β is less than obvious. A simple argument attributed to Savage does imply that the indifference curves should be straight lines. Identify two points (α_0, β_0) and (α_1, β_1) between which you are indifferent. If restricted to select one of these two points, you must mean that you are willing to allow someone else to make the selection. Well, I will make the selection for you in the following way. Making use of a random device, I select (α_0, β_0) with probability π and (α_1, β_1) with probability $(1-\pi)$. As a result, your type I and type II error probabilities are, in fact, $(\alpha_0\pi + \alpha_1(1-\pi), \beta_0\pi + \beta_1(1-\pi)$, and you have revealed your indifference between this point and the two original points. By varying π, all probability couples on the line joining (α_0, β_0) to (α_1, β_1) can be shown to be on the same indifference curve. Needless to say, Bayesian indifference curves are straight lines.

The hypothesis-testing problem can be described more formally as follows. A sample outcome $\mathbf{z}$ (the testimony) is assumed to come from a sample space Z of all possible samples. The sample space is partitioned into a region of acceptance A and a region of rejection R, where R is the set of all $\mathbf{z} \in Z$ that would lead to rejection of the null hypothesis (jailing the defendant); and A is the complement of R. The corresponding error probabilities are

$$\alpha_R = P(\mathbf{z} \in R | H_0)$$

$$\beta_R = P(\mathbf{z} \in A | H_1).$$

In discussing the theory of hypothesis testing, we may first consider the purely mathematical problem of defining the set of admissible tests; we must be sure that the partition of the sample space into A and R leads to error probabilities on the line of enlightened judgments. Second, we must choose a particular test from among the set of admissible tests. A failure of the theory of classical inference is that it offers no meaningful comment on this second problem. And the rule "set $\alpha = .05$" regardless of sample size seems undesirable under close examination.

4.2 Testing a Point-Null Hypothesis Against a Point Alternative

This section deals briefly with testing a point-null hypothesis against a point alternative. The purpose of this material is to explain how the discussion in the previous section applies to a formal problem. The test depends on a sample of size T, $(Y_1, Y_2, \ldots, Y_T)$ from a normal distribution with mean μ and variance one. The null hypothesis is $H_0: \mu=0$, and the alternative is $H_1: \mu=1$. In terms of the distribution of the mean $\bar{Y}=\Sigma Y_t/T$, the hypotheses are

$$H_0: \bar{Y} \sim N(0, T^{-1})$$

$$H_1: \bar{Y} \sim N(1, T^{-1}).$$

These two distributions are graphed in Figure 4.5 for $T=1$.

Large values of $\bar{Y}$ favor the alternative hypothesis, and a typical decision function is

$$\text{if } \bar{Y} \geqslant c, \text{ reject } H_0$$

$$\text{if } \bar{Y} < c, \text{ accept } H_0$$

where c is some preassigned cutoff point.

The probabilities of error depend on the cutoff point c:

$$\alpha(c) = P(\bar{Y} \geqslant c \,|\, \mu=0)$$

$$\beta(c) = P(\bar{Y} \leqslant c \,|\, \mu=1).$$

Using a table of normal distribution, we may compute $\alpha(.5)=\beta(.5)=.31$, as depicted in Figure 4.5. If a smaller value of α is desired, c may be increased, say, to 1, which lowers α to .16, but this change in c also

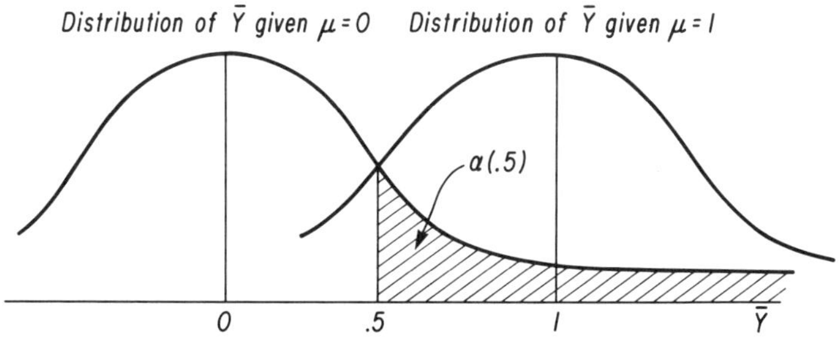

Fig. 4.5 Distributions given null and alternative hypotheses.

increases β to .5. As c is varied from $-\infty$ to $+\infty$, $\alpha(c)$ and $\beta(c)$ thus trace out a line of "enlightened judgments" as in Figure 4.1.

It is also easy to see that the line of enlightened judgments shifts in toward the origin as more evidence is accumulated, that is, as T grows. If T is four instead of one, the standard errors of the distributions become one-half. For $c=.5$, it is easy to calculate that $\alpha=\beta=.16$, compared to .31 before. Similar calculations apply to all values of c.

Traditionally, the cutoff point c is chosen so that $\alpha(c)=.05$. This implies a value of c equal to $1.65/\sqrt{T}$, which shifts toward zero as T increases. This contrasts with a Bayesian cutoff point, which shifts toward .5 as T increases. A Bayesian test selects H_0 if the expected loss from acting as if H_0 were true is less than the expected loss from other action. The probability that H_0 is true given the data $\bar{Y}$ is by Bayes' rule

$$P(H_0|\bar{Y})=\frac{f(\bar{Y}|H_0)P(H_0)}{f(\bar{Y}|H_0)P(H_0)+f(\bar{Y}|H_1)P(H_1)}$$

where $f(\bar{Y}|H_0)=(T/2\pi)^{1/2}\exp[-T\bar{Y}^2/2]$ and $f(\bar{Y}|H_1)=(T/2\pi)^{1/2}\exp[-T(\bar{Y}-1)^2/2]$. The expected loss from proceeding as if H_0 were true is $l_1P(H_1|\bar{Y})$, where l_1 is the loss if H_1 is true and action H_0 is taken; similarly, for H_1. Thus H_0 is to be rejected if

$$l_1P(H_1|\bar{Y})>l_0P(H_0|\bar{Y}),$$

that is, if

$$l_1f(\bar{Y}|H_1)P(H_1)>l_0f(\bar{Y}|H_0)P(H_0),$$

or if $T[\bar{Y}^2-(\bar{Y}-1)^2]>2\log[l_0P(H_0)/l_1P(H_1)]$ or if $2\bar{Y}-1>2T^{-1}\log[l_0P(H_0)/l_1P(H_1)]$. The term on the right-hand side converges to zero as T increases, and the region of rejection becomes $\bar{Y}>.5$.

4.3 Testing a Point-Null Hypothesis Against a Composite Alternative

The more difficult problem of testing a point-null hypothesis against a composite alternative is discussed in this section. A composite hypothesis is a set of values of the parameter vector, each of which determines a different data distribution. In testing a point-null hypothesis against a composite alternative, we ask the question "was the data more likely to have come from the null distribution or from one distribution selected from the set of distributions which comprise the alternative hypothesis?" Except in certain trivial cases, this question cannot admit an unambiguous answer.

An examination of a likelihood function illustrates the difficulty of testing composite hypotheses. Based on a sample from a normal distribution with unknown mean μ and known variance σ^2, we would like to test the null hypothesis $H_0:\mu=0$ against the alternative $H_1:\mu\neq 0$. Letting $\mathbf{1}_T$ be a vector of ones and $\mathbf{Y}$ the vector of T independently drawn observations, the sampling distribution can be written as

$$f_N(\mathbf{Y}|\mathbf{1}\mu,\sigma^2\mathbf{I})=(2\pi\sigma^2)^{-T/2}\exp\left[-\frac{1}{2\sigma^2}(\mathbf{Y}-\mathbf{1}_T\mu)'(\mathbf{Y}-\mathbf{1}_T\mu)\right].$$

If μ were equal to zero, the density function of the data would be $f_N(\mathbf{Y}|\mathbf{0},\sigma^2\mathbf{I})$. The likelihood function

$$L(\mu;\mathbf{Y})=\frac{f_N(\mathbf{Y}|\mathbf{1}\mu,\sigma^2\mathbf{I})}{f_N(\mathbf{Y}|\mathbf{0},\sigma^2 I)}$$

formally summarizes the evidence in favor of some other value of μ in comparison with the hypothesized value $\mu=0$ by indicating how much more likely it is that the data were drawn from a distribution located at μ than from a distribution located at zero. By a simple manipulation we may write it as

$$L(\mu;\mathbf{Y})=\exp\left[-\frac{T}{2\sigma^2}(\mu-\bar{Y})^2\right]\exp\left[\frac{T}{2\sigma^2}\bar{Y}^2\right]$$

where $\bar{Y}$ is the sample mean $\mathbf{Y}'\mathbf{1}/T$.

This is a function that is symmetric around its maximum point $\mu=\bar{Y}$ where it assumes the value $\exp[\bar{Y}^2T/2\sigma^2]$. An example is graphed in Figure 4.6.

We need now to indicate whether the data favor or cast doubt on the null hypothesis $\mu=0$. In the ideal situation the likelihood function $L(\mu;\mathbf{Y})$ is either zero at $\mu=0$ or zero everywhere else, and we could unambiguously conclude in the former case against the value $\mu=0$, and in the latter case in favor of it. Less precise information that nonetheless incontrovertibly favors one hypothesis or the other is implied by a likelihood function that attains either its minimum or its maximum at $\mu=0$. Unhappily, the probability of these unambiguous outcomes is zero, and we are forced to deal almost always with the sort of ambiguous situation depicted in Figure 4.6, in which the null hypothesis looks better than the alternative at some values of μ but worse at others.

We argue subsequently that there is, in fact, no solution to this dilemma. Corresponding to three different statistical schools are three different approaches, each of which is discussed, each of which has apparent shortcomings.

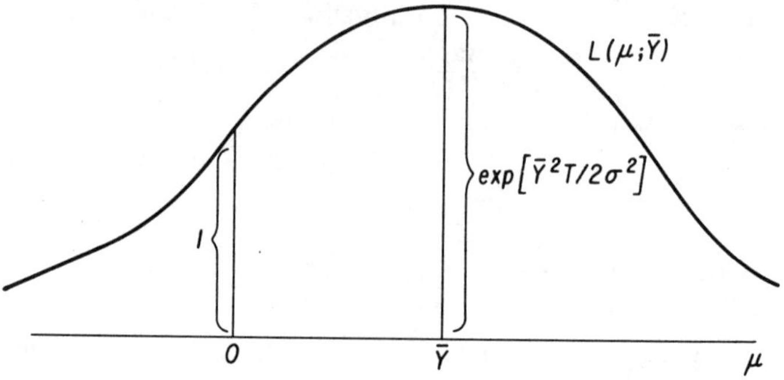

Fig. 4.6 Likelihood function $f_N(\mathbf{Y}|\mathbf{1}\mu, \sigma^2\mathbf{I})/f_N(\mathbf{Y}|\mathbf{0},\sigma^2\mathbf{I})$.

THE LIKELIHOOD SCHOOL

The likelihood approach is apparently straightforward. It is hypothesized that $\mu=0$. Suppose there were another value of μ for which the data were 100 times more probable than for $\mu=0$. Would not this shake your faith in $\mu=0$? This suggests that the right number to report against the null hypothesis is the likelihood ratio at the value of μ that makes the alternative appear best, namely, at $\mu=\bar{Y}$:

$$L_1=\max_{\mu} L(\mu;\mathbf{Y})=\exp\left[\frac{\bar{Y}^2T}{2\sigma^2}\right],$$

which by a series expansion truncated after the first term can be written approximately as

$$L_1 \sim 1+\frac{\bar{Y}^2T}{2\sigma^2}$$

$$=1+\frac{z^2}{2}$$

where $z^2=\bar{Y}^2T/\sigma^2$. Thus for $z^2=2$ we can say that the data cast doubt on H_0 in the sense that there is a value of μ that is approximately twice as likely to have generated the given data.

SAMPLING THEORY SCHOOL

Sampling theory is concerned not with the likelihood function of μ given the data $\mathbf{Y}$ but rather with the sampling distribution of $\mathbf{Y}$ given the null hypothesis $\mu=0$. Statements such as the following are usually made. If the observed value of $\mathbf{Y}$ is in the tail of the distribution $f(\mathbf{Y}|\mu=0,\sigma^2)$ the data cast doubt on H_0 in the sense that something unlikely would have had to occur if H_0 were true.

Most commonly, tests are based on the sufficient statistic $\bar{Y}$ which is distributed normally with mean μ and variance σ^2/T. An indication of whether $\bar{Y}$ came from the tail of its distribution given $\mu=0$ is

$$S=\frac{\max_{\bar{Y}} f_{\bar{Y}}\left(\bar{Y}|\mu=0,\sigma^2/T\right)}{f_{\bar{Y}}\left(\bar{Y}|\mu=0,\sigma^2/T\right)}$$

$$=\exp\left[\frac{\bar{Y}^2T}{2\sigma^2}\right]=\exp\left[\frac{z^2}{2}\right]$$

which is seen to be identical to L_1 before. Thus for $z^2=\bar{Y}^2T/\sigma^2$ large we may say that the data cast doubt on $\mu=0$ both in the sense that there are other values of μ that are more likely to have generated the data and also in the sense that if μ were zero, an unlikely event occurred.

A more traditional description of the tail of the distribution is expressed in terms of the probability mass rather than in terms of the relative density. That is, before observing $\bar{Y}$, it is decided that a value of $\bar{Y}^2$ in excess of some arbitrary number, say, c^2, will cast doubt on H_0. The probability mass in the tail of the distribution beyond c^2 is called the significance level of the test,

$$\alpha(c)=P\left(\bar{Y}^2>c|\mu=0\right),$$

and if $\bar{Y}$ falls in the described region, the null hypothesis is said to be rejected at level α. It is customary to select c such that $\alpha(c)=.05$, and the familiar "region of rejection" is

$$|\bar{Y}|\geqslant\frac{1.96\sigma}{\sqrt{T}}$$

or in terms of z^2, $z^2\geqslant 1.96^2$. The corresponding probability statement is

$$P\left(z^2\geqslant 1.96^2|\mu=0,\sigma^2\right)=.05.$$

Thus z^2 in excess of 1.96 casts doubt on H_0 in the sense that given such a z we would be led to reject H_0 at the .05 level.

There is yet another "metric" for measuring the evidence against H_0. The P-value of a test is the level at which the data is "just significant"; at any significance level less that P the null hypothesis would not be rejected, whereas at any larger significance level it would be. Both the P value and S are graphed in Figure 4.7. Both indicate whether the observed value of $\bar{Y}$ is in the tail of its distribution or not, and both are (increasing) functions of z^2 only.

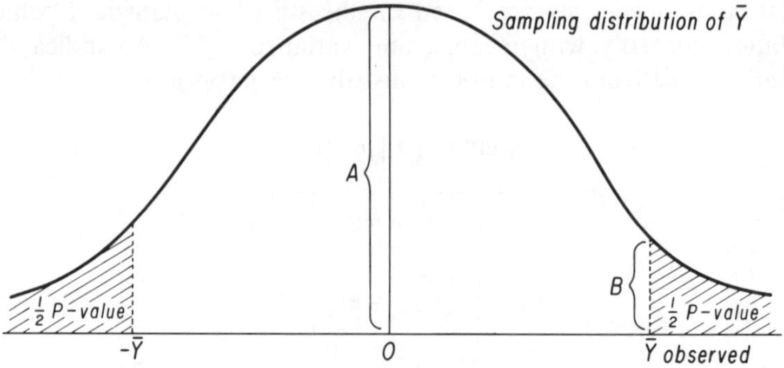

Fig. 4.7 Sampling distribution: P-value and $S=A/B$.

THE BAYESIAN SCHOOL

Bayesians, of course, apply their single commandment, Bayes' rule, and conclude that the null hypothesis is favored by the data if its posterior probability exceeds its prior probability (ignoring the loss function, for now). If the prior for μ is continuous, both the prior and the posterior probability of the point null hypothesis are zero, and the problem is uninteresting. One prior distribution that allocates probability π to the null hypothesis is

$$P(\mu)=\begin{cases}\pi & \mu=0\\ (1-\pi)f_N(\mu|0,h^{*-1})\,d\mu & \mu\neq 0,\end{cases}$$

where $f_N(\mu|0,h^{*-1})$ indicates a normal distribution located at the origin with variance h^{*-1}. In words, there is a spike of mass π at $\mu=0$, and the rest is allocated normally over the line. By a straightforward application of Bayes' rule we have the posterior distribution of μ given the data mean $\bar{Y}$ as

$$P(\mu|\bar{Y})\propto P(\bar{Y}|\mu)P(\mu),$$

and the odds ratio in favor of the alternative hypothesis is

$$\frac{P(H_1|\bar{Y})}{P(H_0|\bar{Y})}=\frac{\int_{\mu\neq 0}f_N(\bar{Y}|\mu,\sigma^2/T)P(\mu)}{f_N(\bar{Y}|\mu=0,\sigma^2/T)\pi}=\int L(\mu;\bar{Y})P(\mu)\,d\mu$$

$$=(1-\pi)\pi^{-1}\left(1+\frac{T}{\sigma^2h^*}\right)^{-1/2}\exp\left[z^2\left(1+\frac{h^*\sigma^2}{T}\right)^{-1}\Big/2\right]$$

where $z^2=\bar{Y}^2/(\sigma^2/T)$ and where we have made use of some results to follow (alternatively, it requires some straightforward numerical integration). That is, the posterior odds ratio is the prior odds ratio times what is called the "Bayes factor"

$$B(h^*)=\left(1+\frac{T}{\sigma^2 h^*}\right)^{-1/2}\exp\left[z^2\left(1+\frac{h^*\sigma^2}{T}\right)^{-1}\Big/2\right]. \tag{4.6}$$

This, like the previous data summaries, is an increasing function of $z^2=\bar{Y}^2T/\sigma^2$; the larger z^2 is the more the data cast doubt on H_0. However, and most importantly, B is a function of sample size also. Let z^2 take on some arbitrarily large value so that, classically, we would say the data cast doubt on H_0. For sufficiently large $T/\sigma^2, B(h^*)$ can take on any arbitrarily small value, and rather than concluding that the data cast doubt on H_0, we would claim that they quite strongly favor H_0.

This is a version of the Lindley (1957) paradox. It represents a sharp disagreement between classicists and Bayesians over the interpretation of evidence. Classicists claim for this problem that the evidence against H_0 can be fully summarized in terms of z^2 alone. Bayesians would be in general agreement that the sample size matters, as well, in the sense that the larger is the sample size the greater must be z^2 to constitute convincing evidence against H_0. Is there a resolution to this controversy? I think there is, and I think it is quite clear that the Bayesians are right.

Consider first the significance level approach. Corresponding to a .05-level test is a power curve $\phi(\mu)=P(z^2\geqslant c^2(.05)|\mu)$ indicating the probability of rejecting the null hypothesis for particular values of μ. An example is graphed in Figure 4.8. The reader may convince himself that ϕ is a symmetric function around $\mu=0$, converging to one as μ increases.

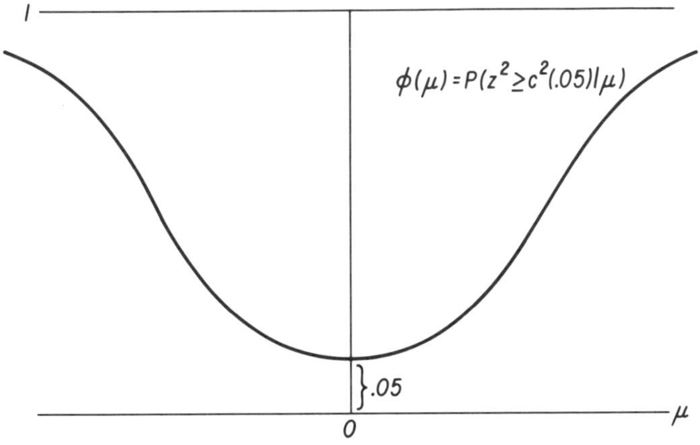

Fig. 4.8 Power curve.

As sample size increases, this power function gets steeper and steeper around $\mu=0$, and $[1-\phi(\mu)]$ which is the probability of making an "error of the second kind"—accepting a false null hypothesis—converges to zero at any value of $\mu\neq 0$. In contrast, the significance level—the probability of making an error of the first kind—stays forever at .05. The result is perfect error control against an error of the second kind but "mediocre" (.05) error control against an error of the first kind. The null hypothesis, however, is meant to be the favored hypothesis in the sense that incorrectly rejecting it is a more serious error than incorrectly accepting it. Fixed-level hypothesis testing, in contrast, clearly favors the alternative hypothesis—more and more so as sample size grows. The cure for this problem is obvious—the significance level must be made a decreasing function of sample size. Thus the conclusion we reached previously from the Bayesian viewpoint (that the interpretation of z^2 as evidence against H_0 depends on sample size) can also be reached within the confines of classical hypothesis testing. This still leaves arbitrary the particular function of sample size that the significance level should be set to. Although the Bayes factor $B(h^*)$ is a precise function of sample size, it depends on a somewhat arbitrary prior distribution. What is clear from both viewpoints is the fact that the interpretation of z^2 should depend on sample size.

As should be expected, a similar argument can be made from the likelihood standpoint. Suppose that the likelihood function of Figure 4.6 were discontinuous at μ_1 as in Figure 4.9. If we let the evidence against H_0 be summarized by $L(\mu_1;\bar{Y})$ we would conclude against H_0, even though for every other value of μ the null hypothesis is favored. It seems doubtful that we would really conclude against H_0 in this instance—some reference would be made to the a priori probability of μ_1, particularly to the fact that the function approximates its modal value on a "zero volume" set. But, essentially, the same thing happens to the likelihood function as sample size increases—it gets steeper and steeper around its mode and approximates the modal value in an ever-decreasing region. This suggests that instead of using the modal value as an indicator of the evidence against H_0 we might use the average value within a small fixed-size region around the mode, say, $\bar{Y}\pm\varepsilon$

$$L_2=\int_{\mu=\bar{Y}-\varepsilon}^{\bar{Y}+\varepsilon}(2\varepsilon)^{-1}L(\mu;\mathbf{Y})\,d\mu$$

$$=(2\varepsilon)^{-1}\exp\left(\frac{z^2}{2}\right)\int_{x=-\varepsilon}^{\varepsilon}\exp\left[-\frac{1}{2\sigma^2}\frac{x^2}{T}\right]dx.$$

The integral in this expression is the area under a normal curve. Making use of the fact that as σ^2/T gets small, essentially all of the probability is

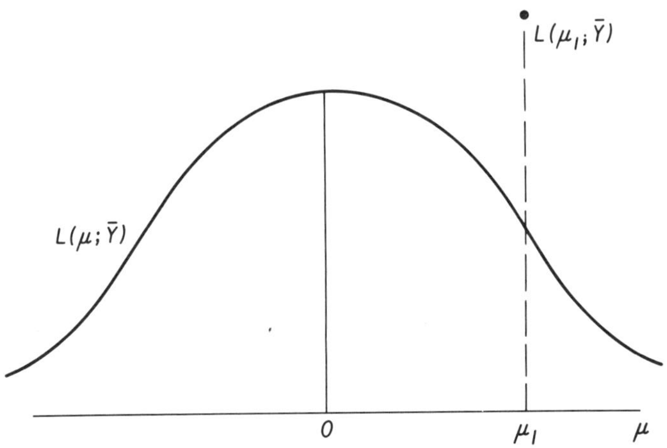

Fig. 4.9 Likelihood ratio: a peculiar example.

within an ε of its mean, we may write L_2 approximately as

$$L_2 \propto \frac{1}{2\varepsilon}\left(\frac{2\pi\sigma^2}{T}\right)^{\frac{1}{2}} \exp\left[\frac{z^2}{2}\right], \tag{4.7}$$

a decreasing function of sample size T. Incidentally, as sample size increases, the Bayes factor $B(h^*)$ converges to

$$B(h^*) = h^{*\frac{1}{2}}\left(\frac{\sigma^2}{T}\right)^{\frac{1}{2}} \exp\left(\frac{z^2}{2}\right),$$

not unlike L_2.

Before proceeding, it may be useful to point out that the testing of a composite hypothesis is an index number problem. An index is a single-valued function of a set of numbers $(x_1,\ldots,x_n)$ that in some sense captures the essential aspects of the entire set. Weighted averages are commonly used as indexes, say, $I(x_1,\ldots,x_n)=\Sigma_{i=1}^{n} w_i x_i$, $\Sigma^i w_i = 1, w_i \geqslant 0$. Theories of index numbers usually imply weight values w_i that are known only approximately, but this has not inhibited the practical construction of indexes. For example, consumer price indexes have weights that depend theoretically on unobservable marginal utilities.

An alternative index—the maximum of the class $I^*(x_1,\ldots,x_n)=\max_i x_i$—has to my knowledge never been proposed in the context of the theory of index numbers. It clearly misrepresents the class of numbers except in the special case when most other members of the class attain or nearly attain the maximum.

The problem of characterizing the evidence in favor of the foregoing hypotheses is precisely an index number problem. Instead of having a single number for each hypothesis, we have a continuous set of numbers. Interestingly enough, a classical analysis implicitly solves this index number problem by taking the maximum of the class, since a test of H_0 versus H_1 is based on the likelihood ratio $L_1 = \max_\mu L(\mu;\mathbf{Y})$. This solution to the index number problem seems inappropriate in general, but it is increasingly inappropriate as sample size increases, since as sample size increases the likelihood function becomes steeper and steeper and approximates its maximum in an ever decreasing region. In contrast, for a small sample size the likelihood function is relatively flat and the maximum *is* representative of wide regions.

A natural alternative is a weighted-average index, $\int L(\mu;\mathbf{Y})w(\mu)\,d\mu$, $w(\mu) \geq 0$, $\int w(\mu) = 1$. A prior distribution suggests itself as a weight function $w(\mu)$. It assigns low weight to relatively implausible values, thereby indicating that the performance of the "model" with a priori unlikely parameters is less important than the performance of the model with a priori likely parameters.

To put this another way, the hypotheses become hypotheses *cum* distributions; $H_1 : \mu \sim w(\mu)$. Each composite hypothesis is thereby mixed into a simple hypothesis, in the sense that each specifies *one* data distribution; $H_1 : \mathbf{Y} \sim \int f_N(\mathbf{Y}|\mathbf{1}\mu, \sigma^2\mathbf{I})w(\mu)\,d\mu$. Data are interpreted to favor one hypothesis relative to another if the data were more likely to have come from one of these distributions than from another.

This way of putting it highlights the fact that the Bayesian approach, rather than solving the composite hypothesis-testing problem, in fact transforms it into the simple hypothesis-testing problem of discriminating among completely specified distributions. It can be a useful approach only if we can identify personally and publicly acceptable weight functions and/or if these weight functions do not matter "too much." The difficulty of selecting weights has not inhibited the use of weighted-average indices in other problems, nor shall it necessarily inhibit their use for this problem.

4.4 Weighted Likelihoods: Conjugate Priors

Bayes factors for testing regression models are now discussed. The Bayes factor in favor of the ith hypothesis relative to the jth hypothesis is $P(\mathbf{Y}|H_i)/\ \mathrm{P}(\mathbf{Y}|H_j)$. It is convenient here to compute the marginal data density $P(\mathbf{Y}|H_i)$ for some hypothesis, and the subscript i may be suppressed. The model is taken to be

$$\mathbf{Y} = \mathbf{X}\boldsymbol{\beta} + \mathbf{u}, \qquad \mathbf{u} \sim N(\mathbf{0}_T, h^{-1}\mathbf{I}_T), \tag{4.8}$$

where $\mathbf{Y}$ is a $(T\times 1)$ observable vector, $\mathbf{X}$ is a $(T\times k)$ observable matrix, $\boldsymbol{\beta}$ is a $(k\times 1)$ unobservable parameter vector, and $\mathbf{u}$ is a $(T\times 1)$ unobservable error vector distributed normally with mean vector zero and variance-covariance matrix $h^{-1}\mathbf{I}_T$ with h a scalar precision parameter.

As discussed in the previous section, a Bayesian approach requires this hypothesis to be complemented with a prior distribution for the parameters $(\boldsymbol{\beta},h)$. This distribution is used to marginalize out the parameters, and the hypothesis is treated as if it specified a unique distribution

$$f(\mathbf{Y})=\int\int f(\mathbf{Y}|\boldsymbol{\beta},h)f(\boldsymbol{\beta},h)\,d\boldsymbol{\beta}\,dh. \tag{4.9}$$

The posterior probability of this hypothesis is, by Bayes' rule, proportional to the prior probability times the density (4.9) evaluated at the observed value of $\mathbf{Y}$. For obvious reasons, this is called a marginal likelihood.

Beginning with the simplest case, suppose the process variance h^{-1} is known, and let the prior for $\boldsymbol{\beta}$ be normal with mean $\mathbf{b}^*$ and variance $\mathbf{H}^{*-1}$. By inspection of Equation (4.8), $\mathbf{Y}$ is a linear combination of normals which is itself normal with mean $E(\mathbf{Y})=\mathbf{X}\mathbf{b}^*$ and variance $V(\mathbf{Y})=\mathbf{X}\mathbf{H}^{*-1}\mathbf{X}'+h^{-1}\mathbf{I}_T$:

$$f(\mathbf{Y})=(2\pi)^{-T/2}|V(\mathbf{Y})|^{-1/2}\exp\left[-\frac{1}{2}hQ\right] \tag{4.10}$$

where

$$hQ=(\mathbf{Y}-\mathbf{X}\mathbf{b}^*)'V^{-1}(\mathbf{Y})(\mathbf{Y}-\mathbf{X}\mathbf{b}^*).$$

This expression can be rewritten by defining $\mathbf{N}^*=h^{-1}\mathbf{H}^*$ and by observing that

$$V^{-1}(\mathbf{Y})=\left(\mathbf{X}\mathbf{H}^{*-1}\mathbf{X}'+h^{-1}\mathbf{I}_T\right)^{-1}=h\left(\mathbf{I}_T-\mathbf{X}(\mathbf{X}'\mathbf{X}+\mathbf{N}^*)^{-1}\mathbf{X}'\right)$$
$$|V(\mathbf{Y})|^{-1}=h^T|\mathbf{N}^*||\mathbf{N}^*+\mathbf{X}'\mathbf{X}|^{-1}.$$

Letting $\mathbf{N}=\mathbf{X}'\mathbf{X}$ and $\mathbf{b}=(\mathbf{X}'\mathbf{X})^{-1}\mathbf{X}'\mathbf{Y}$, and after some manipulation, the quadratic form Q can be written as either

$$Q=(\mathbf{Y}-\mathbf{X}\mathbf{b})'(\mathbf{Y}-\mathbf{X}\mathbf{b})+(\mathbf{b}-\mathbf{b}^*)(\mathbf{N}^{*-1}+\mathbf{N}^{-1})^{-1}(\mathbf{b}-\mathbf{b}^*) \tag{4.11}$$

or,

$$Q=(\mathbf{Y}-\mathbf{X}\mathbf{b}^*)'(\mathbf{Y}-\mathbf{X}\mathbf{b}^*)-(\mathbf{b}-\mathbf{b}^*)'\mathbf{N}(\mathbf{N}^*+\mathbf{N})^{-1}\mathbf{N}(\mathbf{b}-\mathbf{b}^*). \tag{4.12}$$

By inspection of (4.10), the data favor the hypothesis in question if Q is small. Q is written in (4.11) as the minimum error sum of squares plus a factor that depends on the difference between the least-squares estimate $\mathbf{b}$ and the prior location $\mathbf{b}^*$. It is apparent that a model is to be judged in terms of not only its R^2 but also by the "plausibility" of its estimates. In fact, (4.12) describes Q as an error sum of squares using the prior location as the estimate minus a factor also depending on the difference between $\mathbf{b}$ and $\mathbf{b}^*$. Thus Q is a number between $(\mathbf{Y}-\mathbf{X}\mathbf{b})'(\mathbf{Y}-\mathbf{X}\mathbf{b})$ and $(\mathbf{Y}-\mathbf{X}\mathbf{b}^*)'(\mathbf{Y}-\mathbf{X}\mathbf{b}^*)$, which suggests that a model's performance should be judged not in terms of the acceptability of $\mathbf{b}$ per se, but rather in terms of the R^2 when a more acceptable coefficient vector is employed.

Of course, the case of known process variance, h^{-1}, is of little practical interest. If we use a conjugate prior with uncertain h, these results generalize straightforwardly. Let $(\boldsymbol{\beta}, h)$ have the distribution $f_N(\boldsymbol{\beta}|\boldsymbol{b}^*, h^{-1}\mathbf{N}^{*-1})\, f_\gamma(h|s_1^2, \nu_1)$ where f_N is a multivariate normal distribution with mean $\mathbf{b}^*$ and variance matrix $h^{-1}\mathbf{N}^{*-1}$ and f_γ is a gamma distribution with location and scale parameters s_1^2 and ν_1. The predictive density then becomes a multivariate Student function

$$f(\mathbf{Y}) = \int\int f(\mathbf{Y}|\boldsymbol{\beta},h) f(\boldsymbol{\beta},h)\, d\boldsymbol{\beta}\, dh$$

$$= k(\upsilon_1, T)\left|\frac{\mathbf{M}}{s_1^2}\right|^{1/2}\left(\nu_1 + \frac{Q}{s_1^2}\right)^{-(\nu_1+T)/2} \qquad (4.13)$$

where

$$\mathbf{M} = \mathbf{I}_T - \mathbf{X}(\mathbf{N}^* + \mathbf{X}'\mathbf{X})^{-1}\mathbf{X}'$$

$$|\mathbf{M}| = |\mathbf{N}^*||\mathbf{N}^* + \mathbf{X}'\mathbf{X}|^{-1}$$

$$k(\nu_1, T) = \nu_1^{\nu_1/2}\left(\frac{\nu_1}{2} + \frac{T}{2} - 1\right)! \Big/ \pi^{T/2}\left(\frac{\nu_1}{2} - 1\right)!$$

Note that Equations (4.13) and (4.10) are somewhat similar monotonic functions of Q.

Leamer (1970) provides an approximation to the marginal likelihood when the normal prior for $\boldsymbol{\beta}$ is independent of h; Dickey (1971), for the more general case of a Student prior for $\boldsymbol{\beta}$, writes the marginal likelihood as a one-dimensional integral.

4.5 Weighted Likelihoods: Diffuse Priors

The critical defect of a Bayesian analysis of data is that prior distributions are both personally difficult to specify and also subject to variation among interested people. As a consequence, a Bayesian analysis based on any particular prior is of little interest. Instead, a researcher is obligated to report as well as possible the mapping from priors to posteriors, thereby servicing a wide readership and also highlighting those features of the prior that are critical in this sense. Posterior probabilities based on a particular prior are of interest *if* the particular prior is characteristic of a class of interesting priors and if the posterior implied by this prior is "essentially the same as" the posterior for all priors in the class.

For example, in the usual analysis of the linear-regression model the posterior distribution corresponding to the diffuse prior ($\boldsymbol{\beta}$ uniform) is of interest because there is a wide, easily identifiable class of priors that are dominated by the data and that, therefore, lead to posteriors essentially the same as the posterior corresponding to the diffuse prior. Unfortunately, this is not also true for posterior probabilities of composite hypotheses, since various priors that are all relatively diffuse and "noninformative" lead to radically different posterior probabilities.

Making use of Equation (4.13), for example, the Bayes factor in favor of H_i relative to H_j is

$$\frac{f(\mathbf{Y}|H_i)}{f(\mathbf{Y}|H_j)}=\frac{k(\nu_{1i},T)}{k(\nu_{1j},T)}\,\frac{|\mathbf{N}_i^*|^{1/2}}{|\mathbf{N}_j^*|^{1/2}}\,\frac{|\mathbf{N}^*{}_i+\mathbf{X}_i'\mathbf{X}_i|^{-1/2}}{|\mathbf{N}_j^*+\mathbf{X}_j'\mathbf{X}_j|^{-1/2}}\,\frac{s_{1i}^{-T}}{s_{1j}^{-T}}\,\frac{\left(\nu_{1i}+Q_i/s_{1i}^2\right)^{-(\nu_{1i}+T)/2}}{\left(\nu_{1j}+Q_j/s_{1j}^2\right)^{-(\nu_{1j}+T)/2}} \tag{4.14}$$

This formula involves the ratio $|\mathbf{N}_i^*|^{1/2}/|\mathbf{N}_j^*|^{1/2}$. As we let the matrices $\mathbf{N}^*$ become small to reflect decreasing information about the coefficients, this ratio converges to the indeterminate ratio of two zeroes, which can take on any number between zero and infinity, depending on the assumed speeds of convergence.

We may also set ν_1 to zero, to let the prior for h be diffuse. Three obvious ways of making our prior for $\boldsymbol{\beta}$ diffuse ($\mathbf{N}^*\rightarrow\mathbf{0}$) are to define $\mathbf{N}^*=\delta\mathbf{I}_k$ or $\mathbf{N}^*=\sqrt[k]{\delta}\,\mathbf{I}_k$ or $\mathbf{N}^*=(\sqrt[k]{\delta}\,)\mathbf{X}'\mathbf{X}$ and let the scalar δ go to zero. As the reader may verify, these lead to three different limiting expressions:

$$f(\mathbf{Y}|H_i)/f(\mathbf{Y}|H_j)=\begin{cases}0 & \text{if } k_i>k_j\\ \left(\dfrac{|\mathbf{X}_j'\mathbf{X}_j|}{|\mathbf{X}_i'\mathbf{X}_i|}\right)^{1/2}\left(\dfrac{ESS_j}{ESS_i}\right)^{T/2} & \text{if } k_i=k_j\\ \infty & \text{if } k_i<k_j\end{cases}$$

$$\frac{f(\mathbf{Y}|H_i)}{f(\mathbf{Y}|H_j)}=\left(\frac{|\mathbf{X}_j'\mathbf{X}_j|}{|\mathbf{X}_i'\mathbf{X}_i|}\right)^{1/2}\left(\frac{ESS_j}{ESS_i}\right)^{T/2}$$

$$\frac{f(\mathbf{Y}|H_i)}{f(\mathbf{Y}|H_j)}=\left(\frac{ESS_j}{ESS_i}\right)^{T/2}$$

where $ESS_i = \mathbf{Y}'(\mathbf{I} - \mathbf{X}_i'(\mathbf{X}_i'\mathbf{X}_i)^{-1}\mathbf{X}_i')\mathbf{Y}$. Each of these expressions is representative of the posteriors corresponding to a mathematically well-defined class of diffuse priors, but no class has an unambiguous claim to representing "vague" prior information. Thus the class of posteriors corresponding to diffuse priors is not well defined, and the posterior corresponding to a particular kind of diffuseness is of reduced interest. This contrasts with a posterior for the coefficient vector $\boldsymbol{\beta}$, which is the same under all (limiting) definitions of diffuseness.

We can see from another viewpoint the problem with the diffuse priors by examining the distribution of $\mathbf{Y}$. The variance of $\mathbf{Y}$ is

$$V(\mathbf{Y}) = \mathbf{X}V(\boldsymbol{\beta})\mathbf{X}' + V(\mathbf{u}).$$

Those linear combination $\boldsymbol{\psi}'\mathbf{Y}$ that are orthogonal to the columns of $\mathbf{X}$, $\boldsymbol{\psi}'\mathbf{X} = \mathbf{0}$, have variance independent of $V(\boldsymbol{\beta})$

$$\begin{aligned} V(\boldsymbol{\psi}'\mathbf{Y}) &= \boldsymbol{\psi}'\mathbf{X}V(\boldsymbol{\beta})\mathbf{X}'\boldsymbol{\psi} + \boldsymbol{\psi}'V(\mathbf{u})\boldsymbol{\psi} \\ &= \boldsymbol{\psi}'V(\mathbf{u})\boldsymbol{\psi}. \end{aligned}$$

Thus as $V(\boldsymbol{\beta})$ explodes, the predictive density on the $T-k$ dimensional subspace of $\mathbf{Y}$ defined by $\boldsymbol{\psi}'\mathbf{X} = \mathbf{0}$ maintains a finite variance. An interpretation of this might be that predictions about joint events in the full T-dimensional space are called off, but predictions on certain subspaces are still on. The subspace over which predictions are proper clearly depends on $\mathbf{X}$. Since we can compare the predictive performance of two models in a nonarbitrary way only if the two models are predicting the same events, it will prove impossible to choose models without informative priors. This statement holds, incidentally, even when the number of explanatory variables is the same in all models, since although the dimensionalities of the predictions are the same, the prediction spaces are different unless $\mathbf{X}$ is the same for all hypotheses.

This dilemma does have a potential escape. Instead of seeking diffuse priors, let us find dominated priors, that is, let us explore the behavior of the marginal likelihoods as sample evidence accumulates. As T, Q, and $\mathbf{X}'\mathbf{X}$ grow, Equation (4.14) is well approximated by

$$\frac{f(\mathbf{Y}|H_i)}{f(\mathbf{Y}|H_j)} \cong c\left(\frac{|\mathbf{X}_j'\mathbf{X}_j|}{|\mathbf{X}_i'\mathbf{X}_i|}\right)^{1/2}\left(\frac{ESS_j}{ESS_i}\right)^{T/2} \tag{4.15}$$

where c is a constant that does not change with sample size. Like the prior odds ratio, this constant will come to be dominated by the other terms. By the same argument, if the explanatory variables come from a stationary process, $\mathbf{X}'\mathbf{X}/T$ will converge to a constant, and a useful approximation is

$$\frac{f(\mathbf{Y}|H_i)}{f(\mathbf{Y}|H_j)} \cong cT^{(k_j-k_i)/2}\left(\frac{ESS_j}{ESS_i}\right)^{T/2}. \tag{4.16}$$

Equation (4.15) has one obvious defect. It is not invariant to scale transformations of the explanatory variables. By a suitable change in the units of measurement, $|\mathbf{X}_j'\mathbf{X}_j|/|\mathbf{X}_i'\mathbf{X}_i|$ can be made to favor any hypothesis (if each has at least one explanatory variable not found in any of the others). On the other hand, an argument can be made that some adjustment should be made for the variability of the explanatory variable set, as is done by this term. If there is a constant term in the regression, then $|\mathbf{X}'\mathbf{X}|$ is proportional to the determinant of the matrix of moments about the means of the nonconstant variables or the generalized variance of the explanatory variable set. Thus, by Equation (4.15), a model that enjoys a richly variable set of explanatory variables is expected to yield a smaller error sum of squares than a model with a poorer set of explanatory variables.

A formula that is both invariant to scale transformations and adjusts for the variability of the explanatory variable set is

$$\frac{f(\mathbf{Y}|H_i)}{f(\mathbf{Y}|H_j)} = c\left(\frac{|\mathbf{R}_j|}{|\mathbf{R}_i|}\right)^{1/2} T^{(k_j - k_i)/2}\left(\frac{ESS_j}{ESS_i}\right)^{T/2} \tag{4.17}$$

where $\mathbf{R}_i$ is the matrix of correlation coefficients of the explanatory variables. Note that $|\mathbf{X}'\mathbf{X}| = T^k|\mathbf{R}|\Pi_{l=1}^k \hat{\sigma}_l^2$ where $\hat{\sigma}_l^2$ is the sample estimate of the variance of the lth explanatory variable. Equation (4.17) thus involves the assumption that $|\mathbf{N}^*|/\Pi_l \hat{\sigma}_l^2$ is constant across models.

Which of these many formulas are we then to choose? The following properties do seem desirable:

a. There must be no arbitrary constants.
b. The posterior probability of a model should be invariant to linear transformations of the data.
c. There should be a degrees-of-freedom adjustment; of two models that both yield the same *ESS*, the one with the fewer explanatory variables should have the higher posterior probability.
d. A model with a richly variable explanatory variable set should be expected to yield a smaller *ESS* than one with highly collinear data.

Of these properties only (d) seems to be open to serious question, because by a linear transformation any set of explanatory variables can be made to be orthogonal. As a consequence, (d) is in conflict with (b). The formula that satisfies (a), (b), and (c) is Equation (4.16), currently this author's favorite. If there is an obvious parameterization such that $|\mathbf{N}^*|/\Pi_l \hat{\sigma}_l^2$ is constant across models, than Equation (4.17), which does adjust for the variability of the explanatory variable set, is preferred.

It should be emphasized that the dominance notion is not really a solution to the problem, even though it leads unambiguously to formulas such as Equation (4.16), since in order to apply these formulas we need to know whether the sample does, in fact, dominate the prior. For any proper prior, no matter how diffuse, there are observations that make Equation (4.16) a very poor approximation to (4.14). The only way to know if the particular sample does, in fact, dominate the prior is to specify fully the prior, but having done that Equation (4.14) applies.

For other discussions of improper priors for this problem see Jeffreys (1961), Thornber (1966), Geisel (1969), Lempers (1971), Zellner (1971), and Dagenais (1972).

To conclude this section, we have seen that various reasonable definitions of diffuseness lead to rather different posterior probabilities of composite hypotheses. This makes posterior probabilities computed from any particular formula less interesting. This author has the personal opinion that the problem is of academic interest only, since a prior that allocates positive probability to subspaces of the parameter space but is otherwise diffuse represents a peculiar and unlikely blend of knowledge and ignorance. Parenthetically, what often appears to be choice among potentially true models is, in fact, the choice of a simple model that works well for some decisions. In other cases, hypothesis testing is used to introduce into the analysis uncertain prior information about parameters. These as well as other specification searches are discussed in detail in subsequent chapters.

4.6 Conclusion

To conclude, let us reconsider the five problems mentioned in the introduction, this time providing answers:

Answer 1. Classical hypothesis testing at a fixed level of significance increasingly distorts the interpretation of the data against a null hypothesis as the sample size grows. *The significance level should consequently be a decreasing function of sample size.*

Under one definition of diffuseness we saw that the posterior odds ratio in favor of the alternative hypothesis is the prior odds times the factor

$$B=\left(\frac{ESS_0}{ESS_1}\right)^{T/2}T^{(k_0-k_1)/2}.$$

We say that the evidence favors the alternative hypothesis if $B>1$, which can be written in terms of the F value defined in (4.2) as

$$F>\frac{T-k}{p}(T^{p/T}-1)$$

where $p=k_1-k_0$ is the number of restrictions. Table 4.1 provides these critical values as a function of sample size T, the degrees of freedom $T-k$ and the number of restrictions p. For comparison, the critical value of the F test at the .05 level are also included, and we observe the general result that a Bayesian with this kind of prior requires much larger F values as sample size increases.

Answer 2. There is nothing special about complex, nonnested structures of hypotheses. The posterior probability of an hypothesis H_i is $P(H_i|Y)=P(Y|H_i)P(H_i)/\Sigma_j P(Y|H_j)P(H_j)$ regardless of the structure of the hypotheses.

Answer 3. The existence of prior information about the parameters influences hypothesis testing in the sense that a hypothesis is to be judged at a priori likely values of the parameters as well as at those values favored by the data.

However, there does not seem to me to be a proper prior distribution that would lead to the common procedure of discounting an R^2 in four increasing steps, depending on whether a parameter is significantly different from zero and the right sign, insignificant and the right sign, insignificant and the wrong sign, and (worst of all) significant and the wrong sign. Although the normal priors discussed previously inadequately capture information about signs, they lead one to hope that his estimates are insignificantly different from his prior mean. Take the case when $\boldsymbol{\beta}$ is known to be $\mathbf{b}^*$ and the prior for h is diffuse, $f(h)\propto h^{-1}$. The marginal likelihood is then proportional to

$$f(\mathbf{Y}|H)\propto\int h^{T/2-1}\exp\left[-\frac{1}{2}h\big(ESS+(\mathbf{b}^*-\mathbf{b})(\mathbf{X}'\mathbf{X})(\mathbf{b}^*-\mathbf{b})\big)\right]dh$$

$$=\left[ESS+(\mathbf{b}-\mathbf{b}^*)'(\mathbf{X}'\mathbf{X})(\mathbf{b}-\mathbf{b}^*)\right]^{-T/2}$$

$$=ESS^{-T/2}\left(1+F\frac{k}{T-k}\right)^{-T/2}$$

where F is the F statistic for testing $\boldsymbol{\beta}=\mathbf{b}^*$. The larger is this F value, the less is $f(\mathbf{Y}|H)$ and the less likely is the data to have been generated by this model. Thus you are hoping for insignificance, not significance.

The "bigger is better" philosophy embedded in the usual procedure would seem to require an improper prior that says "bigger is more likely." For example, given the sample mean $\overline{Y}$ which is distributed normally with mean μ and variance σ^2/T and a prior for μ that is uniform between zero and M, the Bayes factor in favor of the hypothesis $\mu=0$ versus the

Table 4.1

Bayesian and Classical Critical Values of the F test

	$T-k=$	1	2	3	4	5	10	50	100	1000
	$k=1$	0.41	0.88	1.24	1.52	1.74	2.44	4.01	4.68	6.93
	2	0.44	0.83	1.14	1.39	1.60	2.30	3.95	4.64	6.92
	3	0.41	0.76	1.04	1.28	1.48	2.18	3.89	4.60	6.91
$p=1$	4	0.38	0.70	0.96	1.19	1.38	2.07	3.83	4.57	6.91
	5	0.35	0.64	0.89	1.11	1.29	1.98	3.78	4.53	6.90
	10	0.24	0.46	0.65	0.83	0.99	1.62	3.53	4.37	6.87
	20	0.16	0.30	0.44	0.57	0.69	1.20	3.13	4.07	6.81
	5% point[b]	161	18.5	10.1	7.8	6.6	4.96	4.03	3.94	3.85
	$k=1$	0.50	1.08	1.50	1.81	2.04	2.73	4.17	4.78	6.95
	2	0.54	1.00	1.36	1.63	1.86	2.57	4.10	4.75	6.94
	3	0.50	0.90	1.23	1.49	1.70	2.42	4.04	4.71	6.94
$p=2$	4	0.45	0.82	1.12	1.36	1.57	2.29	3.98	4.67	6.93
	5	0.41	0.74	1.02	1.26	1.46	2.17	3.92	4.63	6.93
	10	0.27	0.51	0.73	0.92	1.09	1.75	3.66	4.46	6.90
	20	0.17	0.32	0.47	0.61	0.73	1.27	3.23	4.15	6.84
	5% point[a]	200	19.0	9.55	6.94	5.79	4.10	3.18	3.09	3.00
	$k=1$	0.61	1.33	1.83	2.17	2.42	3.08	4.34	4.90	6.97
	2	0.67	1.22	1.63	1.93	2.17	2.87	4.27	4.86	6.97
	3	0.61	1.08	1.45	1.74	1.97	2.69	4.20	4.82	6.96
$p=3$	4	0.54	0.97	1.30	1.57	1.80	2.53	4.13	4.78	6.96
	5	0.48	0.87	1.18	1.44	1.66	2.40	4.07	4.74	6.95
	10	0.31	0.57	0.81	1.01	1.20	1.89	3.79	4.56	6.92
	20	0.18	0.35	0.51	0.65	0.79	1.35	3.33	4.24	6.86
	5% point[b]	216	19.2	9.28	6.59	5.41	3.71	2.79	2.70	2.61
	$k=1$	0.75	1.66	2.25	2.62	2.88	3.48	4.52	5.01	7.00
	2	0.83	1.50	1.97	2.30	2.55	3.22	4.44	4.97	6.99
	3	0.75	1.31	1.73	2.04	2.29	3.00	4.37	4.93	6.99
$p=4$	4	0.66	1.15	1.53	1.83	2.07	2.81	4.30	4.89	6.98
	5	0.58	1.02	1.37	1.66	1.89	2.65	4.23	4.85	6.97
	10	0.35	0.64	0.90	1.13	1.32	2.05	3.92	4.66	6.94
	20	0.20	0.38	0.54	0.70	0.84	1.43	3.43	4.33	6.88
	5% point[b]	225	19.25	9.12	6.39	5.19	3.48	2.56	2.46	2.38
	$k=1$	0.93	2.10	2.79	3.20	3.45	3.95	4.70	5.13	7.02
	2	1.05	1.86	2.40	2.76	3.01	3.63	4.62	5.09	7.02
	3	0.93	1.60	2.07	2.41	2.67	3.36	4.54	5.05	7.01
$p=5$	4	0.80	1.38	1.81	2.13	2.39	3.13	4.47	5.00	7.00
	5	0.69	1.21	1.60	1.91	2.16	2.93	4.40	4.96	7.00
	10	0.39	0.73	1.01	1.25	1.47	2.23	4.07	4.76	6.97
	20	0.21	0.41	0.59	0.75	0.90	1.53	3.55	4.42	6.91
	5% point[b]	230	19.30	9.01	6.26	5.05	3.33	2.40	2.30	2.22

[a] T = number of observations, k = number of parameters, p = number of restrictions being tested.

[b] Classical critical value at the .05 level of significance.

hypothesis $\mu>0$ is

$$\frac{f(\bar{Y}|\mu=0)}{\int_0^M f(\bar{Y}|\mu,\sigma^2)M^{-1}d\mu}$$

$$=\frac{(2\pi\sigma^2/T)^{1/2}\exp\left[-\bar{Y}^2/(2\sigma^2/T)\right]}{\int_0^M(2\pi\sigma^2/T)^{1/2}\exp\left[-(\bar{Y}-\mu)^2/(2\sigma^2/T)\right]M^{-1}d\mu}$$

$$=\frac{(2\pi\sigma^2/T)^{1/2}\exp\left[-z^2/2\right]}{M^{-1}P^*(0\leqslant\mu\leqslant M)}$$

where z^2 is the square of the normal statistic for testing $\mu=0$, $z^2=\bar{Y}^2/(\sigma^2/T)$, and $P^*(0\leqslant\mu\leqslant M)$ is the posterior probability that $0\leqslant\mu\leqslant M$ given an improper prior for μ that is uniform on the whole line. The numerator of this Bayes factor is unambiguously a decreasing function of z^2. The denominator, however, may either increase or decrease with z^2 depending on whether $\bar{Y}$ is in the interval $0\leqslant\bar{Y}\leqslant M$ or not. It may, nonetheless, be approximately true that the Bayes factor in favor of $\mu>0$ is greatest if $\bar{Y}$ is positive and z^2 large, relatively great for $\bar{Y}$ positive, small for $\bar{Y}$ negative, and especially small if $\bar{Y}$ is significantly negative.

Answer 4. Measures of location and dispersion of a parameter vector β_i follow necessarily from the probability function

$$f(\boldsymbol{\beta}_i|\mathbf{Y})=\sum_j f(\boldsymbol{\beta}_i|\mathbf{Y},H_j)f(H_j|\mathbf{Y}).$$

The interpretation of this simple formula is not entirely trivial. We have yet to identify the slightly confusing p.d.f. $f(\boldsymbol{\beta}_i|\mathbf{Y},H_j)$ for $i\neq j$. It summarizes opinions about a parameter in the ith model given that the jth model generated the data. Conditional on the jth model the data come from the distribution $f(\mathbf{Y}|\boldsymbol{\beta}_j,H_j)$ and therefore contain information only about $\boldsymbol{\beta}_j$. It may, nonetheless, be the case that $\boldsymbol{\beta}_i$ and $\boldsymbol{\beta}_j$ are a priori correlated, and we would then obtain information about $\boldsymbol{\beta}_i$, or in terms of probability functions $f(\boldsymbol{\beta}_i|\mathbf{Y},H_j)\neq f(\boldsymbol{\beta}_i|H_j)$.

For example, consider the hypotheses $H_1:\mathbf{Y}=\mathbf{x}\gamma+\mathbf{z}\alpha+\mathbf{u}$ and $H_2:\mathbf{Y}=\mathbf{x}\gamma+\mathbf{w}\delta+\boldsymbol{\mu}$, with parameter vectors $\boldsymbol{\beta}_1=(\gamma,\alpha)$, $\boldsymbol{\beta}_2=(\gamma,\delta)$, and with the first parameters thus perfectly correlated. In this case the probability function for $\gamma=\beta_{11}=\beta_{12}$ is the mixture $f(\gamma|H_1,\mathbf{Y})f(H_1|\mathbf{Y})+f(\gamma|H_2,\mathbf{Y})f(H_2|\mathbf{Y})$ where $f(\gamma|H_i,\mathbf{Y})$ is the usual posterior distribution for γ given model H_i. Letting $\pi_i=P(H_i|\mathbf{Y})$, $m_i=E(\gamma|\mathbf{Y},H_i)$ and $V_i=V(\gamma|\mathbf{Y},H_i)$

the mean and variance of γ are

$$E(\gamma|\mathbf{Y}) = \sum \pi_i m_i$$

$$V(\gamma|\mathbf{Y}) = E(\gamma^2|\mathbf{Y}) - \left(\sum \pi_i m_i\right)^2$$

$$= \sum \pi_i (V_i + m_i^2) - \left(\sum \pi_i m_i\right)^2$$

$$= \sum \pi_i V_i + \left[\sum \pi_i m_i^2 - \left(\sum \pi_i m_i\right)^2\right]$$

$$= \sum \pi_i V_i + \sum \pi_i \left(m_i - \sum \pi_i m_i\right)^2$$

where the last term in the brackets is the variance of a discrete probability function that allocates probability π_i at location m_i. The point we wish to draw attention to is that although the mean is a mixture of the means from each of the regression equations, the variance exceeds a weighted average of the variances by an amount that depends on the variability of the estimates across equations. Thus although the several regressions may individually yield highly accurate estimates, if those estimates are very different and if given the data there remains considerable ambiguity about the model, the result may be considerable uncertainty about the parameter.

Next consider hypotheses that have no common parameters, and assume furthermore that the parameters are completely independent across hypotheses. Thus if the data are generated by the first model, no information can be gathered about coefficients in other models; in the foregoing notation we must have $f(\boldsymbol{\beta}_i|\mathbf{Y}, H_j) = f(\boldsymbol{\beta}_i|H_j)$. In order to apply the formula we must, of course, also determine $f(\boldsymbol{\beta}_i|H_j)$. The reader may verify that without saying so in the previous example we have set $f(\boldsymbol{\beta}_i|H_j) = f(\boldsymbol{\beta}_i)$ for all j. In words, our prior information about the coefficient $\boldsymbol{\beta}_i$ is independent of the hypothesis that applies. It is more natural in this second situation to assume that $f(\boldsymbol{\beta}_i|H_j)$ is a degenerate probability function that assigns all probability to a point. For example, consider the pair of hypotheses $H_1: \mathbf{Y} = \mathbf{x}\gamma + \mathbf{u}$, $H_2: \mathbf{Y} = \mathbf{z}\delta + \mathbf{u}$. We may either say that the distribution of δ given H_1 assigns all probability to the value zero, or we can define a new parameter β called the effect of $\mathbf{z}$ on $\mathbf{Y}$ which is zero given the first hypothesis. In either case, the distribution of the parameter is a mixture of the origin with weight $1 - P(H_i|\mathbf{Y})$ and the conditional posterior $f(\boldsymbol{\beta}_i|H_i, \mathbf{Y})$ with weight $P(H_i|\mathbf{Y})$. If we prefer to have $f(\boldsymbol{\beta}_i|H_j) = f(\boldsymbol{\beta}_i)$ we will obtain a posterior distribution for $\boldsymbol{\beta}_i$ that is a mixture of the prior $f(\boldsymbol{\beta}_i)$ with weight $1 - P(H_i|\mathbf{Y})$ and the conditional posterior $f(\boldsymbol{\beta}_i|H_i, \mathbf{Y})$ with weight $P(H_i|\mathbf{Y})$. The resultant increased uncertainty about the parameter, due to uncertainty about the model, is obvious.

Answer 5. A researcher who uses more than one model can report the overall effectiveness of his research in terms of the average marginal likelihood

$$f(\mathbf{Y}) = \sum_i f(\mathbf{Y}|H_i)P(H_i).$$

Assuming equal prior probabilities of M different models and the diffuse prior result (4.16), this becomes

$$f(\mathbf{Y}) \propto M^{-1} \sum_{i=1}^{M} T^{-k_i/2} (ESS_i)^{-T/2}.$$

We can transform $f(\mathbf{Y}|H_i)$ into an R^2 by the formula

$$f(\mathbf{Y}|H_i) \propto T^{-k_i/2} \left(1 - R_i^2\right)^{-T/2}$$

which solves to

$$\left(1 - R_i^2\right) \propto \left[\frac{T^{-k_i/2}}{f(\mathbf{Y}|H_i)} \right]^{2/T}.$$

The same transformation may be applied to $f(\mathbf{Y})$ to obtain a "grand" R^2

$$(1 - R^{*2}) = \left[\frac{T^{-k^*/2}}{f(\mathbf{Y})} \right]^{2/T}$$

$$= \left[M^{-1} \sum_{i=1}^{M} T^{(k^* - k_1)/2} \left(1 - R_i^2\right)^{-T/2} \right]^{-2/T}$$

where k^* is a (fictitious) average k and R^{*2} is the grand R^2. Assuming equal k_i or ignoring the $T^{k^* - k_i}$ terms we have the "overall" R^2 as

$$R^{*2} = 1 - \left[\frac{\sum_{i=1}^{M} \left(1 - R_i^2\right)^{-T/2}}{M} \right]^{-2/T}.$$

This formula is intended to penalize specification searches, since estimated models with low R^2s tend to lower the grand R^2. The penalty is not as great as you might imagine, however. Grand R^2s are reported in Table 4.2 which makes use of the assumptions that the best model yields an R^2 equal to .9, and that all the others yield identical R^2s. The lowest grand R^2 in the table is .43, which requires 99 models with zero R^2s and a very small

Table 4.2

Grand R^2's

		$M=2$	4	10	100
	$T=5$	.87	.64	.50	.43
$R_2^2=0$	10	.89	.87	.84	.75
	100	.9	.9	.9	.89
	$T=5$	.89	.85	.83	.8
$R_2^2=.8$	10	.89	.87	.85	.81
	100	.9	.9	.9	.89

M = number of models, T = number of observations
$R^2=.9$; $R_2^2=R_3^2=\cdots=R_M^2$

sample size, $T=5$. For reasonably large sample sizes compared with the number of models, the grand R^2 is almost equal to the maximum R^2. This result derives from the fact that for reasonably large T, the marginal likelihoods are extremely sensitive to R^2. Imperceptible differences in R^2s translate into large differences in posterior probabilities, and all but the best model will be assigned nearly zero posterior probability.

CHAPTER 5

INTERPRETIVE SEARCHES

In this chapter we discuss searches designed to interpret multi-dimensional evidence. A completely specified model usually contains a large number of collinear explanatory variables, and the least-squares estimates that result are rarely "acceptable." Various constraints on the parameters may be imposed to "improve" the estimate, and one among many constrained least-squares estimates is usually selected to convey the data evidence.

A pair of fictitious examples illustrate the phenomenon.

Example 1. The demand for oranges D is thought to depend negatively on the price of oranges P, positively on the price of grapefruit π, and positively on money income Y. The following regression is estimated (with standard errors in parentheses)

$$\log D = 7.0 + \underset{(.3)}{.1}\ \log P + \underset{(.4)}{.2}\ \log \pi + \underset{(.2)}{.6}\ \log Y.$$

Unhappily, the direct price elasticity—the coefficient of $\log P$—has the wrong sign. Neither price coefficient is significantly different from zero. In an effort to "improve" these results, the following constrained regression is estimated:

$$\log D = 4.0 - \underset{(.1)}{.2}\log\left(\frac{P}{\pi}\right) + \underset{(.2)}{.5}\log\left(\frac{Y}{P}\right).$$

The constraint that implies this equation is suggested by economic theory; it is not rejected at the 5% level, and it yields estimates that are both the right sign and also statistically significant. The second regression is, therefore, selected to convey the content of the data.

Example 2. Consumption expenditures C_t in quarter t are thought to depend especially on income receipts in that quarter Y_t, but also somewhat on receipts in previous quarters $Y_{t-1}, Y_{t-2}, \ldots$. To estimate this relationship, the explanatory variables $Y_t, Y_{t-1}, \ldots$ are sequentially included in the equation until one of the variables becomes statistically insignificant. This procedure yields the estimated relationship

$$C_t = \underset{(.2)}{4.0} + \underset{(.1)}{.3}Y_t + \underset{(.1)}{.4}Y_{t-1} + \underset{(.1)}{.1}Y_{t-2}.$$

These two examples illustrate, first, a *contracting search*, in which a series of constraints is imposed on a general model and, second, an *expanding search*, in which the assumptions implicit in an initial model are relaxed and a variety of more general models are estimated. In both cases the most general model leads to, or would lead to "unacceptable" results. In the case of the expanding search it is assumed at the outset that the data could not "support" the most general model. A severe set of constraints is initially imposed and then gradually loosened. The contracting search, on the other hand, occurs when the researcher discovers that, in fact, the data would not support the most general model, and he imposes a series of constraints designed to "improve" the results.

The most important feature of these examples is the fact that the constraints are thought to be, a priori, likely. If the constraints were certain, they would have been imposed without testing. The researcher is, in fact, less confident than this. He feels that the constraint may be "approximately true," but he checks with the data to "make sure." If the constraint "works," he will impose it; otherwise he will not. To put this another way, the researcher has a priori knowledge about some parameter or some linear combination. If the sample evidence is sufficiently strong, he will disregard that information. Given weak evidence, he may prefer to use his a priori estimate. By definition, then, the intent of an interpretive search is to integrate into the data analysis uncertain a priori information. In the absence of such information, no interpretive search should be performed.

The Bayesian solution to this problem is quite straightforward. The data evidence is summarized in terms of the unconstrained equation and its sufficient statistics. The evidence is interpreted by bouncing the data off a prior distribution where the word "interpret" refers to the process of updating one's opinions from prior to posterior distribution in response to the data evidence. A Bayesian evaluation of interpretive searches thus amounts to the question: does an interpretive search lead to a description of the uncertainty similar to a posterior distribution corresponding to some prior? A secondary issue is whether anyone actually holds that prior opinion.

The failure of ad hoc interpretive searches is twofold. First, it may be difficult to find a prior distribution that makes a search seem reasonable. But much more important is the fact that the output of an interpretive search is an interpretation of the data evidence built on some implicit prior information. This interpretation is relevant to the reader only to the extent that he accepts the implicit prior information as his own, and only then if he understands that it is already built into the result. Publication of the output of an interpretive search is thus equivalent to publication of a posterior distribution without either the sample result or the prior. Publication of the search process is useful only in simple cases when the procedure simply and unambiguously reveals the prior. Most interpretive searches are terribly complex and would be almost impossible to comprehend even if they were fully reported. An interpretive search is thus an inefficient way to use ill-defined, uncommunicable prior information.

That uncertain prior information is used in the evaluation of nonexperimental evidence is incontrovertible. Nonexperimental models worthy of degrees of belief almost always have large numbers of collinear explanatory variables. The amount we can learn from the data about individual parameters in these models would be almost nil if there were no prior information that effectively constrained the ranges of at least some of the parameters. The mining of data that is common among nonexperimental scientists constitutes prima facie evidence of the existence of prior information. Arguments concerning the use of prior information should thus address the question of how rather than whether prior information should be used.

There are at least three alternative approaches that may be taken with respect to the use of prior information in a regression model.

1. *Complete and understandable description of the sample likelihood function*. We may decide that a researcher should report only the likelihood function. Prior information is difficult to specify personally and may vary considerably among intended readers. We may prefer, therefore, to report the evidence and not to interpret it.
2. *Bayesian analysis*. In one or two dimensions, a likelihood function may be straightforwardly described. In higher dimensions, a likelihood function defies intelligible reporting. Described least pretentiously, a Bayesian analysis is merely a tool for exploring likelihood functions. Difficulties in specifying a personally or a publicly acceptable prior distribution should be dealt with by performing a sensitivity analysis designed to characterize as generally as possible the mapping from prior to posterior distribution. In fact, unless he has a strong reason to believe that his priors are somehow superior to his readers', a researcher's only obligation is to report this mapping as informatively as possible.
3. *Interpretive search*. The unintelligibility of the complete likelihood function has led most researchers to use interpretive searches that involve fitting and refitting the equation with various a priori likely constraints. One of the perhaps hundreds of equations is selected and reported, often as if the others had never been estimated. The resulting estimate involves an unknown and perhaps an undesirable mixture of prior and sample information. It, furthermore, constitutes an interpretation of the evidence built, surprisingly, without a theory of interpretation.

The choice between a complete description of the likelihood function and the Bayesian approach involves *only* a disagreement over how to report results. A Bayesian merely explores and reports the region of the parameter space that is favored by the data by computing how the likelihood function affects various prior distributions. In higher dimensional problems, the Bayesian approach seems viable, but the likelihood approach does not, which is another way of saying that I think it is possible to identify and to choose the critical features of multidimensional priors. The choice between the Bayesian approach and the interpretive search approach is, however, a choice between theory and "ad hockery." Interpretative searches lead to ill-defined use of ill-defined prior information. They are an abuse that has led many to discount completely all statistical analyses with nonexperimental data. It is highly misleading, however, to regard them to be complete evils. Rather, they are a commonsense solution designed by intelligent men to complete an unworkable incomplete theory of inference. As we see in this chapter and again in later chapters, intuition and common sense often lead in a desirable direction. What we are proposing is only a formal structure to police our intuitive instincts and to help avoid judgmental errors. Never do we desire a cessation of common sense.

The rules that are used to direct an interpretive search are rarely sufficiently well defined to be written mathematically. An incomplete list of hypothetical rules will give some flavor of the great menu of search strategies. For *contracting sequential searches*, estimate the complete unconstrained model and do one of the following:

a. Drop all variables that have t values less than some cutoff point.
b. Drop the variable with the lowest t value, refit the equation, and continue until all coefficients have significant t's.
c. Specify an a priori sequence of variables. If the first is insignificant, drop it and refit. Proceed similarly with the second, and terminate the process when a variable is reached that has a significant t.
d. Apply a linear transformation to the explanatory variables that makes them orthogonal, and drop any of the new variables with insignificant coefficients.
e. Proceed as in either (b)–(d) but terminate the search when the coefficient on a particular variable is (1) positive or (2) significantly positive or (3) not significantly negative.

For *expanding sequential searches*, estimate a constrained model and do one of the following:

a. Add sequentially the omitted variables in a predetermined order and terminate when a variable has an insignificant coefficient.
b. Add only one other variable selected to maximize the R^2.
c. Proceed as in either (a) or (b), but seek to find an equation that yields a significantly positive coefficient on a particular variable.

For *nonsequential searches*, select the regression equation that yields:

a. The highest $\bar{R}^2$.
b. The biggest percentage of statistically significant coefficients.
c. Estimated residuals that are not autocorrelated.
d. The most number of coefficients with the "right" signs.
e. Some complex combination of (a)–(d).

Only the simplest of these rules yields to a theoretical classical analysis. In this chapter we explore a few of them from that point of view and many others from the Bayesian view. As far as a Bayesian is concerned the effectiveness of a rule depends on how well it implements prior information and how relevant that prior information may be. We argue that this is, in fact, the only question that should be asked. Since classical inference includes no theory of learning and no prior information, classical analysis instead evaluates rules in terms of their sampling properties. This material is discussed in Sections 5.2 and 5.3, the first dealing with the analysis of several simple *ad hoc* rules, the second dealing with the Stein–James estimators and "ridge" estimators.

This classical approach supposes that the researcher has a formal point estimation problem with a conveniently chosen quadratic loss function. Interpretive search rules are then evaluated in terms of the expected loss they imply. The formal shortcoming of this approach is that, at best, it can determine only whether a search estimator is admissible or not—an estimator being inadmissible if there exists another estimator that yields smaller expected loss regardless of the true parameter value. But since the class of admissible estimators is enormous, ruling out inadmissible estimators is only modestly useful. A classical approach often concludes (rather sheepishly?) that choice among the set of admissible estimators depends in some vague way on prior information.

It makes sense to me to begin the analysis with prior information—Bayes estimators derived from proper prior distributions are always admissible, anyway. More importantly, I think the estimation framework does not capture the essential motivation for interpretive searches, that is the pooling of prior information with the data information. It thereby encourages the arbitrary distortion of the data evidence, since it suggests that a "better" estimator results by adjusting the least-squares estimator toward a location not necessarily related to prior information. Further discussion of these points is reported in Section 5.5.

In the first section the problem of this chapter is described as the choice of constrained least-squares estimates, and it is shown that all constrained least-squares estimates lie on an ellipsoid. One shortcoming of a classical analysis is that it considers the very restricted problem of selecting one of only two (arbitrarily chosen) points on this ellipsoid.

As mentioned before, classical analysis of search rules is reported in Sections 5.2 and 5.3. A Bayesian analysis of the same estimation problem is discussed in Section 5.4, and comments are given in Section 5.5. In Section 5.6 we develop a correspondence between search strategies and prior distributions. It is shown that certain classes of priors are implied by certain search strategies in the sense that a Bayesian with such a prior can (loosely) approximate his posterior distribution with a set of constrained least-squares points.

Multicollinearity is discussed in Section 5.7. A distinction is drawn between the weak evidence problem and the interpretation problem. Multicollinearity implies weak evidence in the sense that coefficient standard errors are large, but nothing can be done about that except getting more data. A more confusing consequence of collinearity is that the apparent sample evidence about one parameter depends on the prior information about other parameters. Collinearity, therefore, creates an incentive to use carefully formulated prior information.

I do not believe that anyone could meaningfully specify a complete multivariate prior distribution. Furthermore, readers are certain to vary in their judgments. For both reasons, it is necessary to perform a sensitivity analysis that determines the sensitivity of features of the posterior distribution to changes in the prior distribution. Local sensitivity analysis is discussed in Section 5.7 and global sensitivity analysis in Section 5.8.

A game of definitions is reported in Section 5.9. The words identifiable, estimable, publicly informative, and the phrase "leads to a consensus" are shown to be equivalent. It is also pointed out that although a parameter θ may not be identified, the experiment may nonetheless yield information about θ (because of prior dependence between θ and some other parameters). Lastly, an example is reported in Section 5.10.

Before proceeding, one shortcoming of this chapter must be acknowledged. Almost exclusively, our attention focuses on the choice of point estimates, and tends to ignore the choice of interval around the estimates or other measures of dispersion. This reflects the state of theoretical developments, not the importance of measures of dispersion.

5.1 The Family of Constrained Estimates

The problem under study in this chapter is the choice of one or more constrained regression estimates that jointly imply an "adequate" interpretation of the data evidence. As a preliminary it is useful identify the set of all constrained estimates. It is trivial to show that if all constraints of the form $\mathbf{R}\boldsymbol{\beta}=\mathbf{r}$ are allowed, then any value of $\boldsymbol{\beta}$ is a constrained regression for some value of $\mathbf{R}$ and $\mathbf{r}$. If, however, we consider only constraints of the form $\mathbf{R}\boldsymbol{\beta}=\mathbf{0}$, the family of constrained estimates is an ellipsoid described in Theorem 5.1 and pictured in Figure 5.1. No interpretation of the choice $\mathbf{r}=\mathbf{0}$ seems possible classically, but from the Bayesian point of view this amounts to assuming a prior distribution that is located at the origin.

THEOREM 5.1 (FEASIBLE ELLIPSOID). *A constrained least-squares estimate computed subject to a set of constraints* $\mathbf{R}\boldsymbol{\beta}=\mathbf{0}$ *lies on the ellipsoid*

$$\left(\boldsymbol{\beta}-\frac{\mathbf{b}}{2}\right)'\mathbf{X}'\mathbf{X}\left(\boldsymbol{\beta}-\frac{\mathbf{b}}{2}\right)=\frac{\mathbf{b}'\mathbf{X}'\mathbf{X}\mathbf{b}}{4} \tag{5.1}$$

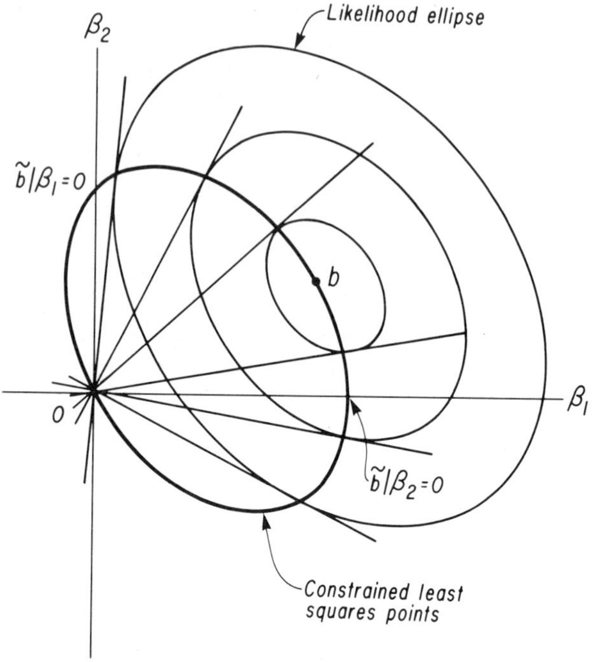

Fig. 5.1 The ellipse of constrained least-squares points.

where **b** *is the unconstrained least-squares vector. Furthermore, any point on this ellipsoid is a constrained estimate for some* **R**.

Proof: Equation (3.14) in Chapter 3 is the formula for computing a constrained estimate. Simply insert that value into (5.1). Conversely, any point on the ellipsoid (5.1), say, $\tilde{\boldsymbol{\beta}}$, is a least-squares estimate subject to the constraint $(\mathbf{b}-\tilde{\boldsymbol{\beta}})'\mathbf{X}'\mathbf{X}\boldsymbol{\beta}=\mathbf{0}$. (Verification left to reader.)

The set of constrained estimates described in Equation (5.1) is merely a translated likelihood ellipsoid. It is located at half the least-squares vector and travels through the origin and **b**. We argue below that if a researcher can select the location but is unable or unwilling to describe more fully his prior, then he should be interested in all points in this ellipsoid, but no other points. Incidentally, any origin $\mathbf{b}^*$ may be selected or, equivalently, constraints of the form $\mathbf{R}(\boldsymbol{\beta}-\mathbf{b}^*)=\mathbf{0}$ may be considered. It is easy to show that (5.1) continues to apply but with $\boldsymbol{\beta}-\mathbf{b}^*$ and $\mathbf{b}-\mathbf{b}^*$ replacing $\boldsymbol{\beta}$ and **b**.

An interpretive search is a procedure for selecting points from among the set of feasible points (5.1). An *interpretive-search estimator* can thus be generally defined as a weighted average of points on the feasible ellipsoid

$$\hat{\boldsymbol{\beta}}^s = \int_{\mathbf{R}} \omega(\mathbf{R})\,\hat{\boldsymbol{\beta}}(\mathbf{R})\,d\mathbf{R}, \qquad \int_{\mathbf{R}} \omega(\mathbf{R})\,d\mathbf{R} = 1,$$

where $\hat{\boldsymbol{\beta}}(\mathbf{R})$ is a constrained estimate and $\omega(\mathbf{R})$ is a weight function. Usually, only a finite subset of the feasible points is considered. Sometimes a set of at most k linearly independent constraints is identified, for example, $\beta_i = 0, i = 1, \ldots, k$, and only estimates that involve subsets of this set of constraints are computed. In two dimensions the constraints $\beta_1 = 0$ and $\beta_2 = 0$ imply the four different estimates illustrated in Figure 5.1: $\mathbf{b}$, $\mathbf{b}_{\{2\}}$, $\mathbf{b}_{\{1\}}$, and the origin, where $\mathbf{b}_{\{i\}}$ is the estimate subject to the ith constraint. Given k constraints, there are 2^k ways of imposing them, and 2^k different constrained estimates.[1] Let I be a subset of the first k integers; let $\mathbf{b}_I$ be a constrained estimate, then a (discrete) interpretative search estimator is

$$\hat{\boldsymbol{\beta}}^s = \sum_I \omega_I(\mathbf{Y}, \mathbf{X})\mathbf{b}_I, \tag{5.2}$$

where we have written the weight function ω_I to indicate its possible dependence on the data, $\mathbf{Y}$ and $\mathbf{X}$.

The interpretive search estimator (5.2) can thus be built in three steps: (1) An *origin* is selected. This restricts the set of constrained estimates to lie on the ellipsoid (5.1). (2) A set of k linearly independent constraints (a *coordinate system*) is chosen. This further restricts the set of interesting constrained least squares points from the ellipsoidal continuum to a set of 2^k points. (3) Last, a weighting function $\omega_I(\mathbf{Y}, \mathbf{X})$ is identified.

Measured in terms of its effect on reducing the set of interesting constrained least-squares points, the choice of origin is the most critical decision. After that, the choice of coordinate system is important. A shortcoming of the classical analysis of interpretative searches now to be discussed is its failure to comment meaningfully on either the choice of origin or the choice of coordinate system.

5.2 Classical Evaluation of Ad Hoc Rules for Interpretive Searches

In this section we report a classical analysis of interpretive searches. Both an origin and a coordinate system for imposing constraints are assumed to be known before the analysis begins. Furthermore, only very special weight functions may be considered.

[1]Computation of the 2^k regressions is discussed by Schatzoff et al. (1968), Garside (1965), and Furnival (1971).

Consider, first, the two variable linear regression

$$\mathbf{Y}=\mathbf{x}\beta+\mathbf{z}\gamma+\mathbf{u}, \qquad E(\mathbf{u})=\mathbf{0}, V(\mathbf{u})=\sigma^2\mathbf{I},$$

where $\mathbf{Y},\mathbf{x},\mathbf{z}$ and $\mathbf{u}$ are $T\times 1$ vectors and β and γ are scalar parameters. The least-squares estimator is

$$\begin{bmatrix} b \\ g \end{bmatrix} = \begin{bmatrix} \mathbf{x}'\mathbf{x} & \mathbf{x}'\mathbf{z} \\ \mathbf{z}'\mathbf{x} & \mathbf{z}'\mathbf{z} \end{bmatrix}^{-1} \begin{bmatrix} \mathbf{x}'\mathbf{Y} \\ \mathbf{z}'\mathbf{Y} \end{bmatrix}$$

which is unbiased with variance-covariance matrix

$$V(b,g)=\sigma^2\begin{bmatrix} \mathbf{x}'\mathbf{x} & \mathbf{x}'\mathbf{z} \\ \mathbf{z}'\mathbf{x} & \mathbf{z}'\mathbf{z} \end{bmatrix}^{-1}. \tag{5.3}$$

An alternative estimator with γ set to zero (i.e., with $\mathbf{z}$ omitted) is

$$\begin{bmatrix} \tilde{\beta} \\ \tilde{\gamma} \end{bmatrix} = \begin{bmatrix} (\mathbf{x}'\mathbf{x})^{-1} & \mathbf{x}'\mathbf{Y} \\ & 0 \end{bmatrix}.$$

The expected value of $\tilde{\beta}$ is

$$\begin{aligned} E(\tilde{\beta}) &= (\mathbf{x}'\mathbf{x})^{-1}\mathbf{x}'E(\mathbf{Y}) \\ &= (\mathbf{x}'\mathbf{x})^{-1}\mathbf{x}'(\mathbf{x}\boldsymbol{\beta}+\mathbf{z}\gamma)=\beta+(\mathbf{x}'\mathbf{x})^{-1}\mathbf{x}'\mathbf{z}\gamma. \end{aligned}$$

The bias of $(\tilde{\beta},\tilde{\gamma})$ is, therefore,

$$E\begin{bmatrix} \tilde{\beta} \\ \tilde{\gamma} \end{bmatrix} - \begin{bmatrix} \beta \\ \gamma \end{bmatrix} = \begin{bmatrix} (\mathbf{x}'\mathbf{x})^{-1} & \mathbf{x}'\mathbf{z}\gamma \\ & -\gamma \end{bmatrix}, \tag{5.4}$$

a linear function of γ, taking on the value of zero only at $\gamma=0$. The variance is straightforwardly calculated as

$$V(\tilde{\beta},\tilde{\gamma})=\sigma^2\begin{bmatrix} (\mathbf{x}'\mathbf{x})^{-1} & 0 \\ 0 & 0 \end{bmatrix}. \tag{5.5}$$

The classical theory of estimation suggests choosing between these two estimators on the basis of their sampling properties. This used to mean discarding $(\tilde{\beta},\tilde{\gamma})$ because of its bias given in (5.4). That counsel has been disregarded in practice; researchers often report $(\tilde{\beta},\tilde{\gamma})$ even when, with essential certainty, γ is not equal to zero. Although the least-squares estimator has minimum variance among linear unbiased estimators, few researchers are willing to accept "peculiar" estimates, and the standard operating procedure is to search for constraints that yield "acceptable" estimates. The fact that the resulting estimator is neither unbiased, linear, nor "best" is no large deterrent to a person whose research project would be dubbed "fruitless" if it were summarized in a nonsensical estimate.

The overwhelming body of nonexperimental data analysis that rests on the obviously shaky foundation of interpretative searches has understandably generated interest among theoretical statisticians. It is currently popular now to discount unbiasedness as an irrelevant criterion conjured up to make the problem soluble. Instead of an unbiased estimator, current wisdom suggests an estimator that yields an estimate close to the true parameter on the average. A tractable distance function for measuring closeness is the squared deviation from the true value of the parameter, and the resulting criterion is the mean squared error. For a one-dimensional parameter θ with estimator $\hat{\theta}$, the mean squared error is defined by

$$MSE(\hat{\theta},\theta)=E\left[(\theta-\hat{\theta})^2|\theta\right]$$

which is informatively rewritten as

$$\begin{aligned} MSE(\hat{\theta},\theta)&=E\left[(\theta-E\hat{\theta}+E\hat{\theta}-\hat{\theta})^2|\theta\right] \\ &=V(\hat{\theta})+(\theta-E[\hat{\theta}|\theta])^2 \end{aligned}$$

the sampling variance plus the square of the bias. An estimator according to this criterion is judged desirable if it has small mean squared error. It is readily seen that an estimator may be deemed desirable even though it is biased, that is if the variance is reduced enough to offset the (square of the) bias.

The multivariate generalization of this criterion is the mean squared error matrix

$$\begin{aligned} MSE(\hat{\boldsymbol{\theta}},\boldsymbol{\theta})&=E\left[(\boldsymbol{\theta}-\hat{\boldsymbol{\theta}})(\boldsymbol{\theta}-\hat{\boldsymbol{\theta}})'|\boldsymbol{\theta}\right] \\ &=V(\hat{\boldsymbol{\theta}})+(\boldsymbol{\theta}-E[\hat{\boldsymbol{\theta}}|\boldsymbol{\theta}])(\boldsymbol{\theta}-E[\hat{\boldsymbol{\theta}}|\boldsymbol{\theta}])' \end{aligned}$$

where $\boldsymbol{\theta}$ and $\hat{\boldsymbol{\theta}}$ are vectors. The reader may verify that the mean-square error of any linear combination $\boldsymbol{\lambda}'\hat{\boldsymbol{\theta}}$ of the estimators is

$$MSE(\boldsymbol{\lambda}'\hat{\boldsymbol{\theta}},\boldsymbol{\lambda}'\boldsymbol{\theta})=\boldsymbol{\lambda}'MSE(\hat{\boldsymbol{\theta}},\boldsymbol{\theta})\boldsymbol{\lambda}$$

and it is desirable to have a mean squared error matrix be small in a matrix sense.

Returning to our problem the mean squared error matrix of the least-squares estimator is

$$MSE(b,g)=\sigma^2\begin{bmatrix}\mathbf{x'x} & \mathbf{x'z} \\ \mathbf{z'x} & \mathbf{z'z}\end{bmatrix}^{-1} \tag{5.6}$$

whereas the mean-square error matrix of constrained least squares is

$$MSE(\tilde{\beta},\tilde{\gamma})=\sigma^2\begin{bmatrix}(\mathbf{x'x})^{-1} & 0 \\ 0 & 0\end{bmatrix}+\gamma^2\begin{bmatrix}(\mathbf{x'x})^{-2}(\mathbf{x'z})^2 & -(\mathbf{x'x})^{-1}(\mathbf{x'z}) \\ -(\mathbf{x'x})^{-1}(\mathbf{x'z}) & 1\end{bmatrix}. \tag{5.7}$$

If interest centers on β, we say the estimator $\tilde{\beta}$ is preferred to b according to the mean squared error criterion if

$$\sigma^2(\mathbf{x}'\mathbf{x})^{-1}+\gamma^2(\mathbf{x}'\mathbf{x})^{-2}(\mathbf{x}'\mathbf{z})^2<\sigma^2\left(\mathbf{x}'\mathbf{x}-\mathbf{x}'\mathbf{z}(\mathbf{z}'\mathbf{z})^{-1}\mathbf{z}'\mathbf{x}\right)^{-1}.$$

Feldstein (1973) and Wallace (1964) write this inequality informatively in terms of the ratio of the mean-square errors

$$\frac{MSE(\tilde{\beta})}{MSE(b)}=1+r_{xz}^2\left(\tau_\gamma^2-1\right) \tag{5.8}$$

where r_{xz} is the correlation between $\mathbf{x}$ and $\mathbf{z}$[2]

$$r_{xz}^2=(\mathbf{x}'\mathbf{z})^2(\mathbf{x}'\mathbf{x})^{-1}(\mathbf{z}'\mathbf{z})^{-1}$$

and τ_γ is the "true t" for testing $\gamma=0$

$$\tau_\gamma^2=\frac{\gamma^2\left[(\mathbf{x}'\mathbf{x})(\mathbf{z}'\mathbf{z})-(\mathbf{x}'\mathbf{z})^2\right]}{\sigma^2(\mathbf{x}'\mathbf{x})}.$$

It is readily seen from (5.8) that the MSE of $\tilde{\beta}$ is less than the MSE of b if and only if

$$\tau_\gamma^2<1. \tag{5.9}$$

This identifies the region in the two-dimensional parameter space within which $MSE(\tilde{\beta})<MSE(\mathrm{b})$. Since the mean squared errors do not depend on β, we may draw a one-dimensional graph of the mean squared error function, Figure 5.2.

The reader should take note of the following features of this figure:

a. Neither $\tilde{\beta}$ nor b dominates the other in the sense of yielding uniformly smaller mean squared error.

b. $\tilde{\beta}$ does best around the origin $\gamma=0$ but since $MSE(\tilde{\beta})$ is a quadratic function of γ whereas $MSE(\mathrm{b})$ is just a constant, the relative inferiority of $\tilde{\beta}$ is unbounded as γ grows. The difference at the origin is a function of r_{xz}^2.

c. The origin of this figure is completely arbitrary. That is, there is a whole class of estimators $\tilde{\beta}_{g*}$ estimated by setting γ to some arbitrary value and calculating

$$\tilde{\beta}_{g*}=(\mathbf{x}'\mathbf{x})^{-1}\mathbf{x}'(\mathbf{Y}-\mathbf{z}g^*).$$

This estimator will have a relatively low MSE in the neighborhood of $\gamma=g^*$ but a relatively high MSE elsewhere. Figure 5.2 applies with g^*

[2]If desired, the reader may consider the variables to have had their means removed.

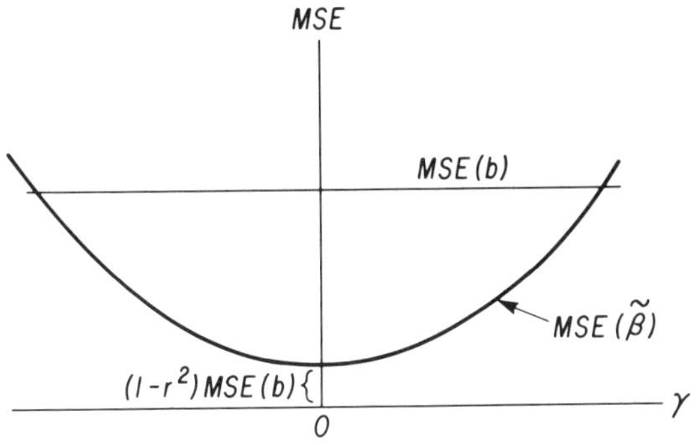

Fig. 5.2 Mean squared error functions.

as the origin and

$$\tau^2_{\gamma,g^*} = \frac{(\gamma - g^*)^2\left[(\mathbf{x}'\mathbf{x})(\mathbf{z}'\mathbf{z}) - (\mathbf{x}'\mathbf{z})^2\right]}{\sigma^2(\mathbf{x}'\mathbf{x})}$$

replacing τ^2_γ.

We have yet to discuss an interpretive search strategy, since neither b nor $\tilde{\beta}_{g^*}$ requires a search over estimates. The simplest interpretive search that has been analyzed involves calculating both b and $\tilde{\beta}_{g^*}$ (for some value of g^*) and picking the "better" estimate. No doubt, every conceivable criterion of choice has been used in practice. Theoretically, one particular criterion has been subject to much analysis: pick b if the least-squares estimate of γ is significantly different from g^*; otherwise pick $\tilde{\beta}_{g^*}$. The t^2 statistic for testing the restriction $\gamma = g^*$ is $t^2_{g^*} = (g - g^*)^2/s^2[\mathbf{z}'\mathbf{z} - \mathbf{z}'\mathbf{x}(\mathbf{x}'\mathbf{x})^{-1}\mathbf{x}'\mathbf{z}]^{-1}$ and the "pretest" estimator more formally is

$$\hat{\beta}^p(g^*,\alpha) = \begin{cases} b & \text{if } t^2_{g^*} > t^2_\alpha \\ \tilde{\beta}_{g^*} & \text{otherwise} \end{cases} \tag{5.10}$$

where t_α is the $\alpha/2$ percentage point of the Student's distribution for some arbitrarily chosen value of α.

It is readily seen that both b and $\tilde{\beta}_{g^*}$ are in the class of pretest estimators with $\alpha = 1$ and $\alpha = 0$, respectively. The mean square error of other members of the class $\tilde{\beta}^p(g^*,\alpha)$ tend to be like the mean squared error of $\tilde{\beta}_{g^*}$ for those values of γ that are highly likely to yield insignificant values of t_{g^*}, that is, for small values of $(\gamma - g^*)^2$. Conversely, $MSE(\hat{\beta}^p)$ approximates

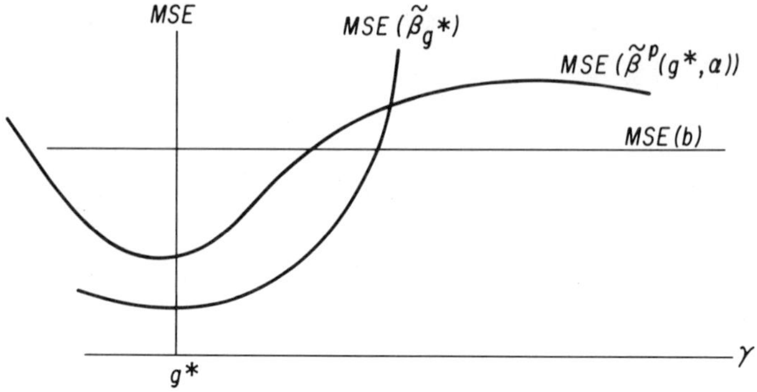

Fig. 5.3 Mean squared error functions.

$MSE(b)$ when b is most likely to be selected, that is, for large values of $(\gamma - g^*)^2$. A typical mean squared error function is depicted in Figure 5.3.

One desirable feature of $MSE(\tilde{\beta}^p)$ is that it is bounded. It exceeds $MSE(\tilde{\beta}_{g^*})$ at $\gamma = g^*$ but attains its minimum there. There is a finite region in which it is the worst of the three estimators. [For other properties of this function, see Wallace and Ashar (1972) and Feldstein (1973).]

We have now identified a two-dimensional infinity of estimates of $\beta, \tilde{\beta}^p(g^*, \alpha)$, $-\infty < g^* < \infty$, and $0 \leqslant \alpha \leqslant 1$. Choice of g^* locates the region in which we will do relatively well, and choice of α determines the amount by which and the region within which this estimator "does better" than ordinary least squares. Which are you to choose? There is simply no answer to this question within the confines of classical inference. That body of statistical theory determines the class of admissible estimators, but does not provide a method of choice from this class. Clearly, however, the choice must depend on where you want to do better than least squares. This surely means the values of γ that you think are most likely and that means prior information about γ. To quote Wallace and Ashar (1972, p. 177):

(1) If one has a strong prior that τ_γ^2 is either greater or less than one half, then no pre-testing is called for. One simply treates $\gamma = 0$ in the latter and $\gamma \neq 0$ in the former if the mean squared error loss function is taken as a guide.

(2) Any intermediate prior about τ_γ^2 can be reflected through choice of α. The stronger the belief that $\gamma \neq 0$, the smaller should be the choice of α and conversely, α should be larger the stronger the prior doubts about the inclusion of z. Of course, if one could cast priors into a precise distributional form, there are both classical and Bayesian procedures to by-pass the pre-testing altogether.

This last sentence brings up the question to be discussed subsequently: "what is the best way of picking a prior: directly or indirectly by picking a search strategy?"[3]

Note that this pretest estimator is a discontinuous function of the data, since a small change in the data **Y** that is large enough to shift the estimate of g from "insignificance" to significance induces a discrete change in the estimate from $\tilde{\beta}$ to b. This discontinuity has been shown by Cohen (1965) to imply that this pretest estimator is inadmissible; there is, necessarily, another estimator that has smaller MSE for all values of (β,γ). We can rid our procedure of this discontinuity by having a continuous mixing scheme, such as the following.

Let a weighted-average estimator be

$$\beta_{g^*}^{w}=\lambda(\mathbf{Y},\mathbf{x},\mathbf{z})b+\left[1-\lambda(\mathbf{Y},\mathbf{x},\mathbf{z})\right]\tilde{\beta}_{g^*}$$

where λ is a continuous function of the data. Feldstein (1970) derives a value of λ independent of $(\mathbf{Y},\mathbf{x},\mathbf{z})$ that minimizes the mean squared error of $\beta_{g^*}^{w}$:

$$\begin{aligned}E(\beta_{g^*}^{w}-\beta)^2&=E\left[\lambda(b-\beta)+(1-\lambda)(\tilde{\beta}_{g^*}-\beta)\right]^2\\&=\lambda^2MSE(b)+(1-\lambda)^2MSE(\tilde{\beta}_{g^*})+2\lambda(1-\lambda)E(b-\beta)(\tilde{\beta}_{g^*}-\beta).\end{aligned}$$

Setting the derivative of this with respect to λ equal to zero yields

$$0=2\lambda MSE(b)-2(1-\lambda)MSE(\tilde{\beta}_{g^*})+2(1-2\lambda)E(b-\beta)(\tilde{\beta}_{g^*}-\beta)$$

which simplifies to

$$\lambda=\frac{\tau_{\gamma,g^*}^2}{1+\tau_{\gamma,g^*}^2}$$

a function of γ. This suggests using a weight proportional to the square of

[3]There are a couple other developments relating to this model that are worth reporting here. Note that the usual α-level test of the hypothesis $H_0:\gamma=g^*$ against the alternative $H_1:\gamma\neq g^*$ is not an α-level test of the hypothesis that $\tilde{\beta}_{g^*}$ is a better estimator than b. If we wanted to test those hypotheses we would test

$$H_0':\tau_{\gamma,g^*}^2=\frac{(\gamma-g^*)^2}{\sigma^2\left(\mathbf{z}'\mathbf{z}-\mathbf{z}'\mathbf{x}(\mathbf{x}'\mathbf{x})^{-1}\mathbf{x}'\mathbf{z}\right)^{-1}}\leqslant 1$$

$$H_1':\tau_{\gamma,g^*}^2>1$$

Toro-Vizcarrondo and Wallace (1968) argue that H_0 and H_1 are irrelevant in this context and propose instead α-level tests of H_0' versus H_1'.

the t statistic used to test $\gamma = g^*$

$$\lambda(\mathbf{Y}, \mathbf{x}, \mathbf{z}) = \frac{t_{g^*}^2}{1 + t_{g^*}^2}$$

where

$$t_{g^*}^2 = \frac{(g - g^*)^2}{s^2\left(\mathbf{z}'\mathbf{z} - \mathbf{x}'\mathbf{z}(\mathbf{z}'\mathbf{z})^{-1}\mathbf{z}'\mathbf{x}\right)^{-1}}.$$

These weights were first suggested by Huntsberger (1955) and have been explored in a Monte Carlo study by Feldstein (1973).

5.3 "Stein" Estimators and Ridge Regression

It is possible to dismiss the "pretest" estimators discussed in the previous section, since they do not, in fact, dominate the least-squares estimator. However, a most provocative result of modern statistical theory is that the least-squares estimator is, in fact, inadmissible when there are more than two coefficients and when the loss function takes a special form.[4] Consider the k-means model

$$Y_i = \xi_i + u_i \qquad i = 1, 2, \ldots, k \tag{5.11}$$

with the u_is having independent normal distributions with mean 0 and variance σ^2. The least-squares estimator and also the maximum likelihood estimator are $\hat{\xi}_{io} = Y_i$, $i = 1, 2, \ldots, k$, or, in vector notation, $\hat{\xi}_0 = \mathbf{Y}$. Assuming the quadratic loss function $\mathrm{L}(\xi, \hat{\xi}) = (\xi - \hat{\xi})'(\xi - \hat{\xi})$ Stein (1956) and James and Stein (1961) have shown that

a. The least-squares estimator $\hat{\xi}_0 = \mathbf{Y}$ is admissible for $k \leqslant 2$. That is, there is no estimator that provides uniformly smaller risk (expected loss) than $\hat{\xi}_0$.
b. For $k \geqslant 3$, an alternative estimator

$$\hat{\xi}_1 = \left(1 - \frac{(k-2)\sigma^2}{\mathbf{Y}'\mathbf{Y}}\right)\mathbf{Y}$$

has uniformly smaller risk than $\hat{\xi}_0$. Thus $\hat{\xi}_0$ is inadmissible.

[4]An excellent summary of this literature is given in Zellner and Vandaele (1975). Our discussion is very abbreviated, and the reader should consult Zellner and Vandaele for a fuller treatment. We discuss here the analysis when σ^2 is known. There are similar developments for σ^2 unknown.

The general linear model $\mathbf{Y}=\mathbf{X}\boldsymbol{\beta}+\mathbf{u}$ with $\mathbf{u}$ distributed normally with mean $\mathbf{0}$ and variance matrix $\sigma^2\mathbf{I}$ can be transformed neatly into the k-means problem as in Sclove (1968). Let us find a matrix $\mathbf{P}$ such that $\mathbf{P}'\mathbf{X}'\mathbf{X}\mathbf{P}=T\mathbf{I}$, where T is the number of observations. The linear model can be written

$$\begin{aligned}\mathbf{Y}&=\mathbf{X}\mathbf{P}\mathbf{P}^{-1}\boldsymbol{\beta}+\mathbf{u}.\\&=\mathbf{W}\boldsymbol{\theta}+\mathbf{u}\end{aligned}$$

where $\mathbf{W}=\mathbf{X}\mathbf{P}$ and $\boldsymbol{\theta}=\mathbf{P}^{-1}\boldsymbol{\beta}$. Premultiplying now by $\mathbf{W}'$, we obtain

$$\begin{aligned}\mathbf{W}'\mathbf{Y}&=\mathbf{W}'\mathbf{W}\boldsymbol{\theta}+\mathbf{W}'\mathbf{u}\\&=T\boldsymbol{\theta}+T\boldsymbol{\varepsilon}\end{aligned}$$

with $T\varepsilon_n \sim N(\mathbf{0},\sigma^2\mathbf{W}'\mathbf{W})=N(\mathbf{0},\sigma^2 T\mathbf{I})$. Dividing by T, we obtain

$$\frac{\mathbf{W}'\mathbf{Y}}{T}=\boldsymbol{\theta}+\boldsymbol{\varepsilon}$$

which is precisely the same form as the k-means problem with variance σ^2/T. Thus the estimator

$$\hat{\boldsymbol{\theta}}_1=\left(1-\frac{(k-2)T\sigma^2}{\mathbf{Y}'\mathbf{W}'\mathbf{W}\mathbf{Y}}\right)\frac{\mathbf{W}'\mathbf{Y}}{T}$$

dominates least squares

$$\hat{\boldsymbol{\theta}}_0=\frac{\mathbf{W}'\mathbf{Y}}{T}$$

when the loss function is $(\boldsymbol{\theta}-\hat{\boldsymbol{\theta}})'(\boldsymbol{\theta}-\hat{\boldsymbol{\theta}})$. The corresponding estimators and loss function in terms of the natural parameters $\boldsymbol{\beta}$ are[5]

$$\hat{\beta}_0=\mathbf{P}\hat{\boldsymbol{\theta}}_0=\frac{\mathbf{P}\mathbf{W}'\mathbf{Y}}{T}=(\mathbf{X}'\mathbf{X})^{-1}\mathbf{X}'\mathbf{Y}=\mathbf{b}$$

$$\hat{\beta}_1=\mathbf{P}\hat{\boldsymbol{\theta}}_1=\left(1-\frac{(k-2)\sigma^2}{\mathbf{Y}'\mathbf{Y}}\right)\frac{\mathbf{P}\mathbf{W}'\mathbf{Y}}{T}=\left(1-\frac{(k-2)\sigma^2}{\mathbf{Y}'\mathbf{Y}}\right)\hat{\beta}_0$$

$$L(\beta,\hat{\beta})=(\beta-\hat{\beta})'\mathbf{P}^{-1\prime}\mathbf{P}^{-1}(\beta-\hat{\beta})=T^{-1}(\beta-\hat{\beta})'\mathbf{X}'\mathbf{X}(\beta-\hat{\beta}).$$

[5]The dependence of the loss function on $\mathbf{X}'\mathbf{X}$ is apparently peculiar. It does make sense for the following prediction problem. Let Y_f be a future outcome of the dependent variable $Y_f=\mathbf{x}_f'\boldsymbol{\beta}+u_f$, and let $\hat{Y}_f$ be a conditional forecast $\hat{Y}_f=\mathbf{x}_f'\hat{\boldsymbol{\beta}}$. Then the squared prediction error is $(Y_f-\hat{Y}_f)^2=(\mathbf{x}_f'(\beta-\hat{\beta})+u_f)^2$. Taking the expected value of this squared error and using $E\mathbf{x}_f\mathbf{x}_f'=\mathbf{X}'\mathbf{X}/T$ we obtain

$$E\left[(Y_f-\hat{Y}_f)^2|\beta,\hat{\beta}\right]=T^{-1}(\beta-\hat{\beta})\mathbf{X}'\mathbf{X}(\beta-\hat{\beta})+\sigma^2.$$

The equation $E\mathbf{x}_f\mathbf{x}_f'=\mathbf{X}'\mathbf{X}/T$ makes sense if the explanatory variable vectors are drawn independently from a fixed multivariate distribution.

The Stein–James estimator $\hat{\boldsymbol{\beta}}_1$ is just least squares times a shrinkage factor; to put it somewhat differently, it is a "weighted average" of the zero vector and the least-squares vector with weights depending on the χ^2 statistics for testing the restriction $\boldsymbol{\beta}=0$, $\chi^2=\mathbf{Y}'\mathbf{Y}/\sigma^2$.[6] The possibility of negative weights has led Baranchik (1964) to propose a "positive part estimator." Except for the possibility of negative weights, the Stein–James estimator is an interpretative search estimator (5.2).

There is another class of estimators mysteriously known as ridge regression estimators that are shown in Section 5.5 to be interpretive search estimators (5.2). Hoerl and Kennard (1970a, b) note that when $\mathbf{X}'\mathbf{X}$ is nearly singular, calculation of the least-squares estimate $\mathbf{b}=(\mathbf{X}'\mathbf{X})^{-1}\mathbf{X}'\mathbf{Y}$ is subject to "a number of 'errors'." They propose the "ridge estimator"

$$\hat{\boldsymbol{\beta}}^r(c)=(\mathbf{X}'\mathbf{X}+c\mathbf{I})^{-1}\mathbf{X}'\mathbf{Y}=(\mathbf{X}'\mathbf{X}+c\mathbf{I})^{-1}(\mathbf{X}'\mathbf{X}\mathbf{b}+c\mathbf{I}\mathbf{0}),$$

where $\mathbf{0}$ is a zero vector and c is some mysteriously chosen scalar. It is readily seen that this is a conditional Bayesian posterior mean with a spherical prior centered at zero and thus has a Bayesian justification.

Hoerl and Kennard's (1970) non-Bayesian arguments in favor of "ridge regression" are difficult for this author to understand. They prove that for any $\boldsymbol{\beta}$ there is a constant c greater than zero such that the mean squared error of $\hat{\boldsymbol{\beta}}^r(c)$ is smaller than the mean squared error of least squares, $\hat{\boldsymbol{\beta}}^r(0)$. Even if we accept the mean-squared-error logic, this result has limited applicability. The constant c is a function of $\boldsymbol{\beta}$, but if $\boldsymbol{\beta}$ is known, there is an even better estimator than the ridge estimator. Hoerl and Kennard also offer informal arguments in favor of ridge regression. They (1970, p. 56) point out that when $\mathbf{X}'\mathbf{X}$ has one or more small eigenvalues "the distance from $\mathbf{b}$ to $\boldsymbol{\beta}$ will tend to be large. Estimated coefficients that are large in absolute value have been observed by all who have tackled live nonorthogonal data problems." It would seem that the average distance between $\mathbf{b}$ and $\boldsymbol{\beta}$ is fully reflected in the sampling variance of $\mathbf{b}$, $\sigma^2(\mathbf{X}'\mathbf{X})^{-1}$, which does indeed have large elements when $\mathbf{X}'\mathbf{X}$ has small eigenvalues. It is hard to imagine that we could cure this disease by shrinking to some arbitrary point. The outcome of such a procedure is, of course, to improve the mean squared error at the arbitrary point, but the cost would usually be worsened mean squared error elsewhere. Whether we want to do this must depend on prior information. This suggests that the last sentence of the quotation in this paragraph might better read: estimated coefficients that are far from a priori likely coefficients have been observed by all who have tackled live nonorthogonal data problems.

[6]The weights need not sum to one. In fact, $\hat{\boldsymbol{\beta}}_1$ and $\hat{\boldsymbol{\beta}}_0$ may be on opposite sides of the origin.

Note also that the origin and the "metric" are arbitrary.[7] "Shrinking" $\mathbf{A}\boldsymbol{\beta}$ to $\mathbf{Ab}^*$ with observations generated by $\mathbf{Y}-\mathbf{Xb}^*=(\mathbf{XA}^{-1})(\mathbf{A}\boldsymbol{\beta}-\mathbf{Ab}^*)+\mathbf{u}$ yields $\mathbf{A}\hat{\boldsymbol{\beta}}-\mathbf{Ab}^*=(\mathbf{A}'^{-1}\mathbf{X}'\mathbf{XA}^{-1}+c\mathbf{I})^{-1}\mathbf{A}'^{-1}\mathbf{X}'(\mathbf{Y}-\mathbf{Xb}^*)=\mathbf{A}(\mathbf{X}'\mathbf{X}+c\mathbf{A}'\mathbf{A})^{-1}\mathbf{X}'(\mathbf{Y}-\mathbf{Xb}^*)$. In terms of $\hat{\boldsymbol{\beta}}$ this is

$$\begin{aligned}\hat{\boldsymbol{\beta}}&=\mathbf{b}^*+(\mathbf{X}'\mathbf{X}+c\mathbf{A}'\mathbf{A})^{-1}(\mathbf{X}'\mathbf{Y}-\mathbf{X}'\mathbf{Xb}^*)\\&=(\mathbf{X}'\mathbf{X}+c\mathbf{A}'\mathbf{A})^{-1}(\mathbf{X}'\mathbf{Xb}^*+c\mathbf{A}'\mathbf{Ab}^*+\mathbf{X}'\mathbf{Y}-\mathbf{X}'\mathbf{Xb}^*)\\&=(\mathbf{X}'\mathbf{X}+c\mathbf{A}'\mathbf{A})^{-1}(\mathbf{X}'\mathbf{Xb}+c\mathbf{A}'\mathbf{Ab}^*)\end{aligned}$$

where $\mathbf{A}'\mathbf{A}$ is an arbitrary symmetric positive definite matrix. Thus the class of ridge estimators is as wide as the class of conditional posterior means (3.28), but the mean-square error logic has nothing to say about either the choice of origin $\mathbf{b}^*$ or the choice of "metric" $\mathbf{A}'\mathbf{A}$.

5.4 Bayes' Decisions and the Admissibility of Bayes' Rules

Although I do not think practical pretesting is intended to solve the estimation problems just discussed, it is important to understand how a Bayesian would solve them, if he had to. It is simple to show that he would estimate the parameters with his posterior mean. An important result is that the posterior mean is necessarily an admissible estimator provided that the prior is proper. That an improper prior may lead to an inadmissible estimator should already be clear from the discussion of Stein's (1956) result on the inadmissibility of least-squares, the posterior mean implied by an improper diffuse prior.

A general quadratic loss function can be written $l(\boldsymbol{\beta},\hat{\boldsymbol{\beta}})=(\boldsymbol{\beta}-\hat{\boldsymbol{\beta}})'\mathbf{Q}(\boldsymbol{\beta}-\hat{\boldsymbol{\beta}})$, where $\boldsymbol{\beta}$ is the $k\times 1$ vector of regression parameters, $\mathbf{Q}$ is a symmetric positive definite $(k\times k)$ matrix, and $\hat{\boldsymbol{\beta}}$ is a $k\times 1$ decision vector representing the estimate of $\boldsymbol{\beta}$.[8] Making use of the data $\mathbf{Y}$, a Bayesian would select an estimate $\hat{\boldsymbol{\beta}}$ that minimizes expected posterior loss

$$\min_{\hat{\boldsymbol{\beta}}} E\left[(\boldsymbol{\beta}-\hat{\boldsymbol{\beta}})'\mathbf{Q}(\boldsymbol{\beta}-\hat{\boldsymbol{\beta}})|\mathbf{Y}\right].$$

It is easy to show that *regardless of the choice of positive definite* $\mathbf{Q}$, this

[7] In Chapter 3 it was shown that a posterior mean is different from the least-squares estimate because it attempts to satisfy the prior that asserts that $(\boldsymbol{\beta}-\mathbf{b}^*)\mathbf{N}^*(\boldsymbol{\beta}-\mathbf{b}^*)$ is small. The word "origin" refers to the prior mean b^*, and the word "metric" refers to the prior precision matrix $\mathbf{N}^*$ that determines the distance function $(\boldsymbol{\beta}-\mathbf{b}^*)'\mathbf{N}^*(\boldsymbol{\beta}-\mathbf{b})$.

[8] For convenience, $\mathbf{Q}$ is not allowed to be positive semi-definite. If $\mathbf{Q}$ were semi-definite, a Bayes estimator is nonunique, and the results reported below would not formally apply. However, even if $\mathbf{Q}$ is semi-definite, the posterior mean can be shown to be admissible.

expression is minimized at the mean of $\boldsymbol{\beta}$

$$\hat{\boldsymbol{\beta}}(\mathbf{Y}) = E(\boldsymbol{\beta}|\mathbf{Y}),$$

where $\hat{\boldsymbol{\beta}}$ is written as a function of $\mathbf{Y}$ to emphasize the fact that the posterior mean of $\boldsymbol{\beta}$ is a function of the data $\mathbf{Y}$.

With a reasonable assumption about the prior for $\boldsymbol{\beta}$ it can be shown that the posterior mean is an admissible estimator, where the word admissible is now to be defined precisely. The *risk function* given the decision rule $\mathbf{a}(\mathbf{Y})$ is the expected loss conditional on $\boldsymbol{\beta}$,

$$R(\boldsymbol{\beta},\mathbf{a}) = E[(\boldsymbol{\beta}-\mathbf{a})'\mathbf{Q}(\boldsymbol{\beta}-\mathbf{a})|\boldsymbol{\beta}].$$

An estimator $\mathbf{a}_1(\mathbf{Y})$ is said to be *inadmissible* if there is another estimator $\mathbf{a}_2(\mathbf{Y})$ such that $R(\boldsymbol{\beta},\mathbf{a}_2) \leqslant R(\boldsymbol{\beta},\mathbf{a}_1)$ for all $\boldsymbol{\beta}$ with strict inequality $R(\boldsymbol{\beta},\mathbf{a}_2) < R(\boldsymbol{\beta},\mathbf{a}_1)$ for at least one value of $\boldsymbol{\beta}$. Otherwise, $\mathbf{a}_1(\mathbf{Y})$ is said to be *admissible.*

The risk function integrated with respect to the prior on $\boldsymbol{\beta}$ is known as the *Bayes risk*:

$$B(\mathbf{a}) = E[R(\boldsymbol{\beta},\mathbf{a})].$$

The Bayes estimator $E(\boldsymbol{\beta}|\mathbf{Y})$ minimizes Bayes risk. To verify this, write the Bayes risk as

$$B(\mathbf{a}) = E[E(\boldsymbol{\beta}-\mathbf{a})'\mathbf{Q}(\boldsymbol{\beta}-\mathbf{a})|\mathbf{Y}]$$

and observe that the expression in the inner brackets is minimized for each value of $\mathbf{Y}$ by setting $\mathbf{a} = E(\boldsymbol{\beta}|\mathbf{Y})$.

The following lemma on the uniqueness of the posterior mean is necessary to prove the admissibility of the Bayes decision.

LEMMA. *If an estimator* $\mathbf{a}(\mathbf{Y})$ *has the same Bayes risk as the estimator* $E(\boldsymbol{\beta}|\mathbf{Y})$, *then* $\mathbf{a}(\mathbf{Y})$ *is identically equal to* $E(\boldsymbol{\beta}|\mathbf{Y})$.

Proof: Write $\mathbf{a}(\mathbf{Y})$ as $\mathbf{a}(\mathbf{Y}) = \mathrm{E}(\boldsymbol{\beta}|\mathbf{Y}) + \mathbf{d}(\mathbf{Y})$. The Bayes risk of $\mathbf{a}$ can be written as

$$\begin{aligned} B(\mathbf{a}) &= E([E(\boldsymbol{\beta}|\mathbf{Y}) + d(\mathbf{Y}) - \boldsymbol{\beta}]'\mathbf{Q}[E(\boldsymbol{\beta}|\mathbf{Y}) + d(\mathbf{Y}) - \boldsymbol{\beta}]) \\ &= E([E(\boldsymbol{\beta}|\mathbf{Y}) - \boldsymbol{\beta}]'\mathbf{Q}[E(\boldsymbol{\beta}|\mathbf{Y}) - \boldsymbol{\beta}]) + E([\mathbf{d}(\mathbf{Y})]'\mathbf{Q}[\mathbf{d}(\mathbf{Y})]) \\ &= B(E(\boldsymbol{\beta}|\mathbf{Y})) + E([\mathbf{d}(\mathbf{Y})]'\mathbf{Q}[\mathbf{d}(\mathbf{Y})). \end{aligned}$$

Thus $B(a)$ exceeds $B(E(\boldsymbol{\beta}|\mathbf{Y}))$ unless the last term vanishes, which can occur only if $\mathbf{d}(\mathbf{Y}) = \mathbf{0}$ if $\mathbf{Q}$ is positive definite.

THEOREM 5.2 (ADMISSIBILITY OF THE POSTERIOR MEAN). *Given a quadratic loss function* $l(\boldsymbol{\beta},\hat{\boldsymbol{\beta}})=(\boldsymbol{\beta}-\hat{\boldsymbol{\beta}})'\mathbf{Q}(\boldsymbol{\beta}-\hat{\boldsymbol{\beta}})$ *with* $\mathbf{Q}$ *positive definite, a proper prior distribution for* $\boldsymbol{\beta}$, *and the normal linear regression model, the posterior mean* $\hat{\boldsymbol{\beta}}(\mathbf{Y})=E(\boldsymbol{\beta}|\mathbf{Y})$ *is an admissible estimator.*

Proof: Assume that $\hat{\boldsymbol{\beta}}(\mathbf{Y})$ is not admissible. Then there is some other estimator $\mathbf{a}$ such that

$$R(\boldsymbol{\beta},\mathbf{a})\leqslant R(\boldsymbol{\beta},\hat{\boldsymbol{\beta}})\quad \text{for all}\quad \boldsymbol{\beta}$$

$$R(\boldsymbol{\beta},\mathbf{a})< R(\boldsymbol{\beta},\hat{\boldsymbol{\beta}})\quad \text{for some}\quad \boldsymbol{\beta}.$$

Integrating these risk functions with respect to the prior yields the inequality

$$E(R(\boldsymbol{\beta},\mathbf{a}))\leqslant E\big(R(\boldsymbol{\beta},\hat{\boldsymbol{\beta}})\big),$$

but by assumption $\hat{\boldsymbol{\beta}}$ minimizes Bayes risk, and this last inequality must be an equality. But by the previous lemma, if the Bayes risks are equal, $\mathbf{a}=\hat{\boldsymbol{\beta}}$.

For a discussion of the admissibility of Bayes decision rules, the reader may consult Ferguson (1967, Section 2.3).

5.5 Comments on Interpretive Searches

There are two other approaches toward the problem of interpreting k-dimensional evidence—the likelihood approach and the Bayesian approach. Comments on interpretive searches from both these viewpoints are now given.

LIKELIHOOD COMMENTS

An examination of the likelihood function reveals that there is only one point that is unambiguously most favored by the data. This is the least-squares point, the maximum likelihood point. Reporting any other point constitutes a distortion of the evidence, an interpretation depending implicitly or explicitly on prior information. A likelihood advocate prefers not to interpret evidence; he prefers only to summarize it. He objects in general to the estimation framework which implicitly or explicitly involves a well-specified decision problem that requires us to summarize our uncertainty about the parameters in a single point or estimate. Ordinarily, no such decision problem is even envisaged. Selecting a tractable loss function in such circumstances is as arbitrary as selecting a tractable prior distribution. In the absence of any loss structure (even a vague one) the estimation framework seems irrelevant. Of course, we tend to use the language of estimation but what we call "estimates" are usually data summaries, not decisions. That is to say, interest centers on the least-squares estimators not because they are best for some decision but rather because the estimates together with confidence ellipsoids provide a useful data summary.

For the simple, two-variable linear-regression problem $\mathbf{Y}=\mathbf{x}_1\beta_1+\mathbf{x}_2\beta_2+\mathbf{u}$ the most straightforward way of summarizing the data evidence is to draw two-dimensional likelihood contours as in Figure 5.4. There are two useful results on the geometrical relationship between a confidence ellipsoid and various confidence intervals. The first is that the projection of a suitably chosen ellipsoid is a confidence interval. In Figure 5.4, the interval $[b_1^-, b_1^+]$ is a 95% interval for β_1. The second result is that the length of a conditional confidence interval given some linear restriction can be found by intersecting a confidence ellipsoid with a suitably chosen linear manifold. In Figure 5.4, the length of a confidence interval for β_1 given a value of β_2 is w^*, found by drawing a line through the center of the ellipsoid perpendicular to the β_2 axis. These two results are stated first, and then further discussed.

THEOREM 5.3 (SUPPORTING HYPERPLANES). *The region between any pair of parallel supporting hyperplanes of the ellipsoid* $(\boldsymbol{\beta}-\mathbf{b})'\mathbf{X}'\mathbf{X}(\boldsymbol{\beta}-\mathbf{b})=\sigma^2\chi^2_\alpha(1)$ *is a* $1-\alpha$ *percent confidence region for* $\boldsymbol{\beta}$.

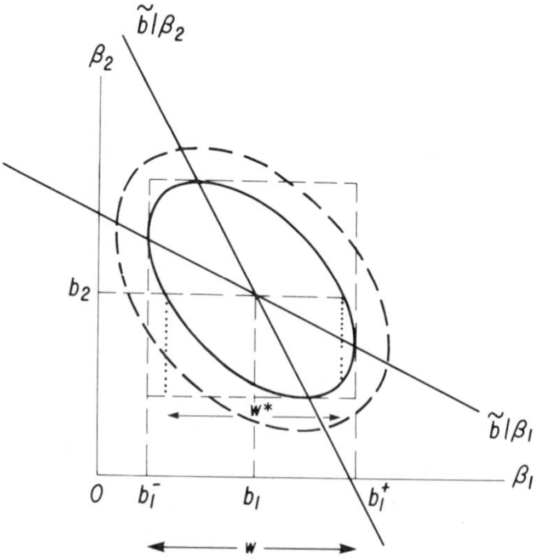

Fig. 5.4 Confidence ellipse.

Proof: A pair of supporting hyperplanes, $\psi'\beta = c_1$ and $\psi'\beta = c_2$, may be found by determining the extreme values of the function $\psi'\beta$ constrained to the ellipsoid. Solving this simple constrained optimization problem leads to the supporting hyperplanes

$$\psi'\beta = \psi'\mathbf{b} \pm \left[\psi'(\mathbf{X}'\mathbf{X})^{-1}\psi\sigma^2\chi^2_\alpha(1)\right]^{1/2}$$

which is a $1-\alpha\%$ interval for $\psi'\beta$.

THEOREM 5.4 (CONSTRAINED CONFIDENCE INTERVAL). *The function $\psi'\beta$ evaluated on the intersection of the ellipsoid $(\beta - \mathbf{b})'\mathbf{X}'\mathbf{X}(\beta - \mathbf{b}) \leqslant \chi^2_\alpha(1)\sigma^2$ and the linear manifold $\mathbf{R}(\beta - \mathbf{b}) = 0$ attains an interval of values equal in length to the conditional $1-\alpha\%$ confidence interval for $\psi'\beta$ given $\mathbf{R}\beta = \mathbf{r}$.*

Proof: This is also a constrained optimization problem left to the reader. Incidentally, Theorem 5.3 is a direct consequence of Theorem 5.4.

These two results illustrate the value of exact prior knowledge of linear combinations of parameters. In Figure 5.4 the outer ellipse is a 95% confidence ellipse for the parameter vector (β_1, β_2). It, furthermore, contains 95% of the volume under the likelihood surface and is a 95% posterior region if the prior distribution is diffuse. The interior ellipse is chosen so that its projection onto the β_1 axis, $[b_1^-, b_1^+]$ is a 95% interval for β_1. By Theorem 5.3, any pair of parallel lines tangent to this ellipsoid defines a 95% region.[9]

The two lines $\tilde{\mathbf{b}}|\beta_1$ and $\tilde{\mathbf{b}}|\beta_2$ are the locus of tangencies between the family of likelihood ellipses and horizontal and vertical lines, respectively. These are the estimates given exact knowledge of one of the parameters. The width of the interval for β_1 given β_2 is, by Theorem 5.4, just w^*. Letting the correlation between x_1 and x_2 be $r^2 = (\mathbf{x}_1'\mathbf{x}_2)^2/(\mathbf{x}_1'\mathbf{x}_1)(\mathbf{x}_2'\mathbf{x}_2)$, the ratio of the two widths is $w^*/w = (1-r^2)^{1/2}$. This is illustrated in Figure 5.5 and the multidimensional analogue of $(1-r^2)^{1/2}$ is suggested in Section 5.7 as a measure of the collinearity problem.

Having made these statements, we may now return to the problem of interpreting the data summarized in Figure 5.4. Assume that interest centers on characterizing the sample evidence about β_1. For simplicity it is desirable to have a one-dimensional description of that evidence. Regrettably, the sample evidence does not allow an unambiguous one-dimensional

[9] A region with minimal area is generated by lines parallel to the major axis of the ellipse. This is a geometrical interpretation of Silvey's (1969) result that a linear combination $\psi'\beta$ can be estimated relatively precisely, if ψ can be expressed as a linear combination of the eigenvectors of $\mathbf{X}'\mathbf{X}$ with the largest eigenvalues.

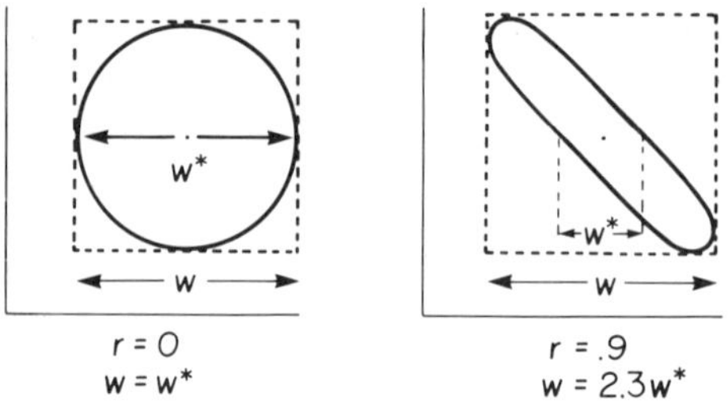

Fig. 5.5 Constrained and unconstrained confidence intervals.

data summary, since the evidence about β_1 depends on the prior information about β_2. For example, if prior information is relatively vague about both β_1 and β_2, the interval $b_1 \pm w$ is a useful summary of the evidence about β_1. If we know something about β_2, the effective sample information about β_1 may change drastically. In particular, if we know $\beta_2=0$, a useful summary of the evidence about β_1 is $(\tilde{b}_1|\beta_2=0)\pm w^*/2$. This interval is both shorter and recentered. The discrepancy between the two intervals will increase with the correlation r (see Figure 5.5). Thus when r is large, the sample evidence cannot be meaningfully interpreted without referring to prior information; to put this somewhat differently, the box implied by the point estimates and their standard errors gives a most inaccurate picture of the region favored by the data, the confidence ellipse. A two-dimensional likelihood function cannot unambiguously be collapsed into two one-dimensional functions.

BAYESIAN COMMENTS

A Bayesian analysis (somewhat magically) implies a one-dimensional summary of the data evidence about β_1, since the marginal posterior distribution of β_1 can be written using Bayes' rule as

$$f(\beta_1|\mathbf{Y}) \propto \left[\int_{\beta_2} f(\mathbf{Y}|\beta_1,\beta_2) f(\beta_2|\beta_1)\, d\beta_2\right] f(\beta_1) = f(\mathbf{Y}|\beta_1) f(\beta_1).$$

The last part of this expression appears to be a one-dimensional application of Bayes' rule with the integrated or marginal (one-dimensional) likelihood function $f(\mathbf{Y}|\beta_1)$ summarizing in the usual way the sample evidence about β_1. The first line indicates, however, that this marginal likelihood depends on the conditional prior $f(\beta_2|\beta_1)$, which is a formal Bayesian way of saying what is already clear: the sample evidence about β_1 depends on the prior information about β_2.

Although a complete Bayesian analysis of this problem requires a fully specified prior distribution, some progress can be made with a crudely specified distribution. Pretesting on β_2 is meaningdul when β_1 is relatively uncertain and when β_2 is thought to be near zero. In this case the prior contours effectively parallel the β_1 axis with the most likely value being $\beta_2=0$. A posterior distribution mixes this information with the sample information according to Bayes' rule $f(\beta_1,\beta_2|\mathbf{Y})\propto f(\mathbf{Y}|\beta_1,\beta_2)f(\beta_2)$. The modes of the posterior must be on the locus of tangencies between the likelihood contours and the prior contours (see Figure 5.6). This locus of points is just a line $\tilde{\mathbf{b}}|\beta_2$. Independent of any further distributional assumptions, the most likely value of the couple (β_1,β_2) after we have seen the data is on the line $\tilde{\mathbf{b}}|\beta_2$ between the least-squares estimate $\mathbf{b}$ and the constrained estimate given $\beta_2=0$.

The position of the mode on this line, as well as the complete joint posterior distribution, depends on a completely specified prior distribution for β_2. The reader may convince himself that there are four posterior distributions that imply the marginal posteriors (and in this case the marginal likelihoods) in Figure 5.7. Cases (a) and (b) occur when there is overwhelming sample or overwhelming prior information. In case (c) the two sources of information are roughly comparable in content and we react by mixing them into a unimodal distribution. This would occur with the conjugate normal prior that is widely discussed. If prior information about β_2 is steeper around zero, we may get the antimixture case (d). This is not an atypical situation with (my?) meaningful prior distributions. Note incidentally that cases (a) and (b) are, in fact, special cases of (c) and (d).

In light of the preceding, a Bayesian might make the following comments:

a. The theory of pretesting is misleading. It suggests that one may implement his prior information without carefully specifying it. It usually leads to pretesting at an arbitrary level of significance, say, .05. This may or may not capture the essential features of your prior.
b. Pretesting works only in the extreme cases when b_1 or $\tilde{b}_1|\beta_2=0$ are appropriate summaries. The mixture and antimixture cases in Figure 5.7 are excluded. (Incidentally, the exclusion of these cases means that small adjustments in the data evidence may imply jumping from the $\mathbf{b}$ summary to the $\tilde{\mathbf{b}}$ summary. This discontinuity ordinarily implies that the estimators are inadmissible. This situation can, of course, be improved by continuous mixing, but the analysis gets quite complex.)
c. Pretesting does not clearly distinguish sample from nonsample information. Ordinarily, our belief in the output of a study should depend on the judgmental inputs. When these inputs are disguised, we have no way of evaluating an empirical study.

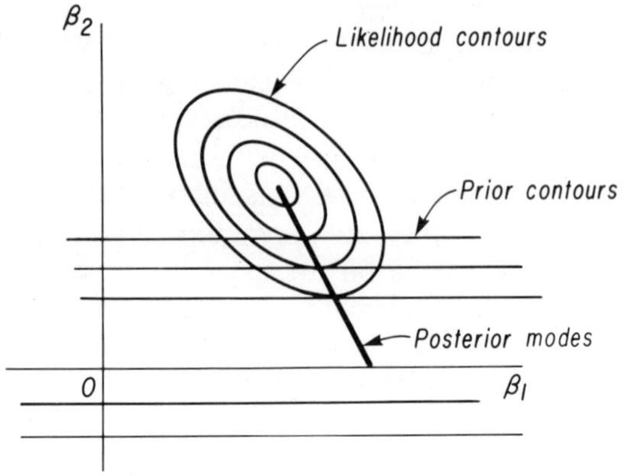

Fig. 5.6 Locus of posterior modes.

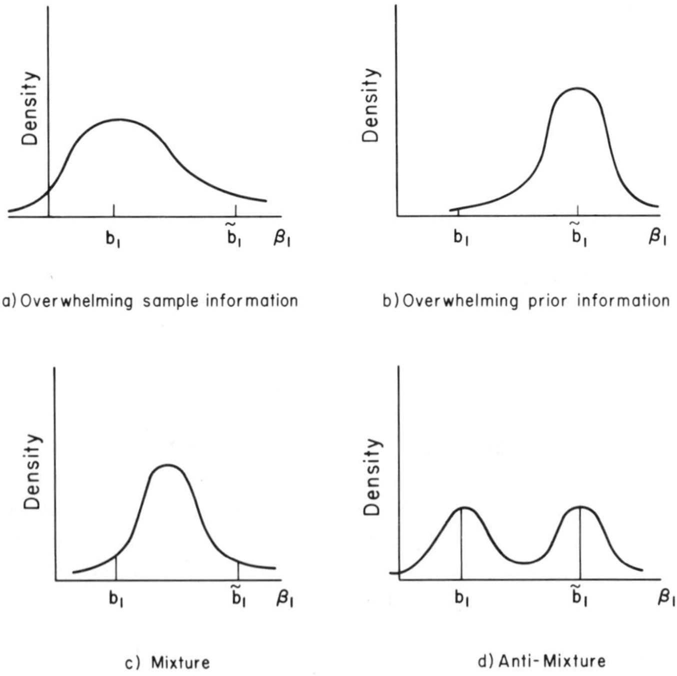

Fig. 5.7 Posterior distributions.

d. The pretesting theory just discussed is inappropriate when there is prior information about the other parameters. It is very difficult to make pretesting meaningful in higher dimensional problems, since perusal of multidimensional risk functions (expected losses) is terribly difficult. More generally, the ambiguities inherent in implementing prior opinion are more easily discussed in terms of alternative prior distributions than in terms of alternative risk functions (or sampling distributions).

In summary, the formal estimation framework does not focus on theoretical questions posed by practical pretesting. Practical pretesting is not designed or should not be designed to improve estimators independently of prior information, as suggested by this literature. In the first place, the estimation framework misstates the problem unless there is a real point decision to be made (a most rare event). Usually, an interpretive search is designed to characterize the information contained in the data set, and the estimation framework suggests the erroneous conclusion that the sample evidence somehow depends on what decisions are to be made. Second, even if you did have such a decision problem, in the context of quadratic loss, a Bayes estimate of one coefficient is just its posterior mean, which does not depend on the weight given another coefficient in the loss function. Third, the resultant estimators rarely dominate least squares and, furthermore, have generally unknown properties (though sampling properties are rarely relevant for interpreting the data). In the fourth place, in the context of the more usual problem of inference, the least-squares point, James and Stein notwithstanding, is the unique point most favored by the data. In the absence of prior information, it is the only point that has an unambiguous claim to be reported.

To put this slightly differently, data analysis can usefully be regarded to include three distinct phases. First, the data evidence is *summarized.* Second, it is *interpreted.* Finally, *decisions* may be made. The first phase requires a theory of sufficient statistics. The second phase requires a theory of learning. And the third phase requires a theory of decision making under uncertainty. The essential differences between Bayesian and classical inference arise in the second or interpretative phase. The Bayesian model of learning is described by Bayes' rule: let data evidence incrementally affect your opinions. The classical theory of learning, however, is either incomplete or is trivially described by: completely believe the data evidence.

The least-squares point and the variance-covariance matrix are the appropriate summary of the data evidence. Practical pretesting occurs at the interpretive stage and constitutes a de facto rejection of the trivial theory of learning implicit in classical inference and/or an ad hoc completion of that incomplete theory. A loss function occurs only in the third (decision-making) phase, and inadmissibility results are relevant to the learning phase only to the extent that we would prefer not to have a model of learning that, when coupled with a *particular* model of decision making under uncertainty leads to decisions that could be improved on, on the average. Inadmissibility results do not, however, have implications for the summarization phase and, for example, the Stein–James result has nothing to do with the adequacy of the least-squares point as a summary of the data evidence. The result *should* be disconcerting to those Bayesians who are fond of improper prior distributions, however, since it suggests that there is a serious inadequacy in their learning model.

5.6 Regression Selection Strategies and Revealed Priors

By definition, an interpretive search is an ad hoc covert method of introducing uncertain prior information. It is hardly remarkable that overt formal Bayesian methods could cast light on these murky goings-on. This section develops a bridge between searchers and Bayesians by answering the question: what kind of prior information does an interpretive search seem to be based on? This discussion is not intended merely to apologize for interpretive searches. It is possible to find priors that justify many procedures. In so doing we hope also to encourage a more careful review of the available prior information, a more overt use of that which is available, and finally, a clearer communication of the prior on which the data analysis rests.

As discussed in Section 5.1, an interpretive search strategy involves three decisions: (a) the choice of an origin, which determines the ellipsoid of constrained estimates, (b) the choice of a coordinate system, which selects 2^k or fewer points on the feasible ellipsoid, and (c) the choice of a weighting function that mixes the 2^k points into a single estimate. The features of the prior distribution that implicitly determine the first two of these choices are the surfaces upon which the prior density is constant. The third choice—the weighting function—is implied by the function that assigns to each prior isodensity surface a particular density value.

In this section, we first explore the relationship between prior isodensities and choice of constraints, and then, for one special case, analyze the relationship between the weighting function and the density labeling function. A correspondence is developed between priors that are uniform on ellipsoids and regression selection strategies that either (a) compute 2^k regressions or (b) compute principal component regressions. Similarly, priors that are uniform on hyperbolas are associated with strategies that omit insignificant variables. Lexicographic priors and priors that are uniform on cones are also discussed, the latter intended to capture the notion that parameters are equal.

5.6.1 Choice of Constraints

The language and method of thinking about this problem is borrowed from the Edgeworth–Bowley analysis of trade between a pair of consumers. In that problem, a fixed quantity of a set of k commodities is to be distributed between two consumers. Let $\mathbf{q}$ be a k-dimensional vector indicating the available quantities of the k commodities; let $\boldsymbol{\beta}$ be a k-dimensional vector indicating the allocation of commodities to the first consumer, who thereby attains utility level $U_1(\boldsymbol{\beta})$. The remainder, $\mathbf{q}-\boldsymbol{\beta}$, is allocated to the second consumer, who thereby attains utility level $U_2(\mathbf{q}-\boldsymbol{\beta})$. An allocation $\boldsymbol{\beta}_a$ is said to dominate another allocation $\boldsymbol{\beta}_b$, if $U_1(\boldsymbol{\beta}_a)\geqslant U_1(\boldsymbol{\beta}_b)$ and $U_2(\mathbf{q}-\boldsymbol{\beta}_a)\geqslant U_2(\mathbf{q}-\boldsymbol{\beta}_b)$ with at least one strict inequality. In words, one person is better off at $\boldsymbol{\beta}_a$ and no one is worse off. The undominated set of allocations is called the Pareto efficient set. Under differentiability and convexity assumptions, the Pareto efficient set is a curve formed by maximizing one of the consumer's utility levels subject to a given utility level of the other. This curve is called a *contract curve*, since given an initial allocation $\boldsymbol{\beta}$ off the contract curve it is likely that the consumers would trade to a suitable point on it, thereby making at least one better off and neither worse off. It is enough that utility be ordinal to define the contract curve. Picking an optimal point on it requires cardinal utility functions and a social welfare function $W(U_1, U_2)$ to be maximized.

Consider now the analogous problem of Bayesian inference with a k dimensional parameter $\boldsymbol{\beta}$. The data communicates its information through a likelihood function, say, $U_1(\boldsymbol{\beta})$, and the researcher communicates his information through a prior density, say, $U_2(\boldsymbol{\beta})$. Posterior modes are found by maximizing the "social information function", $W(U_1, U_2)= U_1U_2$. If the prior density is ordinal, that is, if it is defined only up to the surfaces on which it is constant, all that can be said is that the posterior mode is confined to a curve, which is called the *information contract curve*. In developing the correspondence between regression selection strategies and priors, we hypothesize that a researcher is attempting to approximate the information contract curve with a set of constrained regression estimates, where the word approximates means to find a set of points that contains the curve.

DEFINITIONS: ISODENSITY SURFACES AND LABELING FUNCTIONS

Let $h(\mathbf{x})=z$ be a convenient representation of a family of surfaces indexed by z, and let a family of density functions be written

$$f(\mathbf{x})=g(h(\mathbf{x})),$$

where h is given and g is any monotonically decreasing differentiable function with the restriction that $\int g(h(\mathbf{x}))d\mathbf{x}=1$. The surface $h(\mathbf{x})=z$ is called an *isodensity* surface, and the function g that assigns a density value to each of these surfaces is called a *labeling function*.

Example. A multivariate normal distribution

$$f(\mathbf{x})=(2\pi)^{-k/2}|\mathbf{H}|^{1/2}\exp\left\{-\frac{1}{2}(\mathbf{x}-\bar{\mathbf{x}})'\mathbf{H}(\mathbf{x}-\bar{\mathbf{x}})\right\}$$

is uniform on the ellipsoids

$$h(\mathbf{x})=(\mathbf{x}-\bar{\mathbf{x}})'\mathbf{H}(\mathbf{x}-\bar{\mathbf{x}})$$

with labeling function

$$g(z)\propto \exp-\left(\frac{z}{2}\right).$$

A multivariate Student distribution has the same elliptical isodensity surfaces but has the labeling function

$$g(z|\nu)\propto(\nu+z)^{-(k+\nu)/2}.$$

Johnson and Kotz [1972, p. 296] refer to densities that are uniform on ellipsoids as elliptically symmetric distributions. We call them elliptically uniform distributions to emphasize the fact that the ellipsoids are isodensity surfaces.

Prior-to-posterior analysis of a k-dimensional parameter $\boldsymbol{\beta}$ depends on the data $\mathbf{Y}$ through Bayes' rule

$$f_2(\boldsymbol{\beta}|\mathbf{Y})\propto f_y(\mathbf{Y}|\boldsymbol{\beta})f_1(\boldsymbol{\beta})$$

where the subscripts 1, 2, and y indicate prior, posterior, and sample, respectively. This may be rewritten in terms of isodensity and labeling functions as

$$f_2(\boldsymbol{\beta}|\mathbf{Y})\propto g_y(h_y(\boldsymbol{\beta}))\,g_1(h_1(\boldsymbol{\beta})).$$

Modes of the posterior distribution require the derivatives of this function to be set to zero

$$\mathbf{0}=g_y g_1'\frac{\partial h_1}{\partial\boldsymbol{\beta}}+g_y' g_1\frac{\partial h_y}{\partial\boldsymbol{\beta}}$$

or

$$0=\lambda\frac{\partial h_1}{\partial\beta}+\frac{\partial h_y}{\partial\beta} \tag{5.12}$$

where

$$\lambda(\beta)=\frac{g_1'/g_1}{g_y'/g_y} \tag{5.13}$$

and where λ is assumed to be positive, since $g_y'<0$, $g_1'<0$.

Notice now that as a function of λ, Equation (5.12) defines a k-dimensional space curve that depends on the isodensity functions h_1 and h_y only. In particular, the labeling distributions are irrelevant. To the extent that joint modes of the posterior are of interest, it is, therefore, possible to perform a Bayesian analysis in two steps: (1) select isodensity functions h_1 and h_y and compute the *information contract curve* given by Equation (5.12); (2) specify the labeling functions g_1 and g_y that can be used to compute the modes of the posterior. (All modes necessarily lie on the information contract curve.)

Interest in this two-step approach derives from the following assertion. It is impossible to measure degrees of belief about continuous random variables with enough accuracy that we would be content with a Bayesian analysis based on a single distribution (prior or data). Instead, ambiguities in the choice of distribution are properly dealt with by performing the analysis with many different distributions. A class of distributions that is of interest is the class with fixed isodensity surfaces and with varying labeling functions. This class is an appropriate focus of attention when the choice of isodensity curves is relatively unambiguous, but the choice of labeling functions is relatively ambiguous. In the case of complete ambiguity over choice of labeling function, the Bayesian analysis can at best specify the information contract curve. The value of such a limited statement is not to be underassessed, however. The restriction of modes to a k-dimensional curve is a very significant restriction.

Incidentally, the case of complete ambiguity over choice of labeling distribution represents the continuous analogue of Keynes' (1921) suggestion that probabilities are sometimes only ordinally ordered. In that case values of the random variable may be said to be more likely, less likely, or equally likely to other values, but it is impossible to say how much more likely. More formally, choice of isodensity function with a monotonicity assumption on the labeling function implies an ordinal ranking of points in the outcome space of the random variable. The cardinality of this ordering is determined by the labeling function.

THE DATA INDIFFERENCE SURFACES

The isolikelihood surfaces implied by the normal linear regression model are a family of concentric ellipsoids

$$z=(\boldsymbol{\beta}-\mathbf{b})'\mathbf{X}'\mathbf{X}(\boldsymbol{\beta}-\mathbf{b}),$$

where $\mathbf{b}$ is the least-squares estimate and $\mathbf{X}'\mathbf{X}$ is the design matrix. The data set is indifferent among all values of $\boldsymbol{\beta}$ lying on such an ellipsoid in the sense that each is assigned the same likelihood value. If the residual variance σ^2 were known, the data would assign to each ellipsoidal surface a particular likelihood value. If σ^2 is unknown, relative likelihoods (but not absolute likelihoods) can be computed. For the purposes of this section, only the indifference surfaces are needed, and what is discussed applies to any sampling process that determines elliptical isolikelihood surfaces.

ELLIPTICALLY UNIFORM PRIORS AND PRINCIPAL COMPONENT REGRESSION

The information of the researcher can be packaged in a prior density function. A prior density that is uniform on concentric ellipsoids can be written as[10]

$$f_1(\boldsymbol{\beta})=cg_1\left[(\boldsymbol{\beta}-\mathbf{b}^*)'\mathbf{N}^*(\boldsymbol{\beta}-\mathbf{b}^*)\right],$$

thereby indicating indifference among all points on an ellipsoid $(\boldsymbol{\beta}-\mathbf{b}^*)'\mathbf{N}^*(\boldsymbol{\beta}-\mathbf{b}^*)$.

The information contract curve formed by minimizing the likelihood quadratic form subject to the constraint that the prior quadratic form is a constant is

$$\mathbf{0}=\lambda\mathbf{N}^*(\boldsymbol{\beta}-\mathbf{b}^*)+\mathbf{X}'\mathbf{X}(\boldsymbol{\beta}-\mathbf{b})$$

where λ is a Lagrange multiplier. Solving this for $\boldsymbol{\beta}$, we obtain

$$\boldsymbol{\beta}(\lambda)=(\lambda\mathbf{N}^*+\mathbf{X}'\mathbf{X})^{-1}(\lambda\mathbf{N}^*\mathbf{b}^*+\mathbf{X}'\mathbf{X}\mathbf{b}), \tag{5.14}$$

which should remind you of the curve décolletage defined in Section 3.3. This equation defines the locus of tangencies between the prior ellipsoidal surfaces and the likelihood ellipsoidal surfaces. In two dimensions the curve is a hyperbola (see Figure 5.8).[11]

Points off the information contract curve are inefficient in the sense that

[10]It is interesting to observe that if $\mathbf{N}^*=\mathbf{I}$, and if the coefficients are independent, then f_1 is necessarily a multivariate normal distribution.

[11]The relevant part of the curve consists of the segment between $\mathbf{b}$ and $\mathbf{b}^*$, since any other points on the curve are dominated by points on this segment. It can also be shown that the points on the hyperbola but not on the line segment between $\mathbf{b}$ and $\mathbf{b}^*$ involve negative values of the Lagrange multiplier λ. These are ruled out by the monotonicity assumption on the labeling functions.

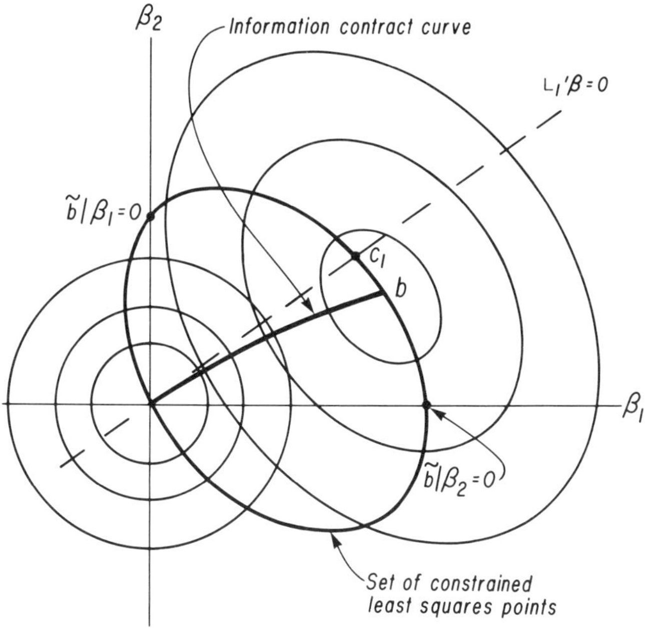

Fig. 5.8 The information contract curve with a spherical prior; $\mathbf{c}_1=$ principal component regression.

both the researcher and the data may be made "happier" by moving to a point on the contract curve. A researcher who imposes constraints on the estimates is trying to make himself happier, but at least if he tests the constraints, he is also keeping in mind the data preferences. He is thus apparently trying to describe the information contract curve, and the question to be addressed now is how accurately a set of constrained estimates approximates the information contract curve.

The first result to be discussed is that the contract curve (5.14) can be written as a weighted average of 2^k constrained least-squares points. The contract curve is also shown to be a weighted average of $k+1$ principal component points. Without loss of generality, the prior precision matrix is assumed to be diagonal $\mathbf{N}^*=\text{diag}\{d_1,d_2,\ldots,d_k\}$, and the prior location is taken as the origin $\mathbf{b}^*=\mathbf{0}$.

THEOREM 5.5 (DIAGONAL PRIOR PRECISION). *A matrix-weighted average can be written as a weighted average of 2^k constrained least-squares points*:

$$\mathbf{b}^{**}=(\mathbf{H}+\mathbf{D}^*)^{-1}\mathbf{H}\mathbf{b}=\sum_I w_I\mathbf{b}_I, \tag{5.15}$$

where $\mathbf{D}^*$ is a diagonal matrix, $\mathbf{D}^* = \text{diag}\{d_1, d_2, \ldots, d_k\}$, $\mathbf{H} = \sigma^{-2}\mathbf{X}'\mathbf{X}$, I is a subset of the first k integers, $\mathbf{b}_I$ minimizes the quadratic form $(\boldsymbol{\beta} - \mathbf{b})'\mathbf{H}(\boldsymbol{\beta} - \mathbf{b})$ subject to the constraints $\beta_i = 0$ for $i \in I$. The weights are

$$w_I = \frac{\left(\prod_{i \in I} d_i\right)|\mathbf{H}_I|}{|\mathbf{H} + \mathbf{D}|} \geqslant 0$$

$$\sum w_I = 1,$$

where $\mathbf{H}_I$ is the square matrix formed by deleting all rows i and columns i of $\mathbf{H}$ for all $i \in I$. By definition, if I is the complete set of k integers, $|H_I| \equiv 1$, and if I is the null set, $\Pi_{i \in I} d_i \equiv 1$.

Proof: See Appendix to Chapter 5.

The implication of this theorem is that a researcher who has elliptically uniform priors with major axes in the directions of the coordinate axes can find the points on his information contract curve by computing the 2^k regressions formed by omitting variables in all different combinations. A stronger result is that the contract curve can be written as a weighted average of $k+1$ principal component points. Before that result is presented, principal component regression is explained.

The intuitive foundation of principal component regression is the assertion that if the explanatory variables were orthogonal, that is, if $\mathbf{X}'\mathbf{X}$ were a diagonal matrix, then a "natural" interpretive search strategy would be to omit the variables with the smallest variance first. Some people have objected that, in fact, it is more "natural" to test the variables to see if they belong in the equation, that is, to omit the variables with the smallest t-values. The point that is being made in this section is that the prior distribution determines what is "natural," and arguments over features of search strategies can only be resolved by explicit reference to features of prior distributions. As it turns out, spherical priors make it "natural" to omit orthogonal variables in the order of their variance. There does not seem to be a prior that would lead one to test the restrictions implicit in principal component regression.

There is, of course, a transformation—in fact many—that make the explanatory variables orthogonal. Write the regression process as

$$\mathbf{Y} = \mathbf{X}\boldsymbol{\beta} + \mathbf{u} = \mathbf{X}\mathbf{L}\mathbf{L}^{-1}\boldsymbol{\beta} + \mathbf{u} = \mathbf{Z}\boldsymbol{\theta} + \mathbf{u} \tag{5.16}$$

where $\mathbf{Z} = \mathbf{X}\mathbf{L}$, $\boldsymbol{\theta} = \mathbf{L}^{-1}\boldsymbol{\beta}$, and where $\mathbf{L}$ is a $(k \times k)$ matrix that diagonalizes $\mathbf{X}'\mathbf{X} : \mathbf{Z}'\mathbf{Z} = \mathbf{L}'\mathbf{X}'\mathbf{X}\mathbf{L} = \boldsymbol{\Lambda} = \text{diag}\{\lambda_1, \lambda_2, \ldots, \lambda_k\}$. The transformed model $\mathbf{Y} = \mathbf{Z}\boldsymbol{\theta} + \mathbf{u}$ thus has orthogonal explanatory variables, and principal component regression would omit these variables (impose the constraints $\theta_i = 0$) as ordered by the variances λ_i.

Among the many matrices that take $\mathbf{X'X}$ into a diagonal matrix, it is necessary to select one. By Theorem 35 in Appendix 1, there is a (unique) matrix $\mathbf{L}$ such that $\mathbf{L'X'XL}$ is a diagonal matrix and $\mathbf{L'N^*L}$ is an identity matrix, where $\mathbf{N^*}$ is an arbitrary $(k \times k)$ symmetric positive definite matrix. Thus choice of $\mathbf{N^*}$ with the restriction on $\mathbf{L}$, $\mathbf{L'N^*L} = \mathbf{I}_k$, determines $\mathbf{L}$ uniquely. Although it is common to choose $\mathbf{N^*} = \mathbf{I}_k$, perhaps with the original explanatory variables standardized to have equal variance, this decision is a critical step in the interpretive search strategy and should not be taken lightly. As is shown, it amounts to choosing a prior with elliptical isodensity surfaces $\boldsymbol{\beta}'\mathbf{N^*}\boldsymbol{\beta} = z$.

When $\mathbf{N^*} = \mathbf{I}_k$, the matrix $\mathbf{L}$ is a matrix of eigenvectors of $\mathbf{X'X}$. The vector of parameters $\boldsymbol{\theta}$ is $\mathbf{L}^{-1}\boldsymbol{\beta} = \mathbf{L}'\boldsymbol{\beta}$, and the constraint $\theta_i = 0$ is equivalent to the constraint $\mathbf{L}_i'\boldsymbol{\beta} = 0$, where $\mathbf{L}_i$ is the ith eigenvector of $\mathbf{X'X}$. The diagonal elements of $\mathbf{L'X'XL}$ are the eigenvalues of $\mathbf{X'X}$. Thus principal component regression sequentially imposes the constraints that the vector $\boldsymbol{\beta}$ is orthogonal to the eigenvectors of $\mathbf{X'X}$, with the constraints ordered from smallest to largest eigenvalue.[12]

In Appendix 1, it is shown that the eigenvectors of $\mathbf{X'X}$ are the principal axes of the ellipsoid $\boldsymbol{\beta}'\mathbf{X'X}\boldsymbol{\beta} = r^2$, and the eigenvalues are ordered the reverse of the ordering of the lengths of the axes. The first restriction is that $\boldsymbol{\beta}$ is orthogonal to the largest axis of the ellipse. For a two-dimensional problem, this restriction is illustrated in Figure 5.8. Notice that $\mathbf{c}_1 - \mathbf{b}$ and $\mathbf{c}_1$ are orthogonal by construction.

The following theorem asserts that the convex hull of the three points $\mathbf{b}$, $\mathbf{c}_1$, and $\mathbf{0}$ contains the contract curve.

THEOREM 5.6 (SPHERICAL PRIOR). *The contract curve* (5.14) *with* $\mathbf{N^*} = \mathbf{I}$ *and* $\mathbf{b^*} = \mathbf{0}$ *can be written as*

$$\boldsymbol{\beta}(\lambda) = (\lambda\mathbf{I} + \mathbf{X'X})^{-1}\mathbf{X'Xb}$$

$$= \sum_{j=0}^{k} w_j(\mathbf{X}, \lambda)\mathbf{c}_j$$

where $\mathbf{c}_j$ *is the jth principal component point formed by "dropping" from the equation the j principal components of* $\mathbf{X'X}$ *with the smallest roots.*

[12]Another way to describe this is that the linear combination of variables $\mathbf{XL}_1$ is selected to have smallest variance $\mathbf{L}_1'\mathbf{X'XL}_1$, given the normalization $\mathbf{L}_1'\mathbf{L}_1 = 1$. The next linear combination, $\mathbf{XL}_2$, is restricted to be orthogonal to $\mathbf{XL}_1$, $0 = \mathbf{L}_2'\mathbf{X'XL}_1$, but is otherwise selected to minimize its variance.

Mathematically, the vectors $\mathbf{c}_j$ *are defined using the eigenvector coordinates*

$$\mathbf{L}'\mathbf{X}'\mathbf{X}\mathbf{L} = \boldsymbol{\Lambda} = \mathrm{diag}(\lambda_1, \lambda_2, \ldots, \lambda_k)$$
$$\lambda_1 \leqslant \lambda_2 \leqslant \cdots \leqslant \lambda_k$$
$$\mathbf{L}\mathbf{L}' = \mathbf{I}$$
$$\mathbf{g} = \mathbf{L}'\mathbf{b}$$

with $\mathbf{h}_j' = (0, 0, \ldots, 0, g_{j+1}, g_{j+2}, \ldots, g_k)$

$$\mathbf{c}_j = \mathbf{L}\mathbf{h}_j.$$

The weights applying to the principal component points are

$$w_0 = \frac{\lambda_1}{(\lambda_1 + \lambda)}$$
$$w_j = \frac{\lambda(\lambda_{j+1} - \lambda_j)}{(\lambda_{j+1} + \lambda)(\lambda_j + \lambda)} \qquad 0 < j < k$$
$$w_k = \frac{\lambda}{(\lambda_k + \lambda)}$$

The proof follows directly by writing $\boldsymbol{\beta}(\lambda)$ *in principal component coordinates,* $\boldsymbol{\beta}(\lambda) = (\lambda\mathbf{I} + \mathbf{L}\mathbf{L}'\mathbf{X}'\mathbf{X}\mathbf{L}\mathbf{L}')^{-1}\mathbf{X}'\mathbf{X}\mathbf{L}\mathbf{L}'\mathbf{b} = \mathbf{L}(\lambda\mathbf{I} + \boldsymbol{\Lambda})^{-1}\boldsymbol{\Lambda}g$.

We can then write $(\lambda\mathbf{I} + \boldsymbol{\Lambda})^{-1}\boldsymbol{\Lambda}\mathbf{g} = \Sigma_{j=0}^{k} w_j h_j$ where w_j is the solution to the triangular system $w_0 = \lambda_1/(\lambda_1 + \lambda), w_0 + w_1 = \lambda_2/(\lambda_2 + \lambda), \ldots, \Sigma_{i=0}^{k-1} w_i = \lambda_k/(\lambda_k + \lambda), \Sigma_{i=0}^{k} w_i = 1$.

These last two results determine a subset of constrained estimates that a Bayesian with an elliptical prior would be interested in. The principal component result is perhaps more useful, since it involves only $k+1$ regressions instead of 2^k and since by a linear transformation any ellipsoid can be taken into a sphere. The principal component result may be used to resolve several questions that trouble users of principal component regression. The first concerns the arbitrary order in which the principal component restrictions are imposed, and the second is the arbitrary normalization rule $\mathbf{l}'\mathbf{l} = 1$. Some writers have suggested that one ought to "test" the restrictions, that is, to order the restrictions by their t values. Theorem 5.6 does not apply if the restrictions are so ordered. They must be ordered by their respective eigenvalues (variances). The arbitrariness in normalization is also resolved, since the researcher is required to use a coordinate system in which his prior is spherical. Equivalently, if his prior is uniform on the ellipsoids $\boldsymbol{\beta}'\mathbf{N}^*\boldsymbol{\beta}$, he should use the normalization $\mathbf{L}'\mathbf{N}^*\mathbf{L} = 1$. The reverse result is also true: if a Bayesian computes principal component points with respect to some normalization, he reveals that he has the prior implicit in that normalization in the sense that otherwise points on the contract curve might not be weighted averages of the points he computes (Leamer, 1977).

Another element of arbitrariness that is not widely recognized arises when the principal component analysis is applied to only a subset of the variables. In that case it is not clear at the outset whether one should attempt to minimize the marginal variance of the components, or the variance conditional on unaffected variables. By the logic in this section, it should be the conditional variance (Leamer, 1977).

HYPERBOLICALLY UNIFORM PRIORS

Although the principal component method of estimation is used occasionally, it is much more common to express restrictions in a predetermined coordinate system, that is, to drop particular variables. In Figure 5.8 the third point that would most often be reported is not $\mathbf{c}_1$ but $\tilde{\mathbf{b}}|\beta_1=0$ or $\tilde{\mathbf{b}}|\beta_2=0$. This is clearly undesirable with elliptically uniform priors, since the information contract curve may be very poorly represented by such a sequence of points. For example, in Figure 5.8 if the variable with the lower t value is dropped first, the convex hull of the three estimates $(\mathbf{b},[\tilde{\mathbf{b}}|\beta_2=0],\mathbf{0})$ does not contain the contract curve. (The t value of the first coefficient exceeds the t value of the second because $\tilde{\mathbf{b}}|\beta_2=0$ is closer in the data metric to $\mathbf{b}$ than is $\tilde{\mathbf{b}}|\beta_1=0$. In other words, the data are "happier" with $\tilde{\mathbf{b}}|\beta_2=0$ than $\tilde{\mathbf{b}}|\beta_1=0$.)

The common procedure of dropping variables with low t values may, therefore, be highly undesirable with elliptically uniform priors. On the other hand, there may be perfectly reasonable priors that do lead to this kind of processing of the data. A completely trivial case in point illustrated in Figure 5.9 occurs when the prior isodensity surfaces are $z=\min(|\beta_1|,|\beta_2|)$. The solid line linking $\mathbf{b}$ to $\tilde{\mathbf{b}}|\beta_2=0$ contains all global modes, although local modes may lie on the dotted line linking $\mathbf{b}$ to $\tilde{\mathbf{b}}|\beta_1=0$. Thus a data analysis that reported $\mathbf{b}$ and $\tilde{\mathbf{b}}|\beta_2=0$ would convey part of the essential features of the data, since global modes are necessarily convex combinations of these two points. Local modes on the segment connecting $\mathbf{b}$ to $\tilde{\mathbf{b}}|\beta_1=0$ are also natural candidates to be reported.

These right-angled prior indifference curves represent a peculiar kind of ordering that seems difficult to approximate with any continuous probability functions. They, furthermore, reflect a peculiar willingness to ignore all but one of the constraints at any but special points in the parameter space. What seems to be a more reasonable family of curves are the hyperbolas

$$z=\prod_i|\beta_i|.$$

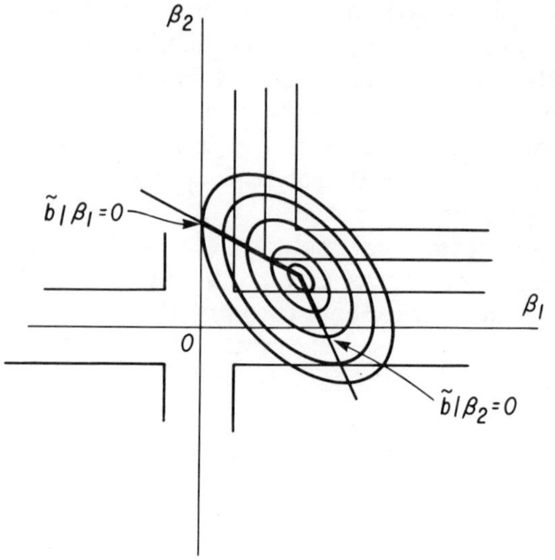

Fig. 5.9 The contract curve with rectangularly uniform priors.

As an example of such a density, products of independent Student functions

$$f(\boldsymbol{\beta}) \propto \prod_i (\nu + \beta_i^2)^{-(\nu+1)/2}$$

are hyperbolically uniform in the degenerate case $\nu = 0$

$$f(\boldsymbol{\beta}) \propto \prod_i |\beta_i|^{-1}.$$

The information contract curve Equation (5.14) with hyperbolically uniform priors becomes

$$\lambda\{\beta_i^{-1}\} + \mathbf{X}'\mathbf{X}(\boldsymbol{\beta} - \mathbf{b}) = \mathbf{0}, \tag{5.17}$$

where $\{\beta_i^{-1}\}$ indicates a vector with elements β_i^{-1}. Solving out the Lagrange multiplier λ yields the system of quadratics

$$\mathbf{n}_1'(\boldsymbol{\beta} - \mathbf{b})\beta_1 = \mathbf{n}_i'(\boldsymbol{\beta} - \mathbf{b})\beta_i, \qquad i > 1$$

where $\mathbf{n}_i'$ is the ith row of $\mathbf{X}'\mathbf{X}$. See Figure 5.10.

This system of equations can also be written as

$$\boldsymbol{\beta} = (\lambda\mathbf{D} + \mathbf{X}'\mathbf{X})^{-1}\mathbf{X}'\mathbf{X}\mathbf{b} \tag{5.18}$$

where $\mathbf{D} = \text{diag}\{\beta_1^{-2}, \beta_2^{-2}, \ldots, \beta_k^{-2}\}$. This equation appears to be a matrix-weighted average with the prior ellipsoids $\boldsymbol{\beta}'\mathbf{D}\boldsymbol{\beta}$ located at the origin, with

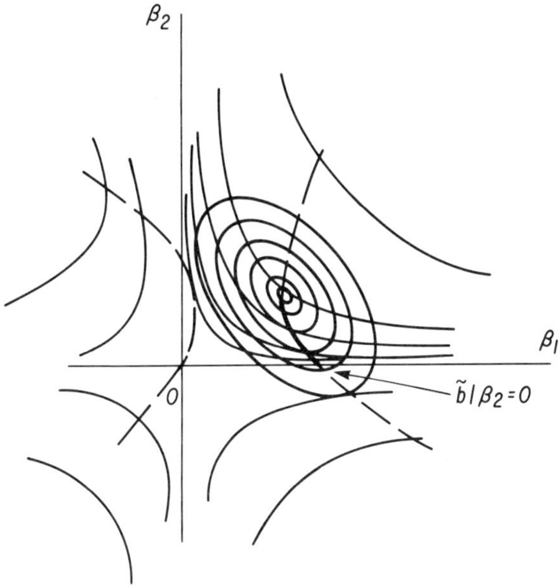

Fig. 5.10 The information contract curve with hyperbolically uniform priors.

principal axes equal to the coordinate axes. The relative lengths of these axes are, however, functions of $\boldsymbol{\beta}$. It is thus useful to think of hyperbolically uniform priors as if they were elliptically uniform priors located at the origin with principal axes equal to the coordinate axes but with the lengths of the principal axes "uncertain." In fact, the Student priors mentioned previously can be written as (elliptically uniform) normals with uncertain variance hyperparameters.

For any arbitrary $\mathbf{D}$ in Equation (5.18), Theorem 5.5 implies that $\boldsymbol{\beta}$ is contained in the hull of the 2^k constrained least-squares points formed by dropping variables. Since $\mathbf{D}$ is not free, it may be possible to further reduce the set of points. Paralleling the previous section, we would like to find a sequence of $k+1$ constrained estimates that contain the relevant curve segment. Although the prior determines the coordinate system, the sequence for imposing constraints is very ambiguous. Many different orderings seem possible. For example, (1) we might delete variables as ordered by the t coefficients in the original equation, or (2) we might recompute t values as constraints are imposed. (3) Proceeding in the other direction, variables may be added to the equation as ordered by their simple correlations with the dependent variable, or (4) as ordered by their partial correlations holding fixed the variables that are already included.

(5) Yet another alternative is to find the set of j variables that maximizes the multiple correlation coefficient, for $j=1,2,\ldots,k$.

It is obvious that the hyperbolic prior cannot induce a change in sign of any coefficient. The sequence of dropping variables as ordered by the sequentially computed t statistics necessarily preserves the orthant of the estimate (Leamer, 1975):

THEOREM 5.7 (ORTHANT PRESERVATION). *The least-squares estimate of β_j subject to a single linear constraint must lie in the interval $(b_j - V_{jj}^{1/2}|t|, b_j + V_{jj}^{1/2}|t|)$, where b_j is the unconstrained least-squares estimate of β_j, $V_{jj}^{1/2}$ is the jth diagonal element of $\mathbf{V}=s^2(\mathbf{X}'\mathbf{X})^{-1}$, and t is the t statistic for testing the restriction.*

Proof: The least-squares estimate subject to the constraint $\mathbf{R}\boldsymbol{\beta}=r$ (where $\mathbf{R}$ is a row vector) is $\hat{\boldsymbol{\beta}}=\mathbf{b}-\mathbf{VR}'(\mathbf{RVR}')^{-1}(\mathbf{Rb}-\mathbf{r})$, with $\mathbf{V}=s^2(\mathbf{X}'\mathbf{X})^{-1}$. The t statistic for testing $\mathbf{R}\boldsymbol{\beta}=r$ is $t=(\mathbf{Rb}-r)/(\mathbf{RVR}')^{1/2}$. Thus we may write the constrained least-squares estimate of β_j as $\hat{\beta}_j=b_j-[\mathbf{V}_j'\mathbf{R}'(\mathbf{RVR}')^{-1/2}V_{jj}^{-1/2}]V_{jj}^{1/2}t$, where $\mathbf{V}_j$ is the jth column of $\mathbf{V}$. The term in square brackets is just the correlation between b_j and $\mathbf{Rb}$, which must be between -1 and $+1$. These two extreme values imply the bound in the statement of the theorem.

A consequence of this theorem is that there can be no change in sign of any coefficient that is more significant than the coefficient of an omitted variable. In particular, if the least significant variable is omitted, all the other coefficients will retain their signs. Thus the sequence of omitting variables in the order of their sequentially computed t statistics necessarily preserves the orthant of the estimate.

This theorem increases my probability of the truthfulness of the conjecture that this sequence of estimates contains the contract curve (5.17), but I have been unable to construct a proof of the proposition. The proposition is true in two dimensions, but in two dimensions all five sequences described in the foregoing paragraph imply estimates that contain the curve. I do have a tedious proof that the curve is contained in the hull of the unconstrained point and the k constrained points formed by omitting a single variable.

LEXICOGRAPHIC ORDERING

Spherically uniform priors have been seen to imply constraints in the coordinate system of the sample. The sequence of imposing these constraints depends on the eigenvalues of $\mathbf{X}'\mathbf{X}$ but not at all on the data $\mathbf{Y}$ or as a result on any test statistics. Hyperbolically uniform priors on the other hand, imply a predetermined coordinate system but with a sequence of imposing constraints that is data dependent. Occasionally, both the coordinate system and the sequence of imposing constraints are predetermined. By this I mean that a researcher decides before looking at either **Y** or **X** first to omit variable one, then variable two, and so on. As an example, in distributed lag analysis a general model involving many lagged explanatory variables is often simplified by omitting sequentially the variable with the longest lag.

In consumer theory, a consumer who first satisfies his desires for good A, then for good B, and so on, is said to have a lexicographic utility ordering. Similarly, a researcher who proceeds this way has a lexicographic information ordering. A probabilistic structure that can effect such an ordering allocates positive probability to a nested sequence of subspaces with a one-dimensional informative prior in each subspace.

CONICALLY UNIFORM PRIORS

A fairly common form of prior information about sets of parameters is expressed in the pair of sentences "I think these coefficients are the same size and sign. I have very little information about their particular magnitudes." This could be translated into one of several families of isodensity surfaces. The traditional degenerate normal distributions would lead to cylindrical surfaces around the vector of ones. Such a density has been used by Lindley and Smith (1972) to produce a modified Stein estimator and by Shiller (1973) to produce a distributed lag estimator. For reasons to be explained below, a better family of isodensity surfaces consists of cones from the origin also around the vector of ones.

Normal priors that reflect this information may be constructed as follows. Begin with a multivariate normal distribution for $\boldsymbol{\beta}$ located at the origin with covariance matrix $\sigma^2\mathbf{I}$. The joint distribution of $\boldsymbol{\beta}$ and the mean of the coefficients $\bar{\beta}$ (a scalar) then has covariance matrix

$$\operatorname{var}\begin{bmatrix} \boldsymbol{\beta} \\ \bar{\beta} \end{bmatrix} = \sigma^2 \begin{bmatrix} \mathbf{I} & \mathbf{1}k^{-1} \\ k^{-1}\mathbf{1}' & k^{-1} \end{bmatrix}$$

where $\mathbf{1}$ is a vector of ones, and $\mathbf{1}'\mathbf{1}=k$. Conditional on $\bar{\beta}$ the moments of $\boldsymbol{\beta}$, therefore, would be

$$E(\boldsymbol{\beta}|\bar{\beta}) = \mathbf{1}\bar{\beta}$$

$$V(\boldsymbol{\beta}|\bar{\beta}) = \sigma^2(\mathbf{I} - \mathbf{1}\mathbf{1}'/k).$$

Retain these conditionals but let $\bar{\beta}$ have variance $v \neq \sigma^2/k$. Marginally, $\boldsymbol{\beta}$

then has mean vector zero and covariance matrix

$$V(\boldsymbol{\beta})=\sigma^2\mathbf{I}+\mathbf{1}\left(v-\frac{\sigma^2}{k}\right)\mathbf{1}'.$$

Isodensity contours take the form

$$\begin{aligned} z&=\boldsymbol{\beta}'V^{-1}(\boldsymbol{\beta})\boldsymbol{\beta}\\ &=\boldsymbol{\beta}'\left(\mathbf{I}-\mathbf{1}(\mathbf{1}'\mathbf{1}+s^{-2})^{-1}\mathbf{1}'\right)\boldsymbol{\beta}\end{aligned}$$

where

$$s^2=v-\frac{\sigma^2}{k}.$$

Last, let v get large to become diffuse on $\bar{\beta}$, and the isodensity contours become

$$z=\boldsymbol{\beta}'(\mathbf{I}-\mathbf{1}k^{-1}\mathbf{1}')\boldsymbol{\beta}.$$

These are cylinders around the vector **1**, since by analogy to the error sum of squares in the least-squares algebra, z is the length of the difference between $\boldsymbol{\beta}$ and the projection of $\boldsymbol{\beta}$ onto **1**.

In other words, a normal probability function for $\boldsymbol{\beta}$ with conditional distribution $f(\boldsymbol{\beta}|\bar{\beta})$ as if $\boldsymbol{\beta}$ were spherical and with $\bar{\beta}$ diffuse implies cylindrical isodensity surfaces. In two dimensions, these are lines parallel to the vector (1,1) in Figure 5.11. To this author, this is an exceedingly poor characterization of the statement "I think β_1 and β_2 have the same sign and magnitude," since it indicates indifference between, for example, the vector (1, −1) and the vector (100,98). The former fails the test of equality of coefficients miserably, and the latter passes it admirably.

A better family of isodensity surfaces to express this kind of prior information is the conically uniform family. This family indicates indifference between all vectors that make the same angle with the vector of ones. That is, isodensity surfaces depend on the cosine of the angle between $\boldsymbol{\beta}$ and **1**

$$z=1-\frac{(\boldsymbol{\beta}'\mathbf{1})^2}{\boldsymbol{\beta}'\boldsymbol{\beta}\mathbf{1}'\mathbf{1}}=1-\cos^2(\boldsymbol{\beta},\mathbf{1})$$

The information contract curve (5.12) can then be written as

$$\mathbf{0}=\left[\frac{(\boldsymbol{\beta}'\mathbf{1})}{\boldsymbol{\beta}'\boldsymbol{\beta}}\mathbf{1}-\frac{(\boldsymbol{\beta}'\mathbf{1})^2}{(\boldsymbol{\beta}'\boldsymbol{\beta})^2}\boldsymbol{\beta}\right]\lambda+\mathbf{X}'\mathbf{X}(\boldsymbol{\beta}-\mathbf{b}) \tag{5.19}$$

or

$$\boldsymbol{\beta}=(\lambda^*\mathbf{I}+\mathbf{X}'\mathbf{X})^{-1}(\lambda^*a^*(\boldsymbol{\beta})\mathbf{1}+\mathbf{X}'\mathbf{X}\mathbf{b}) \tag{5.20}$$

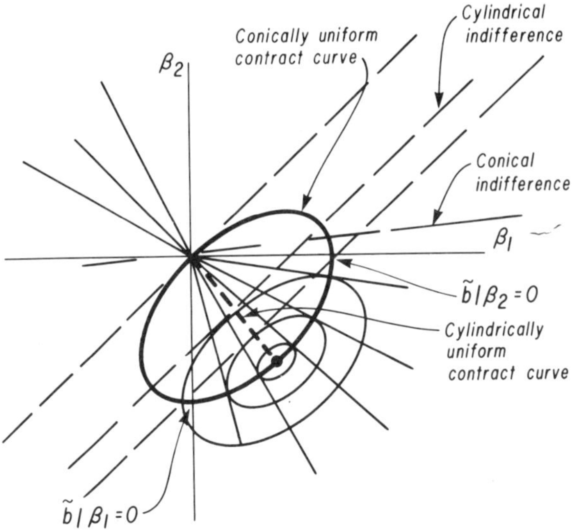

Fig. 5.11 Conically uniform and cylindrically uniform priors.

where

$$a^*(\boldsymbol{\beta})=\frac{(\boldsymbol{\beta}'\boldsymbol{\beta})}{(\boldsymbol{\beta}'\mathbf{1})}=\frac{\bar{\beta}}{\cos^2(\boldsymbol{\beta},\mathbf{1})}$$

$$\lambda^*=\frac{\lambda(\boldsymbol{\beta}'\mathbf{1})^2}{(\boldsymbol{\beta}'\boldsymbol{\beta})^2}.$$

Equation (5.20) appears to be the familiar matrix-weighted average of the least-squares estimate **b** and the vector $a^*(\boldsymbol{\beta})\mathbf{1}$. Thus we can loosely think of the conically uniform prior as inducing a spherically uniform prior located at $a^*(\boldsymbol{\beta})\mathbf{1}$. Note that the location of this vector involves an expansion of $\mathbf{1}\bar{\beta}$ by the amount $\cos^{-2}(\boldsymbol{\beta},\mathbf{1})$. Thus when $\boldsymbol{\beta}$ and $\mathbf{1}$ are orthogonal the effective location of the prior is $\pm\mathbf{1}\infty$, and the estimate may be substantially pulled from its location.

A more accurate understanding of the behavior of this information contract curve can be obtained by premultiplying (5.19) by $\boldsymbol{\beta}'$:

$$0=\left(\frac{(\boldsymbol{\beta}'\mathbf{1})^2}{(\boldsymbol{\beta}'\boldsymbol{\beta})}-\frac{(\boldsymbol{\beta}'\mathbf{1})^2}{(\boldsymbol{\beta}'\boldsymbol{\beta})}\right)\lambda+\boldsymbol{\beta}'\mathbf{X}'\mathbf{X}(\boldsymbol{\beta}-\mathbf{b})$$

$$=\boldsymbol{\beta}'\mathbf{X}'\mathbf{X}(\boldsymbol{\beta}-\mathbf{b}),$$

which can be rewritten as

$$\left(\beta-\frac{\mathbf{b}}{2}\right)'\mathbf{X}'\mathbf{X}\left(\beta-\frac{\mathbf{b}}{2}\right)=\frac{\mathbf{b}'\mathbf{X}'\mathbf{X}\mathbf{b}}{4}.$$

This is the same ellipsoid as the set of all constrained least-squares points (5.1). Since the contract curve is thus a continuous set of points lying on the feasible ellipsoid, it is obvious that the hull of no finite set of constrained least-squares points can contain the curve. In that sense it is undesirable to try to characterize the curve by a set of constrained least-squares points.

An example illustrating the difference between conically and cylindrically uniform priors is depicted in Figure 5.11. The information contract curve with cylindrically uniform priors is a straight line from the least-squares point to the origin, always in the fourth quadrant in this example. In contrast, the information contract curve with conically uniform priors is an ellipse connecting $\mathbf{b}$ to $\mathbf{b}_{\{1\}}$ (the variable with the lower t value is dropped) to the origin. In other words, although imposing the constraint $\beta_1=\beta_2$ leads to the estimate (0,0), this point and the least-squares point alone greatly distort the pooled information. It would be better to report also the estimate computed when the variable with the lower t value is dropped. Even this constitutes a distortion of the pooled information, since the contract curve travels very distinctly through the third quadrant. (The arm of the contract curve that travels into the first quadrant is dominated by the arm that travels into the third quadrant.)

To summarize this section, the vague notion that coefficients are equal in magnitude and sign is not well captured by elliptically uniform priors or their degenerate counterparts—cylindrically uniform priors. A better choice of isodensity surfaces is a family of cones. A family of cones leads to a very different kind of information contract curve than does a family of ellipsoids. Of the contract curves discussed in this section it is the one that is most poorly approximated by a sequence of constrained least-squares estimates.

SUMMARY

The literature on biased estimation of the regression parameter vector can be thought to involve three choices: (a) the choice of a (prior) location, (b) the choice of a distance function for measuring closeness to that location, and (c) the choice of a particular value of the distance function, or in the other language, the shrinkage factor. The "Stein" estimators and both Bayesian and non-Bayesian variants of Hoerl and Kennard's ridge regression presume the existence of (1) the (prior) location and (2) the metric, and they argue over (3) the shrinkage factor. But once the location and metric have been selected, the set of potential estimates has been reduced to a curve, or possibly mixtures of points along a curve. In this section the choice of metric is emphasized, and the choice of shrinkage factor is deemphasized. It is more important to know which curve is appropriate than to pick a particular point on the curve. Furthermore, the choice of a point on the curve depends significantly on features of the prior that are likely to be difficult to select; in that event a useful data analysis tool is a graphical or mathematical representation of the whole curve.

The choice of metric is important from the purist Bayesian viewpoint, since it shrinks the set of potential posterior modes from the feasible ellipsoid to an information contract curve. The choice of metric is equivalently important when only constrained regressions are computable, since it determines the coordinate system, and also the order in which constraints are imposed. All that is left undetermined is the weighting function, which determines a single estimate as a weighted average of the set of constrained estimates.

The usual elliptical metrics are closely associated with principal component regression. The more common regression-selection strategies cannot be justified with elliptical metrics, and we have been forced to consider hyperbolic, lexicographic, and conical metrics. These have been associated, respectively, with strategies that omit insignificant variables, that omit predetermined variables, and that impose equality constraints. Conical metrics might be more appropriate than cylindrical metrics when coefficients are thought to be similar in size, but no regression selection strategy is appropriate with conical metrics. A more careful description of the information contract curve is required.

5.6.2 Choice of Weight Functions

An interpretive search strategy involves, first, a choice of origin, second, a choice of coordinate system for imposing constraints, and third, a weighting function that selects among the set of constrained estimates. Prior isodensities imply a contract curve that can be approximated by a sequence of constrained estimates. In that sense the choice of origin and coordinate system corresponds to the choice of prior isodensities. Selection of one of the constrained estimates or more generally the specification of a weighting function over constrained estimates requires a fully specified prior. In this section we select labeling functions for the elliptically uniform priors and explore the resultant weight functions.

In discussing the weights it is notationally convenient to write the omitted variables as a matrix $\mathbf{Z}$. That is, the regression process may be written

$$\mathbf{Y}=\mathbf{W}\boldsymbol{\delta}+\mathbf{Z}\boldsymbol{\gamma}+\mathbf{u}. \tag{5.21}$$

where $\mathbf{W}$ and $\mathbf{Z}$ are $(T\times k_w)$ and $(T\times k_z)$ observable matrices, and $\boldsymbol{\delta}$ and $\boldsymbol{\gamma}$ are $(k_w\times 1)$ and $(k_z\times 1)$ unobservable vectors. *Letting the prior be normal* with mean vector $\mathbf{0}$ and variance $\mathbf{D}^*=\mathrm{diag}(d_1,d_2,\ldots,d_k)$ the weight to be applied to the estimate $[(\mathbf{W}'\mathbf{W})^{-1}\mathbf{W}'\mathbf{Y},\mathbf{0}]$ is from Theorem 5.5

$$w_x \propto \left(\prod_I d_i\right)|\mathbf{W}'\mathbf{W}|\sigma^{-2k_w} \tag{5.22}$$

where the nonempty set I contains the k_z indices subscripting the left-out variables. These are conditional weights applicable marginally when both the process variance σ^2 is known and when the prior for the coefficients is in the normal family with known variance. Note especially that these weights are independent of the sample result $\mathbf{Y}$ and thus do not depend on any test statistics. This straightforwardly parallels the result that under these assumptions a posterior mean is a fixed (independent of $\mathbf{Y}$) matrix weighted average of the sample mean and the prior mean.

Another result discussed in Chapter 3 is that a conjugate normal-gamma prior with σ^2 uncertain also leads to a posterior mean that is a fixed, weighted average of the sample point and prior mean. Similarly, the weights in (5.22) would be independent of the sample. That is, with $d_i=\sigma^{-2}n_i$ we have

$$w_z \propto \left(\prod_I n_i\right)|\mathbf{W}'\mathbf{W}|$$

since $\sigma^{k_w+k_z}$ is a constant.

The fixed weighted mixing of sample and prior implied by conjugate distributions has justifiably encountered the criticism of Dickey (1975) and others. A fairly tractable analysis that implies variable weighting results when the coefficient vector comes from a multivariate Student distribution. That is, let us write $d_i=\sigma_1^{-2}n_i$, $\mathbf{N}^*=\mathrm{diag}(n_1,n_2,\ldots,n_k)$, $k_1=\mathrm{rank}(\mathbf{N}^*)$, $\sigma_1^{-2}\sim f_\gamma(\ |1,\nu_1^*)$ where $f_\gamma(\ |1,\nu_1)$ indicates a gamma distribution with location and scale parameters 1 and ν_1. Furthermore, let us employ a gamma prior for σ^{-2}, $\sigma^{-2}\sim f_\gamma(\ |s_1^2,\nu_1)$. The resulting posterior is proportional to the product of two multivariate Student distributions and is relatively intractable. The marginal mode, as described in Chapter 3, requires the iterative solution of

$$\mathbf{b}^{**}(\lambda)=(\mathbf{X}'\mathbf{X}+\lambda\mathbf{N}^*)^{-1}\mathbf{X}'\mathbf{Y}$$

$$\lambda=\frac{\left[\nu_1 s_1^2+(\mathbf{Y}-\mathbf{X}\mathbf{b})'(\mathbf{Y}-\mathbf{X}\mathbf{b})+(\boldsymbol{\beta}-\mathbf{b})'\mathbf{X}'\mathbf{X}(\boldsymbol{\beta}-\mathbf{b})\right]/\left[T+\nu_1\right]}{\left[\nu_1^*+\boldsymbol{\beta}'\mathbf{N}^*\boldsymbol{\beta}\right]/\left[\nu_1^*+k\right]}$$

where λ estimates the variance ratio σ^2/σ_1^2. The weights analogous to (5.22) are

$$w_z \propto \left(\prod_I n_i\right)|\mathbf{W}'\mathbf{W}|\lambda^{-k_w}. \tag{5.23}$$

These weights depend on the sample **Y** only through the variance ratio λ. The sample-dependent factor λ^{-k_w} is a constant for all regressions involving the same number of restrictions. Thus *the sample influences only the choice of number of restrictions*. The critical value of λ is one. When λ is less than one, equations with few restrictions are favored, conversely for λ greater than one. The following result makes this point explicit by expressing the posterior mean as a weighted average of $k+1$ rotation invariant average regressions, each of which is a *fixed* weighted average of constrained least squares estimates involving exactly j restrictions ($j=0,1,\ldots,k$).

THEOREM 5.8 (ROTATION INVARIANT AVERAGE REGRESSIONS) *The posterior mean corresponding to a spherical prior can be written as*

$$\mathbf{b}^{**}(\lambda)=(\mathbf{N}+\lambda\mathbf{I})^{-1}\mathbf{N}\mathbf{b}=\sum_{j=0}^{k} w_j(\mathbf{X},\lambda)\mathbf{a}_j \qquad (5.24)$$

where

$$\mathbf{a}_j=\sum_{I\in C_j}|\mathbf{N}_I|\mathbf{b}_I p_j^{-1}$$

$$p_j=\sum_{I\in C_j}|\mathbf{N}_I|$$

$$w_j(\mathbf{X},\lambda)=p_j\lambda^j\Big/\sum_{j=0}^{k}p_j\lambda^j$$

with C_j the set of all subsets of the first k integers taken j at a time and with $\mathbf{N}_I$ a matrix formed by deleting rows i and columns i of $\mathbf{N}=\mathbf{X}'\mathbf{X}$ for all $i\in I$.

Proof: See Appendix 3.

Any prior that is uniform on ellipsoids can, by a linear transformation, be made uniform on spheres. By Theorem 5.8, posterior modes implied by such distributions are weighted averages of $k+1$ rotation invariant average regressions, $\mathbf{a}_j$. Each such point is a weighted average of constrained least-squares points involving exactly j restrictions with weights that cannot depend on the data **Y**. Given an elliptical prior, the choice of labeling function can thus influence *only* the number of restrictions that are imposed.

The rotation invariant average regressions derive their name from the surprising property that they are invariant to rotations of the parameter space. This property is illustrated in the two-dimensional case in Figure

5.12. In the original coordinate system connect with a straight line the two constrained least-squares points given $\beta_1=0$ or $\beta_2=0$. Find the constrained least-squares points in any other (orthogonal) coordinate system and connect them with a straight line. The point of intersection is $\mathbf{a}_1$, because $\mathbf{a}_1$ is a weighted average of constrained least-squares points in any coordinate system.

The weights (5.23) are, in general, very complicated but in the special case when the contract curve is a straight line they do imply familiar test procedures:

ONE RESTRICTION

When prior information is diffuse in all directions but one, the posterior mean is a simple weighted average of the estimates resulting from dropping and not dropping the relevant variable. The weight to be applied to the restricted estimate is given in (5.23)

$$w_z \propto n_z\left(\mathbf{Z'Z}-\mathbf{Z'W}(\mathbf{W'W})^{-1}\mathbf{W'Z}\right)^{-1}\lambda^{-1}. \tag{5.25}$$

The weight given the unrestricted estimator is

$$w_x \propto 1.$$

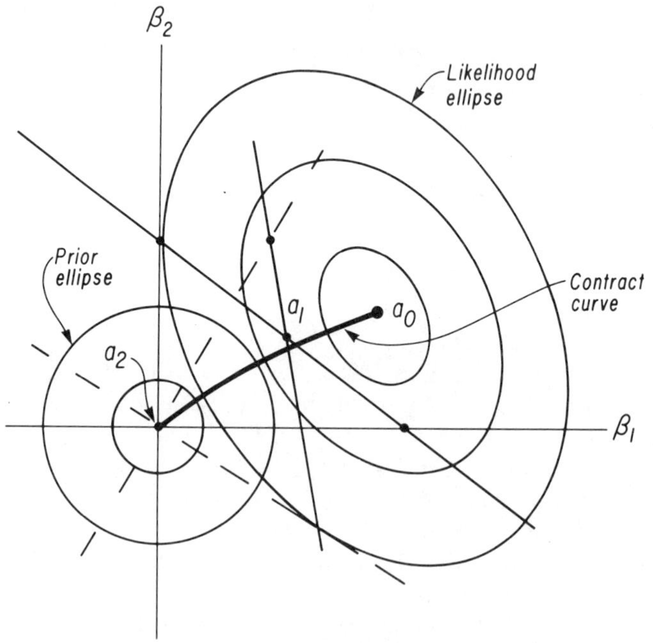

Fig. 5.12 Rotation invariant average regressions.

Note now that for ν_1^* and ν_1 small, the first iteration of the equation for computing λ has

$$\lambda = \frac{\hat{\sigma}^2}{n_z \hat{\gamma}^2}$$

where $\hat{\gamma}$ and $\hat{\sigma}^2$ are the maximum likelihood estimates. Thus the weight becomes approximately

$$w_z \propto t_\gamma^{-2}$$

where t_γ is the t value for testing the restriction $\gamma = 0$. Perhaps more informatively, we may write

$$w_z = \frac{1}{1+t_\gamma^2}, \qquad w_x = \frac{t_\gamma^2}{1+t_\gamma^2}.$$

The estimate of σ^2 in these results is uncorrected for degrees of freedom, this is just the first iteration toward the mode, and the mode is only one aspect of the posterior distribution. Note especially that further iterations to the mode move the estimate closer to the restricted least-squares point. Nonetheless, a somewhat distorted Bayesian analysis with information in one dimension only results in an estimate that is a weighted average of dropping and not dropping the variable with a weight on the unrestricted estimate proportional to the F statistic commonly used to test the restriction.

MULTIPLE RESTRICTIONS: THE STEIN SOLUTION

When information is available in several directions, the weights implied by (5.23) are more complicated. One special case is interesting, however. Suppose $\mathbf{X}'\mathbf{X}$ is proportional to $\mathbf{N}^*$.

$$\mathbf{N}^* = c\mathbf{X}'\mathbf{X}.$$

The posterior modal equation can then be written as a weighted average of the origin and the unrestricted estimator

$$\boldsymbol{\beta} = (1+\lambda c)^{-1}[\mathbf{b} + \lambda c \mathbf{0}]$$

and the variance ratio λ becomes

$$\lambda = \frac{(\nu_1 s_1^2 + (\mathbf{Y}-\mathbf{X}\mathbf{b})'(\mathbf{Y}-\mathbf{X}\mathbf{b}) + (\boldsymbol{\beta}-\mathbf{b})'\mathbf{X}'\mathbf{X}(\boldsymbol{\beta}-\mathbf{b}))/(T+\nu_1)}{(\nu_1^* + c\boldsymbol{\beta}'\mathbf{X}'\mathbf{X}\boldsymbol{\beta})/(\nu_1^*+k)}.$$

The first iteration of these two equations for ν_1^* and ν_1 small yields

$$\lambda = \frac{(\mathbf{Y}-\mathbf{X}\mathbf{b})'(\mathbf{Y}-\mathbf{X}\mathbf{b})/T}{c\mathbf{b}'\mathbf{X}'\mathbf{X}\mathbf{b}/k} \cong (cF)^{-1}$$

where F is the value of the F statistic for testing the restriction $\boldsymbol{\beta}=\mathbf{0}$. Using this in the formula for the mode of $\boldsymbol{\beta}$ and indicating the weight assigned to the zero vector by w_0 and the weight assigned to the unrestricted estimator by w, we obtain approximately

$$w_0=\frac{1}{1+F}, \qquad w=\frac{F}{1+F}.$$

In words, a somewhat distorted Bayesian analysis with prior information structurally equivalent to sample information ($\mathbf{X'X}\propto\mathbf{N^*}$) results in a posterior mode that is approximately a weighted average of the zero vector and the least-squares vector with the weight on the least-squares vector proportional to the F statistic commonly used to test the restriction. The structural equivalence assumption thus appears to be implicit in "Stein estimators." This is a vindication to a Bayesian of such estimators only in the unlikely case that his prior is, in fact, structurally equivalent to the sample.[13]

5.7 Multicollinearity and Local Sensitivity Analysis[14]

An interpretive search that involved omitting variables would be worthless if the data were orthogonal, since the estimates and their standard errors[15] would not change and therefore would not "improve." Thus collinear data is a handmaiden of interpretative searches, and this is a useful point to discuss the "collinearity problem."

The principal claim of this section is that the most important aspects of the collinearity problem derive from the existence of potentially useful, uncertain, prior information, which causes major problems in interpreting the data evidence. It is claimed here that if our a priori knowledge of parameter values were either completely certain or "completely uncertain," the aspects of the collinearity problem that most of us worry about would disappear. As an empirical test of this proposition, consider the situations when collinearity is identified as a culprit. Usually, signs are wrong or point estimates are otherwise peculiar. Occasionally, confidence intervals overlap unlikely regions of the parameter space. Yet to say these things is to say there exists uncertain prior information.

[13]A more general result is that a simple weighted average of imposing and not imposing the restriction is valid (in the sense discussed) if and only if the curve décolletage or contract curve $\mathbf{b}^{**}(\lambda)=(\lambda\mathbf{N}^*+\mathbf{X'X})^{-1}(\lambda\mathbf{N}^*\mathbf{b}^*+\mathbf{X'Y})$ is a straight line. [See Chamberlain and Leamer (1976) for a discussion of the necessary and sufficient conditions.] In this case, the weights are approximated by $F/(1+F)$ and $1/(1+F)$ where F is the usual F statistic for testing the restriction. Weights of this form were originally suggested by Huntsberger (1955). They are also discussed in Alam and Thompson (1964) and Baranchik (1970). Feldstein (1973) uses t^2 weights in a Monte Carlo study.

[14]This section is taken from Leamer (1973).

[15]Conditional on σ^2. See comments at the end of this section.

Classical inference, with the possible exception of the pretesting literature, necessarily excludes undominated uncertain prior information.[16] As a result, most discussions of the collinearity problem miss a critical point. The textbook discussions, including Theil (1971, p. 149), Malinvaud (1970, p. 218), and Goldberger (1964, p. 192), observe that when the design matrix **X′X** becomes singular, the least-squares estimator is nonunique, and the sampling distribution has finite variance only for certain "estimable" functions. Thus extreme collinearity is implicitly defined as total lack of sample information about some parameters.

The case of less extreme collinearity is not dealt with so trivially, since there is nothing in the least-squares theorems that is obviously dependent on the "near noninvertibility" of the design matrix. This fact has led Kmenta (1971, p. 391) to conclude "that a high degree of multicollinearity is simply a feature of the sample that contributes to the unreliability of the estimated coefficients, but has no relevance for the conclusions drawn as a result of this unreliability."

To put this another way, the problem of defining collinearity may be solved by identifying a distance function for measuring the closeness of the design matrix to some noninvertible matrix in which the collinearity problem is unambiguously extreme. Since the extreme case is associated with infinite marginal variances on the parameters, authors such as Theil (1971, p. 152), Malinvaud (1970, p. 218), and Goldberger (1964, p. 193) use a distance function informally related to the sampling variance of the coefficients. Collinearity is defined as large variances. The failure of this definition is that instead of defining a new *problem*, it identifies a new *cause* of an already well-understood problem—weak evidence. Although collinearity as a cause of the weak-evidence problem can be distinguished from other causes, such as small samples or large residual error variances, collinearity as a problem is by this definition indistinguishable from the weak-data problem in general. Thus Kmenta's conclusion that there is really nothing special about the collinearity problem is appropriate. Still, a gnawing confusion remains. Goldberger (1964, p. 201) concludes this discussion with accurate ambiguity, "...when orthogonality is absent the concept of the contribution of an individual regressor remains inherently ambiguous."

The point of this section is that there *is* a special problem caused by collinearity. This is the problem of *interpreting* multidimensional evidence. Briefly, collinear data provide relatively good information about linear combinations of coefficients. The interpretation problem is the problem of deciding how to allocate that information to individual coefficients. This depends on prior information. A solution to the interpretation problem thus involves formalizing and utilizing effectively all prior information.

[16]Dominated prior information is weak relative to the given data information.

The weak-evidence problem, however, remains, even when the interpretation problem is solved. The solution to the weak-evidence problem is more and better data. Within the confines of the given data set there is nothing that can be done about weak evidence.

A Bayesian with a well-defined prior distribution can, of course, have no problem interpreting the sample evidence, since he merely computes his posterior distribution. A Bayesian with poorly defined priors or a wide readership may have extreme difficulties in reporting and interpreting evidence. This suggests the following definition:

Definition. The *collinearity problem* is said to affect a parameter β_i if the apparent sample evidence about β_i depends on ambiguous uncertain prior information about other parameters, where ambiguous means that readers differ in their judgments or that they are not too sure how to select features of their prior. This is made more precise subsequently.

Since classical inference provides no assistance in using uncertain prior information, this definition does not apply directly to everyday "classical" inferences. An easy, ad hoc procedure used when analyzing data is to neglect the off-diagonal terms of $(\mathbf{X}'\mathbf{X})^{-1}$ and to proceed as if the sample evidence were generated by an orthogonal experiment. This may lead to significant misinterpretations of the data and suggests an alternative definition:

Definition. The *collinearity problem* is said to affect β_i if the sample evidence about β_i is distorted by an analysis that proceeds as if the data were orthogonal. This is also made more clear shortly.

MULTICOLLINEARITY: THE WEAK-DATA PROBLEM

The unique problem associated with collinear data is the problem of interpreting multidimensional evidence. Collinear data is also a *cause* of the weak-data problem. In this section we show how collinearity causes weak evidence, where weak evidence is defined as the necessary coincidence of the prior and posterior distribution for some parameter $g(\boldsymbol{\beta})$.

In particular, suppose there is an extreme collinearity problem with the columns of $\mathbf{X}$ being perfectly collinear. Then there exists a vector $\boldsymbol{\psi}$ such that $\mathbf{X}\boldsymbol{\psi}=\mathbf{0}$. We wish to show that there is a function $g(\boldsymbol{\beta})$ that necessarily has the same prior and posterior distribution regardless of the sample outcome $\mathbf{Y}$. This is true, in particular, for $g(\boldsymbol{\beta})=\boldsymbol{\psi}'\mathbf{N}^*\boldsymbol{\beta}$, Malinvaud (1970, pp. 246–249). The prior moments of $\boldsymbol{\psi}'\mathbf{N}^*\boldsymbol{\beta}$ are

$$E(\boldsymbol{\psi}'\mathbf{N}^*\boldsymbol{\beta})=\boldsymbol{\psi}'\mathbf{N}^*\mathbf{b}^*$$

$$V(\boldsymbol{\psi}'\mathbf{N}^*\boldsymbol{\beta})=h_1^{-1}\boldsymbol{\psi}'\mathbf{N}^*\mathbf{N}^{*-1}\mathbf{N}^*\boldsymbol{\psi}=h_1^{-1}\boldsymbol{\psi}'\mathbf{N}^*\boldsymbol{\psi},$$

where the prior variance matrix has been set to $(h_1\mathbf{N}^*)^{-1}$. The posterior moments are

$$\begin{aligned} E(\psi'\mathbf{N}^*\boldsymbol{\beta}|\mathbf{Y}) &= \psi'\mathbf{N}^*\mathbf{b}^{**} \\ &= \psi'(\mathbf{N}+\mathbf{N}^*)\mathbf{b}^{**} \qquad (\text{because } \psi'\mathbf{N}=\mathbf{0}) \\ &= \psi'(\mathbf{N}\mathbf{b}+\mathbf{N}^*\mathbf{b}^*) = \psi'\mathbf{N}^*\mathbf{b}^* \\ V(\psi'\mathbf{N}^*\boldsymbol{\beta}|\mathbf{Y}) &= \psi'\mathbf{N}^*V(\boldsymbol{\beta}|\mathbf{Y})\mathbf{N}^*\psi \\ &= h_1^{-1}\psi'(h\mathbf{N}+h_1\mathbf{N}^*)V(\boldsymbol{\beta}|\mathbf{Y})\mathbf{N}^*\psi \\ &= h_1^{-1}\psi'\mathbf{N}^*\psi \end{aligned}$$

where $h=\sigma^{-2}$. These moments are seen to be the same as the prior moments, and, conditional on h and h_1, there can be no learning about $\psi'\mathbf{N}^*\boldsymbol{\beta}$. The evidence is thus necessarily weak about this linear combination of parameters. (Note, by the way, that this linear combination depends on the prior through $\mathbf{N}^*$. See Section 5.9 also.)

Weaker forms of collinearity imply that this result is almost true. There are functions of the parameters about which we can learn very little. There is, of course, no cure for weak data, except more and better data. If collinearity were only a cause of the weak-data problem it would merit very little mention. The more interesting and more difficult aspect of collinearity is the interpretation problem.

INTERPRETING COLLINEAR DATA: A BAYESIAN ANALYSIS OF AN AD HOC PROCEDURE

Although it is possible to make enlightened use of prior information through interpretive searches, perhaps as suggested by the pretesting literature, we assume that a researcher has before him only the sufficient statistics and no computer, as would be the case of a reader of a technical report. Off-diagonal terms of $(\mathbf{X}'\mathbf{X})^{-1}$ may not be reported and, even if they are, classical inference provides no very clear way of interpreting them. Instead, many of us in this situation would proceed as if $(\mathbf{X}'\mathbf{X})^{-1}$ were diagonal. Furthermore, when prior information on the coefficients is available, we may choose to ignore the a priori covariance terms.

An example can usefully illustrate what I have in mind. A logarithmic regression of a volume index of purchases of meat C_m on money income Y, price of meat P_m, and a general consumer price index P_Y yields a regression that is underreported as

$$\begin{aligned} \log C_m &= \alpha + \beta_1 \log Y + \beta_2 \log P_m + \beta_3 \log P_Y \\ &= \underset{}{5.0} + \underset{(.2)}{.9}\log Y - \underset{(.2)}{.2}\log P_m - \underset{(.2)}{.1}\log P_Y \end{aligned}$$

where standard errors are indicated in parentheses. The researcher then interprets these estimated coefficients by comparing them one by one against his marginal prior for each coefficient. He finds the money-income elasticity β_1 to be a "bit high," the direct-price elasticity β_2 to be "about right," and the cross-price elasticity β_3 "to have the wrong sign but not significantly so." The error that is being made here is, first, to ignore the data tradeoffs implied by the off-diagonal terms of the $(\mathbf{X}'\mathbf{X})^{-1}$ matrix and, second, to ignore the prior tradeoffs implied by a nondiagonal $\mathbf{N}^*$ matrix. For example, the researcher may have independent information about a homogeneity parameter $\beta_1+\beta_2+\beta_3$, a real income elasticity $\beta_1-\beta_3$, and the price elasticity β_2, and he is unlikely to regard β_1, β_2, and β_3 to be a priori independent.

Proceeding as he did, the researcher has made an error in pooling the prior information and the sample information. He has treated a k-dimensional problem as if it were k one-dimensional problems; he will be making misinterpretations of the data evidence unless his prior and data fit together in a special way. Thus the collinearity problem creates a situation in which it is necessary to process prior information carefully.

More formally, inferences about the coefficient vector often proceed as if the posterior mean were on the *diagonalized contract curve*

$$\mathbf{d}(\lambda)=(\mathbf{D}+\lambda\mathbf{D}^*)^{-1}(\mathbf{D}\mathbf{b}+\lambda\mathbf{D}^*\mathbf{b}^*) \tag{5.26}$$

where $\mathbf{D}^{-1}$ and $\mathbf{D}^{*-1}$ are diagonal matrices formed by setting the off-diagonal elements of $\mathbf{N}^{-1}$ and $\mathbf{N}^{*-1}$ to zero. If $\mathbf{N}$ and $\mathbf{N}^*$ are diagonal, Bayes rule may be applied coefficient by coefficient, and the resultant conditional posterior mean is given by (5.26). For $\mathbf{N}$ or $\mathbf{N}^*$ nondiagonal, the true contract curve

$$\mathbf{b}^{**}(\lambda)=(\mathbf{N}+\lambda\mathbf{N}^*)^{-1}(\mathbf{N}\mathbf{b}+\lambda\mathbf{N}^*\mathbf{b}^*) \tag{5.27}$$

may deviate substantially from $\mathbf{d}(\lambda)$ and the ad hoc use of prior information may cause major data misinterpretations and ultimately unnecessary expected losses (see Figure 5.13). Collinearity thus creates an incentive to use prior information more carefully:

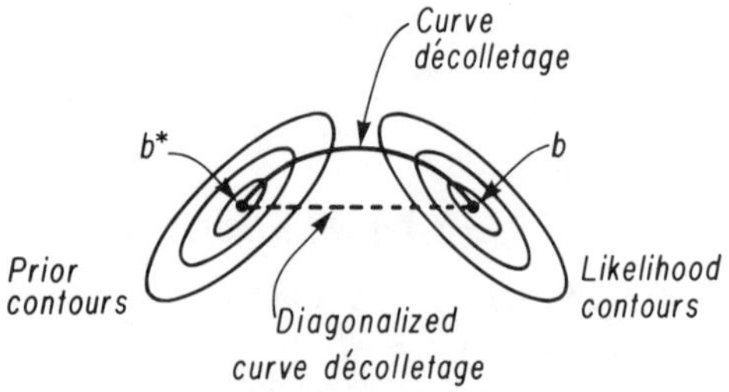

Fig. 5.13 Diagonalized contract curve.

Definition. A coefficient β_i is said to suffer from the collinearity problem if the ith component of the true contract curve $b_i^{**}(\lambda)$ differs substantially from the ith component of the diagonalized contract curve $d_i(\lambda)$.

The difference between $\mathbf{d}(\lambda)$ and $\mathbf{b}^{**}(\lambda)$ may be assessed in several ways. One suggestion of Leamer (1973) is the following:

Rectangle Test. One aspect of a univariate problem is that the posterior mean is necessarily between the prior mean and the sample mean. Thus, the location change induced by the sample evidence has an unambiguous sign and is limited in distance by the sample point. The diagonalized contract curve also has this property, coefficient by coefficient; that is, it lies in the rectangular solid with diagonal $[\mathbf{b}^*, \mathbf{b}]$. The true contract curve need not have this property, and the sign of the elements of $\mathbf{b}^{**}(\lambda) - \mathbf{b}^*$ and $\mathbf{b}^{**}(\lambda) - \mathbf{b}$ may be ambiguous. Thus, it may not be possible to say whether the data suggest positive or negative revisions to opinions about some coefficient or to limit the distance of the revision, until a full prior distribution is specified. This suggests the following definition.

Definition. The collinearity problem in the *rectangle sense* is said to affect a coefficient β_j if the contract curve $\mathbf{b}^{**}(\lambda)$ travels outside the slab $b_j \leqslant b_j^{**}(\lambda) \leqslant b_j^*$. The collinearity problem in the rectangle sense affects no parameters if the contract curve lies everywhere in the rectangular solid with diagonal $[\mathbf{b}, \mathbf{b}^*]$.

The contract curve may, in general, lie anywhere, Theorem (5.10). Even with orthogonal data, that is, with $\mathbf{X}'\mathbf{X}$ diagonal, it need not be restricted to the appropriate rectangular solid. *Thus orthogonal data is not sufficient to prevent the collinearity problem. We must also restrict the class of priors.* If $\mathbf{N}$ and $\mathbf{N}^*$ are both diagonal, there is no collinearity problem in this sense. If $\mathbf{N}$ is proportional to $\mathbf{N}^*$, the contract curve is a straight line, and there is, again, no collinearity problem.[17]

INTERPRETING COLLINEAR DATA: THE BAYESIAN PROBLEMS

Given a prior distribution, the posterior distribution is fully defined, and there is no ambiguity about measures of location and thus no collinearity problem in the sense of the previous section. A Bayesian, therefore, apparently has no special difficulty working with collinear data. This, however, ignores difficulties in selecting an acceptable prior distribution.

[17]Note that even if $\mathbf{N}$ and $\mathbf{N}^*$ are both diagonal, collinearity may be said to affect certain linear combinations of parameters. Collinearity affects no linear combinations only if $\mathbf{N}$ is proportional to $\mathbf{N}^*$.

When collinearity is present, the posterior distribution may be highly sensitive to changes in the prior, and apparently innocuous differences in the prior may be amplified into significant differences in the posterior distribution. Thus the collinearity problem is transformed from a problem of characterizing and interpreting a multidimensional likelihood function into a problem of characterizing and interpreting a multidimensional prior distribution.

In the clearly collinearity-free case with **N** and **N*** diagonal, the posterior distribution of any coefficient is conditionally independent of the prior distributions of the other coefficients. This suggests the following slightly ambiguous definition of the collinearity problem:

Definition. The collinearity problem is said to affect parameter β_i if the interpretation of the sample evidence about β_i depends meaningfully on uncertain prior information about the other parameters. The *interpretation of the sample* evidence about β_i is a mapping of marginal prior distributions for β_i into marginal posterior distributions. The phrase "depends meaningfully" can be interpreted in terms of both the location change and the scale change induced by the sample evidence. As in the previous section, we restrict ourselves to the location change.

In general, the interpretation of the sample evidence about one coefficient is sensitive to the prior about others, because of prior correlations. To make sense out of this definition, we thus have to define meaningful classes of priors within which to perform the sensitivity analysis.

The sensitivity of the posterior mean to variations in the prior mean holding other things constant is indicated by the matrix of derivatives

$$\frac{\partial \mathbf{b}^{**}(\lambda)}{\partial \mathbf{b}^*} = \lambda(\mathbf{N}+\lambda\mathbf{N}^*)^{-1}\mathbf{N}^*.$$

The off-diagonal elements of this matrix indicate the extent to which the conditional posterior mean of one coefficient depends on the prior mean of the others. These are zero for **N** and **N*** diagonal, for **N** proportional to **N***, and for **N*** or λ equal to zero.

We may also be interested in the sensitivity of the posterior mean to changes in the prior variances. Let us write

$$(\mathbf{N}^*)^{-1} = \mathbf{DRD}$$

where **D** is a diagonal matrix with $\sqrt{V(\beta_i)}$ on the diagonal and **R** is the matrix of correlation coefficients. A change in the prior variances induces a change in the prior precision matrix according to the formula

$$d\mathbf{N}^* = -(\mathbf{N}^*\mathbf{D}^{-1}d\mathbf{D} + \mathbf{D}^{-1}d\mathbf{D}\mathbf{N}^*),$$

and a change in $\mathbf{N}^*$ induces a change in $\mathbf{b}^{**}$ according to the formula

$$d\mathbf{b}^{**}(\lambda)=(\mathbf{N}+\lambda\mathbf{N}^*)^{-1}(\lambda d\mathbf{N}^*)(\mathbf{N}+\lambda\mathbf{N}^*)^{-1}\mathbf{N}(\mathbf{b}^*-\mathbf{b}).$$

In the diagonal case, the ith element of $d\mathbf{b}^{**}(\lambda)$ depends only on the ith differential $\{d\mathbf{D}\}_{ii}$. Otherwise, changes in the prior variance of one coefficient induce changes in the posterior means of other coefficients.

Sensitivity analysis can also be performed with respect to the variance ratio λ. In the orthogonal case, the posterior mean (the curve decolletage) lies everywhere in the rectangular solid with diagonal $[\mathbf{b},\mathbf{b}^*]$. In that case the sign and maximum distance of the mapping from prior to posterior mean are unambiguous, and we could say that the sample evidence does not depend meaningfully on prior information about λ. When the curve decolletage travels outside this region, the sample evidence does become ambiguous, and collinearity is the culprit. Note, by the way, that this is mathematically the same as the collinearity problem in the rectangle sense discussed in the previous section, although the interpretations are quite different. In that section the curve decolletage was assumed to lie in the relevant rectangular solid, and when it did not, a major data misinterpretation occurred. Here, we know where the curve décolletage lies, but we are uncertain whether particular points on the curve outside the rectangular solid are relevant, since our prior information about the variance ratio is ambiguous.

The derivatives just reported imply a local sensitivity analysis in which the consequences of small perturbations in the prior are analyzed. Global sensitivity analyses are discussed in the next section, but one result given there yields an especially interesting measure of collinearity. It is shown that if only the prior location is known (taken here to be the origin), then any posterior mean must lie within the feasible ellipsoid (5.1), and any point in the ellipsoid is a posterior mean for some prior. Projection of this ellipsoid on the ith axis yields the set of feasible estimates of β_i. In contrast, if β_i were the only parameter, then the posterior mean would necessarily lie between zero and b_i, the least-squares estimate. The ratio of the lengths of these two intervals is shown in the next theorem to be

$$c_{1i}=\left(\frac{\chi^2}{Z_i^2}\right)^{1/2}\geqslant 1,$$

where χ^2 is the chi-square statistic for testing the multivariate restriction $\boldsymbol{\beta}=\mathbf{0}$ and Z_i is the normal statistic for testing $\beta_i=0$.[18] (See Figure 5.14.) When c_{1i} is one, inferences about β_i are unaffected by the fact that there

[18]Notice that c_{1i} is proportional to the inverse of the square of the t statistic for testing $\beta_i=0$. Thus the ranking of coefficients by t statistics is equivalent to a ranking by this collinearity measure.

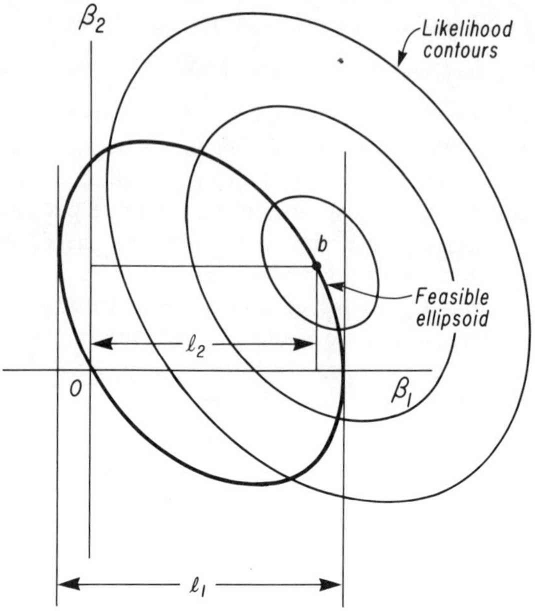

Fig. 5.14 A measure of collinearity $c_1 = l_1/l_2 = \sqrt{\chi^2/Z_1^2}$.

are other parameters, since the posterior location is constrained to lie in the interval $(0, b_i)$. When c_{1i} is greater than one, the existence of other parameters does cause difficulties for interpreting the evidence about β_i in the sense that the set of feasible estimates is enlarged.

THEOREM 5.9 (PROJECTION OF THE FEASIBLE ELLIPSOID). *Given* $\boldsymbol{\beta}$ *constrained to the ellipsoid*

$$(\boldsymbol{\beta}-\mathbf{f})'\mathbf{X}'\mathbf{X}(\boldsymbol{\beta}-\mathbf{f}) = c$$

where $\mathbf{f} = (\mathbf{b}+\mathbf{b}^*)/2$ *and* $c = (\mathbf{b}-\mathbf{b}^*)'\mathbf{X}'\mathbf{X}(\mathbf{b}-\mathbf{b}^*)/4$, *the extreme values of the linear combination* $\boldsymbol{\psi}'\boldsymbol{\beta}$ *occur at the points*

$$\boldsymbol{\beta}^* = \mathbf{f} \pm (\mathbf{X}'\mathbf{X})^{-1}\boldsymbol{\psi}\left(\frac{c}{\boldsymbol{\psi}'(\mathbf{X}'\mathbf{X})^{-1}\boldsymbol{\psi}}\right)^{1/2}.$$

The linear function $\boldsymbol{\psi}'\boldsymbol{\beta}$ *at these points takes on the values*

$$\boldsymbol{\psi}'\boldsymbol{\beta}^* = \boldsymbol{\psi}'\mathbf{f} \pm \left(c\boldsymbol{\psi}'(\mathbf{X}'\mathbf{X})^{-1}\boldsymbol{\psi}\right)^{1/2}.$$

Proof: Setting the derivatives of the Lagrangian to zero yields the vector of equations $\mathbf{0} = \boldsymbol{\psi} + \mathbf{X}'\mathbf{X}(\boldsymbol{\beta}-\mathbf{f})\lambda$, which implies $(\boldsymbol{\beta}-\mathbf{f}) = -(\mathbf{X}'\mathbf{X})^{-1}\boldsymbol{\psi}\lambda^{-1}$. Thus $c = (\boldsymbol{\beta}-\mathbf{f})'\mathbf{X}'\mathbf{X}(\boldsymbol{\beta}-\mathbf{f}) = \lambda^{-2}\boldsymbol{\psi}'(\mathbf{X}'\mathbf{X})^{-1}\boldsymbol{\psi}$, and $\lambda^2 = \boldsymbol{\psi}'(\mathbf{X}'\mathbf{X})^{-1}\boldsymbol{\psi}/c$.

This interval for $\psi'\boldsymbol{\beta}$ *has length equal to* $2(c\psi'(\mathbf{X}'\mathbf{X})^{-1}\psi)^{1/2} = (\chi^2 \operatorname{Var}(\psi'\mathbf{b}))^{1/2}$ *where* χ^2 *is the chi-square statistic for testing* $\boldsymbol{\beta}=\mathbf{b}^*$, $\chi^2 = \sigma^2(\mathbf{b}-\mathbf{b}^*)'\mathbf{X}'\mathbf{X}(\mathbf{b}-\mathbf{b}^*)$, *and* $\operatorname{Var}(\psi'\mathbf{b})$ *is the sampling variance of* $\psi'\mathbf{b}$, $\sigma^2\psi'(\mathbf{X}'\mathbf{X})^{-1}\psi$.

These measures of collinearity have described how the posterior location of one parameter β_i depends on the prior information about the other parameters. One measure mentioned in Section 5.5 describes the relationship between the (posterior) variance of β_i and the prior variance of the other parameters. If the prior for the vector $\boldsymbol{\beta}$ is diffuse, the posterior variance of β_i is proportional to $[(\mathbf{X}'\mathbf{X})^{-1}]_{ii}$, whereas if all the parameters except β_i were known exactly, then the variance of β_i would be proportional to $([\mathbf{X}'\mathbf{X}]_{ii})^{-1}$. The square root of the ratio of these two numbers thus measures the incentive to formulate prior information

$$c_2(\beta_i) = \left\{ \frac{\left[(\mathbf{X}'\mathbf{X})_{ii}\right]^{-1}}{\left[(\mathbf{X}'\mathbf{X})^{-1}\right]_{ii}} \right\}^{1/2}.$$

A value of $c_2(\beta_i)$ equal to one occurs when the ith row of $\mathbf{X}'\mathbf{X}$ has zeroes except in its ith element. A value of one indicates that there can be no gain in information about β_i by gathering more information about the other coefficients. A value of $c_2(\beta_i)=\frac{1}{2}$ indicates that if the other coefficients could be specified exactly, the confidence interval for β_i would be cut in half.

More generally, if we are interested in estimating the linear combination $\psi'\boldsymbol{\beta}$, the incentive to gather information about $\mathbf{R}\boldsymbol{\beta}$ can be measured by

$$c_2^2(\psi, \mathbf{R}) = \frac{\psi' V(\boldsymbol{\beta}|\mathbf{R}\boldsymbol{\beta})\psi}{\psi' V(\boldsymbol{\beta})\psi}.$$

where $V(\boldsymbol{\beta}|\mathbf{R}\boldsymbol{\beta})$ and $V(\boldsymbol{\beta})$ are the conditional and unconditional variances of $\boldsymbol{\beta}$.

These measures of the incentive to gather other information are similar to measures of the intercorrelation of the explanatory variables. Other measures of intercorrelation between a subset $\mathbf{X}_J$, and its complement $\mathbf{X}_{\bar{J}}$ are Hotelling's (1936) "coefficient of alienation"

$$\rho_a^2 = \frac{\det\left[\mathbf{X}_J'(\mathbf{I}-\mathbf{X}_{\bar{J}}(\mathbf{X}_{\bar{J}}'\mathbf{X}_{\bar{J}})^{-1}\mathbf{X}_{\bar{J}}')\mathbf{X}_J\right]}{\det(\mathbf{X}_J'\mathbf{X}_J)}$$

and Hooper's (1962) "trace correlation coefficient"

$$\rho_t^2 = J^{-1}\operatorname{tr}\left[(\mathbf{X}_J'\mathbf{X}_J)^{-1}\left(\mathbf{X}_J'(\mathbf{I}-\mathbf{X}_{\bar{J}}(\mathbf{X}_{\bar{J}}'\mathbf{X}_{\bar{J}})^{-1}\mathbf{X}_{\bar{J}}')\mathbf{X}_J\right)\right].$$

The analogous measure $c^2(\psi,\mathbf{R})$, with $\psi'=(\psi_J,\mathbf{0})$ and $\mathbf{R}=(\mathbf{0},\mathbf{I}_{\bar{J}})$ is

$$c^2=\frac{\psi_J'(\mathbf{X}_J'\mathbf{X}_J)^{-1}\psi_J}{\psi'_J\left[\mathbf{X}_J\left(\mathbf{I}-\mathbf{X}_{\bar{J}}(\mathbf{X}_{\bar{J}}'\mathbf{X}_{\bar{J}})^{-1}\mathbf{X}_{\bar{J}}'\right)\mathbf{X}_J\right]^{-1}\psi_J}$$

Notice that the difference between c^2 and ρ_a^2 is only that c^2 involves variances of a specific linear combination, whereas ρ_a^2 uses the generalized variances, $\det(V(\boldsymbol{\beta}_J))$. They are identical if the set J has only one element.

The discussion to this point has made use of the assumption that the residual variance σ^2 is known. In that event, the conditional confidence interval for $\psi'\boldsymbol{\beta}$ given $\mathbf{R}\boldsymbol{\beta}=\mathbf{r}$ can never be larger than the confidence interval computed without benefit of the restriction. But if σ^2 is unknown, imposition of the constraint may, in fact, lead to a larger confidence interval, since the estimate of σ^2 necessarily changes. The multicollinearity measure, which by definition is the ratio of the length of a conditional to the length of an unconditional interval, may exceed one. This seems to be saying paradoxically that more information is less information; the more information corresponding to knowledge that $\mathbf{R}\boldsymbol{\beta}=\mathbf{r}$, and the less information corresponding to the fact that the interval for $\psi'\boldsymbol{\beta}$ increases in length.

It is, in fact, *not* paradoxical that specific information may make you less certain, especially if that information is greatly at odds with what you currently believe. Knowledge of $\mathbf{R}\boldsymbol{\beta}$ will sometimes increase and sometimes decrease the confidence interval for $\psi'\boldsymbol{\beta}$. But the *expected* variance of $\psi'\boldsymbol{\beta}$ given $\mathbf{R}\boldsymbol{\beta}$, expected with respect to the distribution of $\mathbf{R}\boldsymbol{\beta}$, will always be less than the unconditional variance of $\psi'\boldsymbol{\beta}$, since $V(\psi'\boldsymbol{\beta})=V(E(\psi'\boldsymbol{\beta}|\mathbf{R}\boldsymbol{\beta}))+E(V(\psi'\boldsymbol{\beta}|\mathbf{R}\boldsymbol{\beta}))$.

The incentive to obtain prior information about $\mathbf{R}\boldsymbol{\beta}$ should be measured in terms of the *expected* variance of $\psi'\boldsymbol{\beta}$ given $\mathbf{R}\boldsymbol{\beta}$:

$$c^2(\psi',\mathbf{R})=\frac{E\left[V(\psi'\boldsymbol{\beta}|\mathbf{R}\boldsymbol{\beta})\right]}{V(\psi'\boldsymbol{\beta})}.$$

It is easy to show that this measure is precisely the same as the measures previously suggested when σ^2 was assumed to be known. Thus the results heretofore discussed continue to apply. The proof of this proposition could appeal to properties of multivariate Student distributions, but it is easier merely to observe that, conditional on σ^2, both $V(\psi'\boldsymbol{\beta})$ and $E[V(\psi'\boldsymbol{\beta}|\mathbf{R}\boldsymbol{\beta})]$ are proportional to σ^2. Integrating each with respect to the (posterior) distribution of σ^2, $f(\sigma^2|\mathbf{Y},\mathbf{X})$, will thus multiply each by the *same* constant. Thus the ratio is unaffected by uncertainty in σ^2.

SUMMARY

The most discussed aspect of collinearity is the weak-data problem associated with large standard errors of estimated coefficients and, in a

Bayesian analysis, the coincidence of prior and posterior distributions on certain subspaces. As Kmenta suggests, there is nothing special about this problem in the collinearity context. What is special in the collinearity context is the major problems of interpreting the evidence.

When prior information is fully specified and unique, both personally and publicly, the posterior mean and hence the interpretation of the evidence are unambiguous. The diagonalizations of the data matrix **X′X** and the prior covariance matrix that some of us implicitly perform may, however, lead to very poor approximations to the posterior mean. Qualitative and quantitative summaries of the error of approximation provide one way of assessing the collinearity problem.

The principal implication of collinearity is that data evidence cannot be interpreted in a parameter-by-parameter fashion. The informal use of nondata-based prior information by practicing classical statisticians almost necessarily implies a parameter-by-parameter analysis, and consequently, the data misinterpretation just described. The great benefit of a Bayesian approach is that it provides instruction on how to deal with prior information in a multiparameter problem. For example, the posterior mean is, under suitable assumptions, a *matrix*-weighted average of the prior mean and the sample estimate, not a simple average.

Although the Bayesian approach appropriately spotlights the fundamental source of the collinearity problem—personal prior information—it necessarily leaves the resolution of the problem to the individual. He must "merely" construct his prior distribution. Difficulties in constructing a personal prior and/or variation in opinions among intended readers may cause major difficulties in analyzing and reporting collinear evidence. Thus the problem of collinearity from a Bayesian viewpoint concerns the sensitivity of the posterior distribution to changes in the prior distribution, and quantitative measures of that sensitivity may be used to describe the degree of the problem.

The principal claim of this section is that the collinearity problem concerns the way in which sample evidence fits together with prior information. If prior information dominated sample evidence in all directions, there would be no collinearity problem. When there is a collinearity problem, classical inference, which excludes undominated uncertain prior information, fails as a method of interpreting evidence. Peculiarities in the likelihood surface make the BLUE (least-squares) estimate almost irrelevant. A fuller exploration of the likelihood contours informally directed by prior information is difficult and rarely convincing, especially when the number of dimensions of prior information is more than one. Although a Bayesian approach cannot provide a complete cure, it does indicate the source of the disease.

5.8 Global Sensitivity Analysis: Properties of Matrix-Weighted Averages

The posterior mean of a normal, linear-regression model with normal priors is a matrix-weighted average of a prior location vector and a sample location vector. The prior weight matrix is arbitrary, either because prior distributions are impossible to measure without error or because intended readers may differ in their prior judgements. A Bayesian analysis based on any particular prior distribution, as a result, is of little interest. Practical users of the Bayesian tools necessarily face the difficult reporting problem of characterizing economically the mapping implied by the given data from interesting prior distributions into their respective posterior distributions, thereby servicing a wide readership as well as identifying those features of the prior that critically determine the posterior.

One way of characterizing the mapping from priors into posteriors is a local sensitivity analysis discussed in the previous section that identified the relative sensitivity of aspects of the posterior distribution to infinitesimal changes in the prior. The usefulness of a local sensitivity analysis is somewhat limited, since to have great content it must be performed for many different prior distributions. An alternative is a global sensitivity analysis that constructs a correspondence between classes of priors and classes of posteriors. A correspondence can be constructed by answering questions of the form: "if my prior is a member of this class of priors, what can I say about my posterior?"

Although both the location and the dispersion of the posterior are of interest, we consider here only the location parameter. The location of the prior is taken as given, and a correspondence is developed between classes of prior covariance matrices and regions in the space of the posterior location vector. A great deal can be said about the posterior location without precisely specifying the prior covariance matrix. What is not true, except under special and unlikely circumstances, is that, element by element, the posterior location lies algebraically between the prior location and the sample location. The inappropriateness of this bound is an important reason why a multiparameter problem is fundamentally different from a uniparameter problem.

The three most interesting results of Chamberlain and Leamer (1976) are reported here. The posterior mean, as usual, is written as

$$\mathbf{b}^{**} = E(\boldsymbol{\beta}|\mathbf{Y},\mathbf{H},\mathbf{H}^*) = (\mathbf{H}+\mathbf{H}^*)^{-1}(\mathbf{H}\mathbf{b}+\mathbf{H}^*\mathbf{b}^*) \tag{5.28}$$

where $\mathbf{H}=\sigma^{-2}\mathbf{X}'\mathbf{X}$ and $\mathbf{H}\mathbf{b}=\sigma^{-2}\mathbf{X}'\mathbf{Y}$. The first result is that if only the location vectors $\mathbf{b}$ and $\mathbf{b}^*$ are given, then $\mathbf{b}^{**}$ may lie essentially anywhere. This contrasts with the analogous one-dimensional result, which constrains the scalar b** to lie algebraically between the scalars b^* and b. Next, it is shown that if **X′X** is known as well as **b** and **b***, then the posterior location **b**** must lie in the feasible ellipsoid (5.1). The third result has already been reported in Theorem 5.5: if **b**, **b*** and **X′X** are given and if **H*** is a diagonal matrix, then **b**** is a weighted average of 2^k constrained least-squares points.

The first result of practical interest is that if only the locations **b** and are known, the posterior mean may lie essentially anywhere. The proof requires the following lemma.

LEMMA 5.1. *Given the prior and sample locations,* **b*** *and* **b**, *and an invertible matrix of common conjugate axes* **B** *such that* $\mathbf{B'HB}=\rho\mathbf{I}$ *and* $\mathbf{B'H^*B}=\mathbf{D^*}$, *with* ρ *an arbitrary positive scalar and* **D*** *an arbitrary positive diagonal matrix, the transformed posterior mean*

$$\mathbf{a}^{**}=\mathbf{B}^{-1}\mathbf{b}^{**}=\mathbf{B}^{-1}(\mathbf{H}+\mathbf{H}^*)^{-1}(\mathbf{Hb}+\mathbf{H}^*\mathbf{b}^*)$$

lies everywhere in the orthotope

$$\left|a_i^{**}-\frac{(a_i+a_i^*)}{2}\right|<\frac{|a_i-a_i^*|}{2}$$

where $\mathbf{a}=\mathbf{B}^{-1}\mathbf{b}$ *and* $\mathbf{a}^*=\mathbf{B}^{-1}\mathbf{b}^*$. *The transformation of this region back into natural coordinates takes the axes into the columns of* **B**. *Thus the edges of the bound that are axes in transformed coordinates are columns of* **B** *in natural coordinates. The resultant bound is a parallelotope with* **b****and* **b** *at opposite vertices and with edges parallel to the common conjugate axes.*[19]

Conversely, given any point **b**** *in this bound, there is a unique (up to a factor of proportionality) diagonal matrix* **D*** *such that* **b**** *is a posterior mean. Thus the bound is minimal.*

Proof: This lemma follows trivially by writing

$$\mathbf{a}^{**}=(\mathbf{B}'(\mathbf{H}+\mathbf{H}^*)\mathbf{B})^{-1}(\mathbf{B'HBB}^{-1}\mathbf{b}+\mathbf{B'H^*BB}^{-1}\mathbf{b}^*)$$

$$=(\rho\mathbf{I}+\mathbf{D}^*)^{-1}(\mathbf{a}\rho+\mathbf{D}^*\mathbf{a}^*)$$

where $\mathbf{a}=\mathbf{B}^{-1}\mathbf{b}$, and $\mathbf{a}^*=\mathbf{B}^{-1}\mathbf{b}^*$. Thus, element by element a_i^{**} is a simple weighted average of a_i and a_i^*

$$a_i^{**}=(\rho+d_i)^{-1}(\rho a_i+d_ia_i^*) \tag{5.29}$$

and is constrained to lie in the orthotope described previously.

[19]A parallelotope is an n-dimensional generalization of a parallelogram. It is generated by a pair of points (opposite vertices) and n vectors that define the faces at each of the points. The parallelotope is an orthotope if the vectors are orthogonal to each other.

The converse of this theorem is also true: any point in the orthotope is a posterior mean for some set of eigenvalues d_i. This follows simply by picking ρ and inverting (5.29) to write d_i as a function of a_i^{**}.

THEOREM 5.10 (MATRIX-WEIGHTED AVERAGES CAN LIE ANYWHERE). *Given only the prior and sample locations,* $\mathbf{b}$ *and* $\mathbf{b}^*$, *with* $\mathbf{H}$ *and* $\mathbf{H}^*$ *any symmetric positive definite matrices, the posterior mean may lie on the open-line segment* $(\mathbf{b},\mathbf{b}^*)$ *and anywhere off the line through it. That is, only points on the line through* $\mathbf{b}$ *and* $\mathbf{b}^*$ *exterior to the open line segment* $(\mathbf{b},\mathbf{b}^*)$ *are excepted.*

Proof: Choose any point $\mathbf{b}^{**}$satisfying the foregoing bound, and form a parallelogram in the plane of $\mathbf{b}^{**}$, $\mathbf{b}^*$, and $\mathbf{b}$ that contains $\mathbf{b}^{**}$ and has $\mathbf{b}^*$ and $\mathbf{b}$ at opposite vertices. If we simply choose the edges of this parallelogram as the first two common conjugate axes described in Lemma 1 and further choose $k-2$ additional linearly independent vectors to complete the selection of conjugate axes, by the converse of Lemma 1 there exists a set of d_i such that $\mathbf{b}^{**}$ is a posterior mean. The exception derives from the impossibility of forming such a parallelogram if $\mathbf{b}^{**}$ is on the line through $\mathbf{b}$ and $\mathbf{b}^*$ exterior to the segment $(\mathbf{b},\mathbf{b}^*)$. A tedious algebraic proof is in Leamer (1971).

The following result used to prove Theorem 5.11 is essentially the same as Pratt's (1970); the proof parallels his proof.

LEMMA 2. *Given the sample and prior locations* $\mathbf{b}$ *and* $\mathbf{b}^*$ *and the information that the prior and sample ellipsoids have a complete set of common principal axes (i.e., that the positive definite matrices* $\mathbf{H}$ *and* $\mathbf{H}^*$ *commute,* $\mathbf{HH}^*=\mathbf{H}^*\mathbf{H}$), *the posterior mean is constrained to lie in the hypersphere*

$$(\mathbf{b}^{**}-\mathbf{c})'(\mathbf{b}^{**}-\mathbf{c})<\frac{(\mathbf{b}-\mathbf{b}^*)'(\mathbf{b}-\mathbf{b}^*)}{4} \tag{5.30}$$

where

$$\mathbf{c}=\frac{(\mathbf{b}+\mathbf{b}^*)}{2}.$$

In words, the posterior mean is constrained to lie in the hypersphere with diameter $[\mathbf{b},\mathbf{b}^*]$.

Conversely, any point in this hypersphere is a posterior mean for some choice of $\mathbf{H}$ *and* $\mathbf{H}^*$ *with* $\mathbf{HH}^*=\mathbf{H}^*\mathbf{H}$. *Thus the bound is minimal.*

Proof: We may write

$$\mathbf{b}^{**}-\mathbf{b}^{*}=(\mathbf{H}+\mathbf{H}^{*})^{-1}\mathbf{H}(\mathbf{b}-\mathbf{b}^{*})$$

$$\mathbf{b}^{**}-\mathbf{b}=(\mathbf{H}+\mathbf{H}^{*})^{-1}\mathbf{H}^{*}(\mathbf{b}^{*}-\mathbf{b})$$

and

$$(\mathbf{b}^{**}-\mathbf{b}^{*})'(\mathbf{b}^{**}-\mathbf{b})=-(\mathbf{b}-\mathbf{b}^{*})'\mathbf{H}(\mathbf{H}+\mathbf{H}^{*})^{-2}\mathbf{H}^{*}(\mathbf{b}-\mathbf{b}^{*})<0,$$

since the matrix of this quadratic form is positive definite when $\mathbf{H}$ and $\mathbf{H}^{*}$ commute. After some straightforward rearrangements, this inequality is the same as inequality (5.30).

Conversely, any point in the hypersphere lies in a rectangular solid with vertices at $\mathbf{b}$ and $\mathbf{b}^{*}$. The edges of such a rectangle can be taken as common conjugate axes, and the converse of this lemma follows trivially from the converse of Lemma 1. For an algebraic proof see Pratt (1970).

The following theorem is of considerable practical interest, since it deals with a typical case when the sample precision matrix is known and the prior precision is completely arbitrary.

THEOREM 5.11 (ELLIPSOID BOUND). *Given the sample and prior locations,* $\mathbf{b}$ *and* $\mathbf{b}^{*}$, *and the sample precision* $\mathbf{H}$ *up to a scale factor, the posterior mean is constrained to lie in the ellipsoid*

$$(\mathbf{b}^{**}-\mathbf{c})'\mathbf{H}(\mathbf{b}^{**}-\mathbf{c})<\frac{1}{4}(\mathbf{b}-\mathbf{b}^{*})'\mathbf{H}(\mathbf{b}-\mathbf{b}^{*})$$

where $\mathbf{c}=(\mathbf{b}^{*}+\mathbf{b})/2$. *In words, the posterior mean must lie everywhere within an ellipsoid from the sample family of ellipsoids with center at the midpoint of the line segment joining* $\mathbf{b}$ *to* $\mathbf{b}^{*}$ *and with boundary including* $\mathbf{b}$ *and* $\mathbf{b}^{*}$.

Conversely, any point in this ellipsoid is a posterior mean for some $\mathbf{H}^{*}$. *Thus the bound is minimal.*

Incidentally, the boundary of this ellipsoid is the set of contrained least-squares points described in Section 5.1 and illustrated in Figure 5.1.

Proof: Find the coordinate system that transforms the sample ellipsoid into concentric spheres, $\mathbf{B}'\mathbf{H}\mathbf{B}=\mathbf{I}$. In terms of these coordinates

$$\mathbf{a}^{**}=\mathbf{B}^{-1}\mathbf{b}^{**}=(\mathbf{I}+\mathbf{A})^{-1}(\mathbf{a}+\mathbf{A}^{*}\mathbf{a}^{*})$$

where $\mathbf{A}^* = \mathbf{B}'\mathbf{H}^*\mathbf{B}$, $\mathbf{a} = \mathbf{B}^{-1}\mathbf{b}$, and $\mathbf{a}^* = \mathbf{B}^{-1}\mathbf{b}^*$. By Lemma 2, since $\mathbf{I}$ and $\mathbf{A}^*$ commute, $\mathbf{a}^{**}$ is constrained to the hypersphere

$$(\mathbf{a}^{**} - \mathbf{g})'(\mathbf{a}^{**} - \mathbf{g}) < \frac{1}{4}(\mathbf{a} - \mathbf{a}^*)'(\mathbf{a} - \mathbf{a}^*)$$

where $\mathbf{g} = (\mathbf{a} + \mathbf{a}^*)/2$. Transforming back into natural coordinates we obtain

$$(\mathbf{B}^{-1}\mathbf{b}^{**} - \mathbf{B}^{-1}\mathbf{c})'(\mathbf{B}^{-1}\mathbf{b}^{**} - \mathbf{B}^{-1}\mathbf{c})$$

$$= (\mathbf{b}^{**} - \mathbf{c})'\mathbf{H}(\mathbf{b}^{**} - \mathbf{c}) < \frac{1}{4}(\mathbf{b} - \mathbf{b}^*)'\mathbf{H}(\mathbf{b} - \mathbf{b}^*)$$

where $\mathbf{c} = (\mathbf{b} + \mathbf{b}^*)/2$.

The converse is also true and follows straightforwardly from the converse of Lemma 2 by noting that $\mathbf{A}^*$ is arbitrary.

Knowledge of the sample moment matrix is enough to shrink the bound from essentially complete freedom to a well-defined ellipsoid. We may now consider how various items of information about the prior precision matrix further shrink the bound. One fact we may know is that certain linear combinations of parameters are independent of each other, or equivalently, $\mathbf{H}^*$ may be diagonalized by a known transformation. Theorem 5.5, discussed in Section 5.6.1, implies that the posterior mean is then a weighted average of the 2^k constrained regressions formed by omitting variables in all different combinations.[20] The converse is not true. All weighted averages of the 2^k constrained least-squares points are not necessarily feasible; see Chamberlain and Leamer (1976).

[20]Note, incidentally, that the weighting function w_J in Theorem 5.5 allows us to collapse these 2^k points into a smaller number of points whenever any of the diagonal elements d_i are constrained to be equal. In particular, given k_1 of the diagonals equal to one number, k_2 equal to another..., the 2^k points can be collapsed into $\Pi_j(k_j+1)$ points. The extreme case with all the diagonal elements equal is equivalent to knowing $\mathbf{H}^*$ up to a scale factor, and the resulting minimal bound is the curve decolletage. Theorem 5.5 describes that curve as a convex combination of $k+1$ points.

Theorem 5.5 applies to several familiar models. The exchangeable model of Lindley and Smith (1972) with $\beta_j \sim N(\xi, \sigma_\beta^2)$ and $\xi \sim N(0, \sigma_\xi^2)$ implies a variance matrix for the vector $\boldsymbol{\beta}$ with $\sigma_\xi^2 + \sigma_\beta^2$ on the diagonal and σ_ξ^2 on the off-diagonal. The eigenvalues of this matrix (Rao, 1965, p. 54), are $d_1^{-1} = \sigma_\beta^2 + k\sigma_\xi^2$ with eigenvector $(1, 1, \ldots, 1)$ and $d_j^{-1} = \sigma_\beta^2$ of multiplicity $k-1$ with any set of $(k-1)$ eigenvectors orthogonal to $(1, 1, \ldots, 1)$. Since the eigenvectors are independent of the uncertain parameters (σ_β^2 and σ_ξ^2), there is a known linear transformation that takes the prior variance into a diagonal matrix, and theorem 5.5 applies. The multiplicity of the second eigenvalue implies that the 2^k points can be collapsed into $2k$ points. The constrained least-squares estimates involve one constraint $\Sigma_i\beta_i = 0$ and $k-1$ constraints of the form $\mathbf{l}_j'\boldsymbol{\beta} = 0$ with $\mathbf{l}_j$ a set of eigenvectors orthogonal to the vector of ones. Each of the $2k$ points is a weighted average of constrained least-squares points with or without $\Sigma_i\beta_i = 0$ and with exactly m of the other $k-1$ constraints for $m = 0, 1, \ldots, k-1$. The "ridge regression" special case with $\sigma_\xi^2 = 0$ has only a single distinct eigenvalue, and the number of points is reduced to $k+1$. The limiting degenerate case $\sigma_\xi^2 \to \infty$ has one zero eigenvalue d_1, and all constrained least-squares estimates involving the constraint $\Sigma_i\beta_i = 0$ have zero weight.

Other bounds are discussed in Chamberlain and Leamer (1976). The point of this section is that it is possible to say interesting things about the posterior distribution when the prior is not fully specified. If only the prior location and sample location are known, then the posterior modes may lie essentially anywhere. Knowledge of the sample ellipsoid shrinks the set of feasible points to an ellipsoid; knowledge of the major axes of the prior further restricts the feasible region to the hull of the 2^k restricted least-squares points.

5.9 Identification

If two models imply the same distribution of the data, no observed data can be said to favor one or the other. The posterior odds, defined as the ratio of the posterior probability of one to the posterior probability of the other, necessarily equal the prior odds ratio. A mathematical translation of this statement is, essentially, the classical definition of the identification problem used by Koopman and Riersol (1950). The shortcoming of this definition is that a "model" usually determines a *family* of data distributions indexed by some uncertain parameter vector. The generalization of the definition to deal with families of data distributions is not obvious.

The probability model we use to illustrate the concepts is the usual normal linear-regression model, $\mathbf{Y}=\mathbf{X}\boldsymbol{\beta}+\mathbf{u}$, with $\mathbf{u}$ normally distributed with mean vector $\mathbf{0}$ and covariance matrix $\sigma^2\mathbf{I}$ with σ^2 assumed known. It is assumed that the model suffers from the extreme multicollinearity problem, $\mathbf{X}\boldsymbol{\eta}_i=\mathbf{0}$, for some set of p linearly independent vectors, $\boldsymbol{\eta}_i$, $i=1,\ldots,p$. Where required, we use a prior for $\boldsymbol{\beta}$ that is normal with mean $\mathbf{b}^*$ and variance matrix $(\mathbf{H}^*)^{-1}$. It should be emphasized that the following discussion applies to this linear-regression model, and the definitions and results do not necessarily generalize to other statistical models.

Example. It is useful to have in mind a more specific example. Suppose that the model included only two explanatory variables, $\mathbf{Y}=\mathbf{x}_1\beta_1+\mathbf{x}_2\beta_2+\mathbf{u}$, with the two variables identical, $\mathbf{x}_1=\mathbf{x}_2$. In this case, there is only one $\boldsymbol{\eta}$ vector, $\boldsymbol{\eta}'=(1,-1)$, (or any vector proportional to it).

A special case of the model used here has an $\mathbf{X}$ matrix with zero vectors as the first p columns and $k-p$ linearly independent vectors as its remaining columns. Data generated by such a model pretty clearly provides evidence only about the last $k-p$ parameters. By a linear transformation, any model with p linear dependencies can be taken into this form. Let $\mathbf{C}$ be a $k\times k$ invertible matrix with the p vectors $\boldsymbol{\eta}_i$ as its first p columns, and write the regression process as

$$\mathbf{Y}=(\mathbf{XC})(\mathbf{C}^{-1}\boldsymbol{\beta})+\mathbf{u}.$$

Let the vector $\boldsymbol{\theta}=\mathbf{C}^{-1}\boldsymbol{\beta}$ be partitioned $\boldsymbol{\theta}'=(\boldsymbol{\theta}_1,\boldsymbol{\theta}_2)$ where $\boldsymbol{\theta}_1$ has p elements. By construction, the first p columns of $\mathbf{XC}$ are zero, and the process can be written

$$\mathbf{Y}=\mathbf{X}_2^*\boldsymbol{\theta}_2+\mathbf{u}, \tag{5.31}$$

where $\mathbf{X}_2^*$ is a $T\times(k-p)$ matrix with linearly independent columns.

Example. The two variable regression model with $\mathbf{x}_1=\mathbf{x}_2$ can be written $\mathbf{Y}=\mathbf{x}_1\beta_1+\mathbf{x}_2\beta_2+\mathbf{u}=\mathbf{x}_1(\beta_1+\beta_2)+\mathbf{u}$, and we are led to conclude that the process produces evidence about $\theta=\beta_1+\beta_2$.

The first pair of definitions are the ones most commonly used in the econometric literature.

Definition. A parameter value $\boldsymbol{\beta}^a$ is *observationally equivalent* to a parameter value $\boldsymbol{\beta}^b$ if the data distributions $f(\mathbf{Y}|\boldsymbol{\beta}=\boldsymbol{\beta}^a)$ and $f(\mathbf{Y}|\boldsymbol{\beta}=\boldsymbol{\beta}^b)$ are identical.

Discussion. Let $\boldsymbol{\beta}^b=\boldsymbol{\beta}^a+\boldsymbol{\eta}$ with $\boldsymbol{\eta}$ chosen such that $\mathbf{X}\boldsymbol{\eta}=\mathbf{0}$. Then $\mathbf{X}\boldsymbol{\beta}^b=\mathbf{X}\boldsymbol{\beta}^a$, and both parameter values determine the same data distribution.

Definition. The regression vector $\boldsymbol{\beta}$ is *identifiable* if there exist no two (distinct) observationally equivalent values of $\boldsymbol{\beta}$. The model is *identified* if its parameter vector is identifiable.

The shortcoming of this definition, as stressed by Kadane (1975), is that it tends to give up too soon. Although a model may fail to be identified, it may, nonetheless, provide a great deal of information about some functions of the parameter. The following is a special case of a definition due to Kadane (1975).

Definition. The linear *function* $\boldsymbol{\psi}'\boldsymbol{\beta}$ is *identified* if, whenever a parameter value $\boldsymbol{\beta}^a$ is observationally equivalent to another parameter value $\boldsymbol{\beta}^b$, it is also true that $\boldsymbol{\psi}'\boldsymbol{\beta}^a=\boldsymbol{\psi}'\boldsymbol{\beta}^b$.

Discussion. The logic of this definition is that although two parameter points may be indistinguishable, we do not, in fact, need to distinguish them, if we are interested in functions that assign the same value to both points.

THEOREM 5.12 (IDENTIFICATION OF $\boldsymbol{\psi}'\boldsymbol{\beta}$). *Given the full set of exact linear dependencies among the columns of* $\mathbf{X}$, $\mathbf{X}\boldsymbol{\eta}_i=\mathbf{0}$, $i=1,\dots,p$, *the function* $\boldsymbol{\psi}'\boldsymbol{\beta}$ *is identified if and only if* $\boldsymbol{\psi}'\boldsymbol{\eta}_i=0$ *for* $i=1,\dots,p$.

Proof: It is easily seen that the complete set of values observationally equivalent to $\boldsymbol{\beta}^a$ is $\boldsymbol{\beta}^b = \boldsymbol{\beta}^a + \Sigma_{i=1}^{p} \boldsymbol{\eta}_i w_i$ for any values of w_i. It is also clear that $\boldsymbol{\psi}'\boldsymbol{\beta}^a = \boldsymbol{\psi}'\boldsymbol{\beta}^b$ for all pairs of vectors in this class if and only if $\boldsymbol{\psi}'\boldsymbol{\eta}_i = 0$, $i = 1, \ldots, p$.

A word that is very close in spirit to identifiable is the word estimable. The concept of estimable functions is due to Bose (1944) and is discussed in Scheffé (1959). As is now shown, estimable is mathematically equivalent to identifiable.

Definition. The linear combination of parameters $\boldsymbol{\psi}'\boldsymbol{\beta}$ is *estimable* if there exists an unbiased linear estimator of $\boldsymbol{\psi}'\boldsymbol{\beta}$.

Discussion. A linear estimator is $\mathbf{w}'\mathbf{Y}$, where $\mathbf{w}$ is a vector of constants. It is an unbiased estimator of $\boldsymbol{\psi}'\boldsymbol{\beta}$ if, for all $\boldsymbol{\beta}$, $E(\mathbf{w}'\mathbf{Y}|\boldsymbol{\beta}) = \boldsymbol{\psi}'\boldsymbol{\beta}$, that is, if $\mathbf{w}'\mathbf{X}\boldsymbol{\beta} = \boldsymbol{\psi}'\boldsymbol{\beta}$. This condition is satisfied for all $\boldsymbol{\beta}$ if and only if $\mathbf{w}'\mathbf{X} = \boldsymbol{\psi}'$.

THEOREM 5.13. *Given the full set of exact linear dependencies among the columns of* $\mathbf{X}$, $\mathbf{X}\boldsymbol{\eta}_i = \mathbf{0}$, $i = 1, \ldots, p$, *a necessary and sufficient condition for* $\boldsymbol{\psi}'\boldsymbol{\beta}$ *to be estimable is that* $\boldsymbol{\psi}$ *and* $\boldsymbol{\eta}_i$ *be orthogonal,* $\boldsymbol{\psi}'\boldsymbol{\eta}_i = 0$, *for all* i.

Proof: Post-multiplying the condition $\mathbf{w}'\mathbf{X} = \boldsymbol{\psi}'$ by $\boldsymbol{\eta}_i$ produces the equality $0 = \boldsymbol{\psi}'\boldsymbol{\eta}_i$. This establishes the necessity of $\boldsymbol{\psi}'\boldsymbol{\eta}_i = 0$. A constructive proof of the sufficiency of the condition is useful. We can make use of the same notation as used in Equation (5.31). Observe that $\hat{\boldsymbol{\theta}}_2 = (\mathbf{X}_2^{*\prime}\mathbf{X}_2^*)^{-1}\mathbf{X}_2^{*\prime}\mathbf{Y}$ is an unbiased estimator of $\boldsymbol{\theta}_2$. Let $\hat{\boldsymbol{\theta}}$ have arbitrarily chosen values for its first p elements and have $\hat{\boldsymbol{\theta}}_2$ for its last $k-p$ elements, $\hat{\theta}' = (\mathbf{a}', \hat{\theta}_2')$. An estimator of $\boldsymbol{\psi}'\boldsymbol{\beta}$ is $\boldsymbol{\psi}'\mathbf{C}\hat{\boldsymbol{\theta}}$ with expected value

$$E\boldsymbol{\psi}'\mathbf{C}\hat{\boldsymbol{\theta}} = \boldsymbol{\psi}'\mathbf{C}\begin{bmatrix}\mathbf{a}\\ \boldsymbol{\theta}_2\end{bmatrix} = \boldsymbol{\psi}'\mathbf{C}\boldsymbol{\theta} + \boldsymbol{\psi}'\mathbf{C}\begin{bmatrix}\mathbf{a}-\boldsymbol{\theta}_1\\ \mathbf{0}\end{bmatrix}$$

$$= \boldsymbol{\psi}'\boldsymbol{\beta} + \boldsymbol{\psi}'\mathbf{C}\begin{bmatrix}\mathbf{a}-\boldsymbol{\theta}_1\\ \mathbf{0}\end{bmatrix}.$$

The last term in this expression is zero if the first p elements of $\boldsymbol{\psi}'\mathbf{C}$ are zero, that is, if $\boldsymbol{\psi}'\boldsymbol{\eta}_i = 0$; and then $\boldsymbol{\psi}'\mathbf{C}\hat{\boldsymbol{\theta}}$ is an unbiased estimator of $\boldsymbol{\psi}'\boldsymbol{\beta}$.

Example. In the two-variable model with $\boldsymbol{\eta} = [1, -1]$, the function $\beta_1 + \beta_2$ is estimable.

A feature of a set of observationally equivalent parameter values is that they are all assigned the same value by the likelihood function regardless of the data **Y**. The likelihood function is

$$L(\boldsymbol{\beta};\mathbf{Y})\propto\exp\left[-\frac{1}{2\sigma^2}(\mathbf{Y}-\mathbf{X}\boldsymbol{\beta})'(\mathbf{Y}-\mathbf{X}\boldsymbol{\beta})\right].$$

Given two observationally equivalent values, $\boldsymbol{\beta}^a$ and $\boldsymbol{\beta}^b$, it is easy to see that $L(\boldsymbol{\beta}^a;\mathbf{Y})=L(\boldsymbol{\beta}^b;\mathbf{Y})$ regardless of the value of **Y**. This suggests some alternative definitions.

Definition. A parameter value $\boldsymbol{\beta}^a$ is *observationally equivalent* to a parameter value $\boldsymbol{\beta}^b$ if the likelihood function $L(\boldsymbol{\beta};\mathbf{Y})$ satisfies $L(\boldsymbol{\beta}^a;\mathbf{Y})=L(\boldsymbol{\beta}^b;\mathbf{Y})$ for all **Y**.

Definition. The model is said to be *identified* if the likelihood function attains its maximum at a single point.

Example. Given the two-variable regression model with $\mathbf{x}_1=\mathbf{x}_2$, the likelihood function is maximized along the line $(\beta_1+\beta_2)=(\mathbf{x}_1'\mathbf{x}_1)^{-1}\mathbf{x}_1'\mathbf{Y}$ (see Figure 5.15).

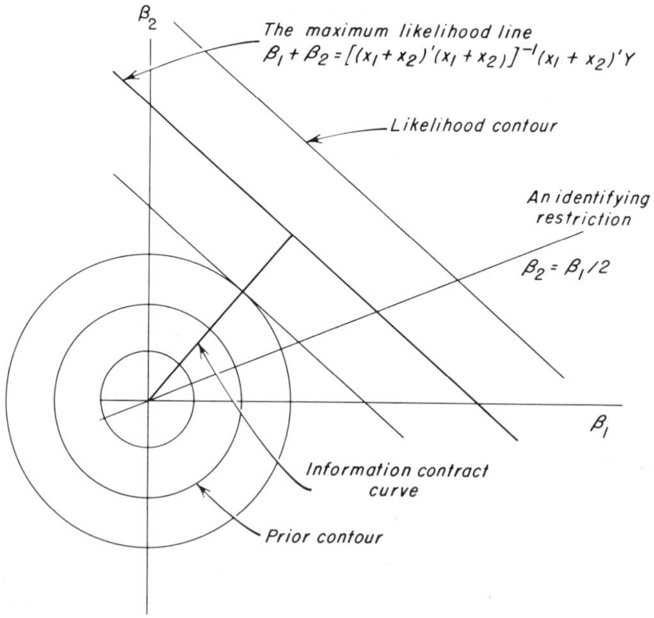

Fig. 5.15 The likelihood function of an unidentified model.

We now consider the role that prior information plays in identifying a model. We first consider exact linear restrictions and then probabilistic prior information.

Definition. A set of q restrictions $\mathbf{R}\boldsymbol{\beta}=\mathbf{r}$ where $\mathbf{R}$ is a $q\times k$ matrix of rank q is set of *identifying restrictions* if on the set $\{\boldsymbol{\beta}|\mathbf{R}\boldsymbol{\beta}=\mathbf{r}\}$ the likelihood function attains its maximum at a single point. The set of restrictions is said to *overidentify* the model if a subset of the restrictions identifies the model.

Discussion. Using the notation of Equation (5.31), the set of restrictions $\boldsymbol{\theta}_1=\mathbf{a}$ is a set of identifying restrictions (see Figure 5.15).

THEOREM 5.14. *Given a set of identifying restrictions any function $\boldsymbol{\psi}'\boldsymbol{\beta}$ is identified and (equivalently) estimable.*

Proof: Left to reader.

Dreze (1962) and (1975) comments that exact restrictions are unlikely to hold with probability one and suggests using probabilistic prior information. Whereas a model is defined before to include the restrictions, a model is now defined to include any probabilistic prior information:

Definition. A model is said to be *identified in probability* if the posterior distribution for $\boldsymbol{\beta}$ is proper regardless of the data $\mathbf{Y}$.

Discussion. It is enough to have a proper prior distribution for $\boldsymbol{\theta}_1$ to have a proper posterior distribution for $\boldsymbol{\theta}$ and hence for $\boldsymbol{\beta}$. If the prior is proper for $\boldsymbol{\beta}$, any model is identified in probability. This leads one to conclude, erroneously, that the concept of identification is uninteresting to a Bayesian. More discussion follows.

Another observation due to Dreze (1962) is that if it is known in advance of the data $\mathbf{Y}$ that the likelihood function will necessarily be uniform on some subspace, then, conditional on this subspace, the prior and posterior distributions will coincide:

Definition. A model is not identified if there exist a set of restrictions $\mathbf{A}\boldsymbol{\beta}=\mathbf{a}$ such that $f(\boldsymbol{\beta}|\mathbf{A}\boldsymbol{\beta}=\mathbf{a})=f(\boldsymbol{\beta}|\mathbf{A}\boldsymbol{\beta}=\mathbf{a},\mathbf{Y})$ for all $\mathbf{Y}$.

Discussion. The reader may verify that $f(\boldsymbol{\beta}|\boldsymbol{\theta}_2)=f(\boldsymbol{\beta}|\boldsymbol{\theta}_2,\mathbf{Y})$ regardless of the prior $f(\boldsymbol{\beta})$.

Without saying so explicitly, these definitions of the identification problem describe deficiencies in the information afforded by the experiment. The concept of estimability, however, does make explicit reference to the sample information. An alternative is to compare the prior distribution with the posterior distribution:

Definition. Given a particular prior distribution, the experiment is *personally uninformative* about a linear function $\boldsymbol{\psi}'\boldsymbol{\beta}$ if the posterior distribution of $\boldsymbol{\psi}'\boldsymbol{\beta}$ is equal to the given prior distribution for all data values $\mathbf{Y}$.

Example. If $\mathbf{H}^*$ is the identity matrix and $\eta=(-1,1)$, then from Section 5.7 the data is personally uninformative about $\boldsymbol{\eta}'\mathbf{H}^*\boldsymbol{\beta}=\beta_2-\beta_1$. Notice in Figure 5.15 that the information contract curve is $\beta_2-\beta_1=0$, regardless of the data $\mathbf{Y}$. Notice also that this curve changes if you choose a different prior metric, $\mathbf{H}^*$.

Discussion. With $\mathbf{H}^*$ as the prior precision matrix, the experiment is personally uninformative about $\boldsymbol{\eta}_i'\mathbf{H}^*\boldsymbol{\beta}$, $i=1,\ldots,p$ and any linear combination of these (see Section 5.7), but the experiment is personally informative (i.e., contains information) about all other linear combinations. Notice the sharp contrast between this notion and the notion of identifiability. Given some linear dependencies among the columns of $\mathbf{X}$, almost all functions are unidentified. Nonetheless, the data are personally informative about almost all functions. The following definition *is* the analogue of identifiable.

Definition. The experiment is *publicly informative* about the linear combination $\boldsymbol{\psi}'\boldsymbol{\beta}$ if there exists no positive definite, prior precision matrix $\mathbf{H}^*$ such that the experiment is personally uninformative about $\boldsymbol{\psi}'\boldsymbol{\beta}$.

THEOREM 5.15. *Given the full set of exact linear dependencies among the columns of* $\mathbf{X}$, $\mathbf{X}\boldsymbol{\eta}_i=0$, $i=1,\ldots,p$, *a necessary and sufficient condition for the experiment to be publicly informative about* $\boldsymbol{\psi}'\boldsymbol{\beta}$ *is* $\boldsymbol{\psi}'\boldsymbol{\eta}_i=0$ *for all i.*

Proof: The experiment is personally uninformative about the function $\Sigma_i w_i\boldsymbol{\eta}_i'\mathbf{H}^*\boldsymbol{\beta}$, for arbitrary w_i. The experiment is publicly informative about $\boldsymbol{\psi}'\boldsymbol{\beta}$ if we cannot find a positive definite $\mathbf{H}^*$ and constants w_i such that $\Sigma_i w_i\boldsymbol{\eta}_i'\mathbf{H}^*=\boldsymbol{\psi}'$. Write this equation as $(\mathbf{H}^*)^{-1}\boldsymbol{\psi}=\Sigma w_i\boldsymbol{\eta}_i$, and postmultiply it by $\boldsymbol{\psi}'$, $\boldsymbol{\psi}'\mathbf{H}^{*-1}\boldsymbol{\psi}=\Sigma w_i\boldsymbol{\psi}'\boldsymbol{\eta}_i$. Given the conditions $\boldsymbol{\psi}'\boldsymbol{\eta}_i=0$, this equation amounts to $\boldsymbol{\psi}'(\mathbf{H}^*)^{-1}\boldsymbol{\psi}=0$ which cannot hold for positive definite $\mathbf{H}^*$. Thus $\boldsymbol{\psi}'\boldsymbol{\eta}_i=0$, $i=1,\ldots,$p is a sufficient condition. Conversely, suppose there is a $\boldsymbol{\eta}$ such that $\mathbf{X}\boldsymbol{\eta}=0$ and $\boldsymbol{\psi}'\boldsymbol{\eta}\neq 0$. Then let c_j, $j=2,\ldots,k$ be a set of orthonormal vectors orthogonal to $\boldsymbol{\psi}$, and let $\mathbf{H}^*=\boldsymbol{\psi}\boldsymbol{\psi}'(\boldsymbol{\psi}'\boldsymbol{\eta})^{-1}+\Sigma_j c_j c_j'$. With this choice of $\mathbf{H}^*$ we have $\mathbf{H}^*\boldsymbol{\eta}=\boldsymbol{\psi}$, which is the required condition.

Observation of an experiment is personally valuable in the sense that it changes the observer's opinions. Observation of an experiment is socially valuable in the sense that it leads to a consensus, thereby eliminating the need for the unresolvable argument over whose prior is best. The last set of definitions are aimed at this social-learning phenomenon.

Definition. An experiment *leads to a consensus* if, given a sufficient number of independent replications of the experiment, all observers with nondogmatic priors have essentially the same posterior distribution, regardless of their prior.

Definition. An experiment *leads to consensus* about the linear function $\boldsymbol{\psi}'\boldsymbol{\beta}$ if, given a sufficient number of independent replications of the experiment, all observers with nondogmatic priors have essentially the same posterior distribution for $\boldsymbol{\psi}'\boldsymbol{\beta}$.

THEOREM 5.16. *An experiment leads to consensus if and only if the model is identified.*

THEOREM 5.17. *An experiment leads to consensus about $\boldsymbol{\psi}'\boldsymbol{\beta}$ if and only if the function $\boldsymbol{\psi}'\boldsymbol{\beta}$ is identified.*

Proof: Left to reader.

An experiment cannot lead to consensus if it is impossible to distinguish one prior from another. The following definition is due to Zellner (1971).

Definition. A prior distribution $f^a(\boldsymbol{\beta})$ is observationally equivalent to another prior distribution $f^b(\boldsymbol{\beta})$ if the marginal data distributions $f^a(\mathbf{Y}) = \int f(\mathbf{Y}|\boldsymbol{\beta}) f^a(\boldsymbol{\beta})\, d\boldsymbol{\beta}$ and $f^b(\mathbf{Y}) = \int f(\mathbf{Y}|\boldsymbol{\beta}) f^b(\boldsymbol{\beta})\, d\boldsymbol{\beta}$ are identical.

Observation. Translate the location of a prior $\mathbf{b}^*$ to $\mathbf{b}^* + \boldsymbol{\eta}$ where $\boldsymbol{\eta}$ satisfies $\mathbf{X}\boldsymbol{\eta} = 0$ to construct an observationally equivalent prior.

In summary, the words identifiable, estimable, and publicly informative, and the phrase "leads to a consensus" are interchangeable. If a model implies a likelihood function that attains its maximum on a (linear) set of points, the model is not identifiable, and, conditional on that set of points (or certain other sets), the prior and posterior distribution coincide.

The concept of personal informativeness is quite different from the concept of public informativeness. No individual would want to discard the information generated by a model just because the model is not identified, or even because the linear function of interest is not identified. Given this prior information, the model may, nonetheless, imply useful information.

5.10 Examples

Two examples of a sensible Bayesian analysis of data are reported in this section. The adjective "Bayesian" refers to the admission that subjective nonsample information is used to interpret the data. The modifier "sensible" refers to the fact that it is unlikely that anyone could with confidence select a particular prior, and as a result we explore the implications of many different prior distributions. These results were computed by a program entitles SEARCH: *S*eeking *E*xtreme and *A*verage *R*egression *C*oefficient *H*ypotheses, which is available on request.

SEARCH assumes that the prior distribution is built in three steps. The prior location is first selected, then the prior "metric" (isodensities), and lastly a particular density value is assigned to each of the isodensity surfaces. If only the prior location is known, the data support an ellipsoid of estimates described alternatively by Theorem 5.11 as a hull of contract curves or by Theorem 5.1 as a set of constrained least-squares estimates. The choice of prior isodensity surfaces further limits the set of estimates to a curve within this feasible ellipsoid. Lastly, the labeling of the prior isodensities selects from this curve a point or a set of points as posterior modes.

SEARCH describes the ellipsoid of constrained estimates in terms of extreme values of coefficients of interest. The information contract curve is described in terms of the "rotation invariant average regressions" (Theorem 5.8) and also in terms of a set of points on the curve.

The priors we are about to discuss are not informative on all the coefficients, and the feasible ellipsoid is suitably adjusted. The prior has implicit in it a set of q uncertain constraints, $\mathbf{R}\boldsymbol{\beta}=\mathbf{r}$. The feasible set of estimates can be described as the set of constrained regressions subject to constraints $\mathbf{M}\mathbf{R}\boldsymbol{\beta}=\mathbf{M}\mathbf{r}$, with $\mathbf{R}$ and $\mathbf{r}$ taken as given and $\mathbf{M}$ free to vary. If $\mathbf{R}$ is invertible, then as $\mathbf{M}$ is varied any point on the "feasible" ellipsoid of constrained estimates is attainable. But when $\mathbf{R}$ is not invertible, only a subset of constrained estimates is attainable. This is made more clear in the following examples.

DOUBTFUL VARIABLES

It is very common to have a model with a few explanatory variables that are known to belong in the equation and a longer list of "doubtful" explanatory variables. The first set of variables is likely to be the focus of the analysis, and the second set is used to "control" for other influences. If the list of doubtful variables is long, estimation with all the doubtful variables included in the equation will produce large standard errors on the coefficients of the "focus" variables. In this situation, it is typical to try different subsets of the doubtful variables, and it is hoped that the coefficients of the focus variables will not change much as the list of doubtful variables is changed. But this search is both haphazard and nonexhaustive. Furthermore, if the coefficients of the focus variables change very much, this ad hoc search does not suggest how to average the many computed estimates into a single number.

SEARCH is ideally suited to deal with this problem. The interesting bounds that the program can report are the extreme estimates of the focus coefficients with ideally chosen doubtful variables included in the equation. There is no way of "fiddling" with the doubtful variables to get an estimate outside the reported range. The points on the contract curve reported by the program are mixtures of the 2^q regressions that could be computed using subsets of the q doubtful variables. Thus the program both searches exhaustively the set of possible regressions and also suggests weighted averages of the regressions, the latter being important when the bounds are wide.

The following example has eight "doubtful" regional dummy variables. The dependent variable is the wage rate, and the focus variables are the education of the wage earner, his age, and the square of his age. A dummy variable for a region is necessary if the labor market in the given region is "separated" from the markets in other regions. To say that the dummy variables are doubtful is to say that in the absence of evidence to the contrary, we should view the labor market as a national market.

The estimated model with all the dummy variables included is (standard errors in parentheses):

$$
\begin{array}{llll}
W=.041\,D_1 & +.098\,D_2 & +.051\,D_3 & -.019\,D_4 \\
\quad(.34) & \quad(.32) & \quad(.46) & \quad(.34) \\
+.004\,D_5 & -.178\,D_6 & +.086\,D_7 & +.060\,D_8 \\
\quad(.46) & \quad(.43) & \quad(.50) & \quad(.35) \\
\multicolumn{4}{l}{+.05\ \mathrm{EDUC}+.137\ \mathrm{AGE}-.0015\ (\mathrm{AGE})^2+5.737} \\
\quad(.030) & \quad(.047) & \quad(.0006) & \quad(.96)
\end{array}
$$

where

D_1 = Mid-Atlantic
D_2 = East North Central
D_3 = West North Central
D_4 = South Atlantic
D_5 = East South Central
D_6 = West South Central
D_7 = Mountain
D_8 = Pacific

(New England omitted)

The bounds for the coefficients of the three focus variables are reported in the table below. The numbers in parentheses are the standard errors of these coefficients if the model that implied the estimate could be taken as given. (Remember that these bounds include regressions subject to constraints such as $\beta_1 = \beta_2$, which says the Mid-Atlantic and East North Central regions can be aggregated. They also include constraints of the form $\beta_i = 0$.)

Bounds for the Focus Coefficients

EDUC	AGE	$(\text{AGE})^2$
.0577 (.0177)	.139 (.029)	−.00147 (.00035)
.0446 (.0178)	.131 (.029)	−.00155 (.00035)

Each of these coefficients is quite insensitive to the choice of regional dummy variables.

Choice of points within these (narrow) bounds requires a more completely specified prior. Suppose that the coefficients of the doubtful variables are thought to be small in the sense that $\Sigma_{i=1}^{8} \beta_i^2$ is likely to be small. This prior "metric" implies the contract curve incompletely reported in Table 5.1. On the contract curve the extremes of all coefficients occur at the end points. One end point is least squares with all the dummies included; the other is least squares with all the dummies excluded. The extremes for the focus variables are:

EDUC	AGE	$(\text{AGE})^2$
.0521	.1332	−.001489
.0502	.1336	−.001535

These bounds are almost points and it hardly seems necessary to select a particular point on the contract curve. But notice from Table 5.1 that the equation with the dummies omitted has a low likelihood ratio (equivalently a large F) and the data have a distinct preference for an estimate close to the unconstrained least-squares points.

Table 5.1

Points on Contract Curve

Likelihood Ratio	EDUC	AGE	$-(\text{AGE})^2$
.14	.0521	1.33	.00148
.31	.0517	1.34	.00150
.48	.0514	1.34	.00150
.66	.0511	1.35	.00151
.83	.0507	1.35	.00152
1.0	.0502	1.37	.00153

To conclude, for this particular problem the ambiguity in the specification does not translate into substantial ambiguity in the focus coefficients. The specification error implies, for example, an interval of estimates for the education coefficient from .0446 to .0577. But the sampling standard error of this coefficient in the unconstrained model is .03, which is large compared to the specification range .0577 − .0446 = .0131. To put it briefly, the sampling error is more important than the specification error.

DISTRIBUTED LAG ESTIMATION

Another common problem in economics is the estimation of distributed lag processes. Consider the import demand function estimated by ordinary least squares

$$\begin{aligned} M_t = {} & \underset{(.42)}{.13\,Y_t} \; \underset{(.48)}{+2.0\,Y_{t-1}} \; \underset{(.48)}{-.91\,Y_{t-2}} \\ & \underset{(.50)}{+.56\,Y_{t-3}} \; \underset{(.39)}{-.33\,Y_{t-4}} \underset{(.50)}{-.42P_t} \; \underset{(.55)}{-.53P_{t-1}} \\ & \underset{(.50)}{+.33P_{t-2}} \; \underset{(.51)}{-.72P_{t-3}} \underset{(.48)}{+.23P_{t-4}} - .15 + .96e_{t-1} \end{aligned}$$

where standard errors are in parenthesis and where

M_t = logarithm (United States imports in the tth quarter divided by a price index of imports)
Y_t = logarithm (United States GNP in quarter t divided by the GNP price index)
P_t = logarithm (import price index divided by GNP price index)
t = 1951 first quarter to 1967 fourth quarter

Economists would generally expect to see the coefficients on the income variables positive and the coefficients on the price variables negative. The peculiar saw-tooth pattern of coefficients would be regarded as highly unlikely, and some constraint on the coefficients would undoubtably be used to "improve" or to smooth the estimates. One possibility is to constrain the coefficients of each of the distributed lag patterns to lie on a line. The resulting estimates are

$$\begin{aligned} M_t = \alpha & + .83\,Y_t \; + .55\,Y_{t-1} + .27\,Y_{t-2} - .01\,Y_{t-3} \\ & - .29\,Y_{t-4} - .56P_{t-1} - .38P_{t-2} - .19P_{t-3} \\ & - 0P_{t-4} \; + .19P_{t-5}. \end{aligned}$$

Although this constraint does eliminate the wild pattern of coefficients, it does not produce coefficients that are all the same sign for each variable. Perhaps we should constrain them all to be equal, yielding the estimated equation

$$M_t = \alpha + .28 \sum_{\tau=0}^{4} Y_{t-\tau} - .31 \sum_{\tau=0}^{4} P_{t-\tau}.$$

Each of these three estimated equations is appropriate for one extreme form of prior information about the coefficients. Since the researcher in fact holds as his opinions neither one of these three forms of prior information, he may informally mix together the three results. He notes that the sum of the coefficients on the income variable is either 1.29, 1.35, or 1.21, and the sum of the coefficients on the price variable is either -1.04, $-.94$, or -1.29. Neither of these estimates is particularly sensitive to the form of prior information. The shape of the lag distribution does seem to be highly sensitive to the form of the prior, but it does seem that the biggest coefficient is "probably" either the first or second.

The point of much of the discussion in Section 5.6, especially Theorem 5.5, is that a Bayesian can do nothing more than compute sensible weighted averages of constrained estimates. The value of the Bayesian approach is that it provides instruction concerning both the choice of constraints and the choice of sensible weight functions.

The analysis now to be discusses makes us of three different prior distributions. These priors make use of the assumption that $\mathbf{R}\boldsymbol{\beta}$ has spherical normal distribution with $\mathbf{R}$ defined below:

PRIOR 1 SMALL DIFFERENCES

$$\mathbf{R} = \begin{bmatrix} 1 & -1 & 0 & 0 & 0 & 0 & 0 & 0 & 0 & 0 & 0 \\ 0 & 1 & -1 & 0 & 0 & 0 & 0 & 0 & 0 & 0 & 0 \\ 0 & 0 & 1 & -1 & 0 & 0 & 0 & 0 & 0 & 0 & 0 \\ 0 & 0 & 0 & 1 & -1 & 0 & 0 & 0 & 0 & 0 & 0 \\ 0 & 0 & 0 & 0 & 0 & 1 & -1 & 0 & 0 & 0 & 0 \\ 0 & 0 & 0 & 0 & 0 & 0 & 1 & -1 & 0 & 0 & 0 \\ 0 & 0 & 0 & 0 & 0 & 0 & 0 & 1 & -1 & 0 & 0 \\ 0 & 0 & 0 & 0 & 0 & 0 & 0 & 0 & 1 & -1 & 0 \end{bmatrix}$$

This prior reflects the fact that the first five coefficients are likely to be similar, $\beta_1 = \beta_2 = \beta_3 = \beta_4 = \beta_5$, and the next five coefficients are likely to be similar, $\beta_6 = \beta_7 = \beta_8 = \beta_9 = \beta_{10}$. (This is one of Shiller's (1973) proposals, first proposed by Whitakker and Robinson, 1940).

PRIOR 2

$$\mathbf{R} = \begin{bmatrix} 1 & -1 & 0 & 0 & 0 & 0 & 0 & 0 & 0 & 0 & 0 \\ 0 & 1 & -1 & 0 & 0 & 0 & 0 & 0 & 0 & 0 & 0 \\ 0 & 0 & 1 & -1 & 0 & 0 & 0 & 0 & 0 & 0 & 0 \\ 0 & 0 & 0 & 1 & -1 & 0 & 0 & 0 & 0 & 0 & 0 \end{bmatrix}$$

This prior will smooth only the income coefficient pattern.

PRIOR 3

$$\mathbf{R}=\begin{bmatrix} 0 & 0 & 0 & 0 & 0 & 1 & -1 & 0 & 0 & 0 & 0 \\ 0 & 0 & 0 & 0 & 0 & 0 & 1 & -1 & 0 & 0 & 0 \\ 0 & 0 & 0 & 0 & 0 & 0 & 0 & 1 & -1 & 0 & 0 \\ 0 & 0 & 0 & 0 & 0 & 0 & 0 & 0 & 1 & -1 & 0 \end{bmatrix}$$

This prior will smooth only the price coefficients.

The bounds over all regressions which make use of the prior constraints are reported in Table 5.2. Note, for example, that the prior 1 bound for the sum of the income coefficients is the fairly small interval [1.2, 1.65], whereas the bounds for the individual coefficients are rather wide. This indicates that the choice of smoothness prior does not have much impact on the estimated long-run effect but does have a substantial effect on how that impact is allocated over the individual coefficients.

The set of constrained regressions just discussed includes those that use constraints of the form $\beta_1=\beta_2$; but the set of constraints also includes the unlikely constraint $\beta_1-\beta_2=\beta_7-\beta_6$. Recall that the set of constraints are of the form $\mathbf{MR}\boldsymbol{\beta}=0$ for any $\mathbf{M}$. To avoid constraints that involve jointly the price coefficients and the income coefficients we would have to restrict $\mathbf{M}$ to a block diagonal. As it turns out, this is a difficult computational task, and instead we use priors 2 and 3, which are diffuse, respectively, on the price coefficients and the income coefficients. The prior 2 bound for the sum of the income coefficients is [1.24, 1.61] which indicates the set of estimates if only the income coefficients are smoothed. But the set of

Table 5.2
Bounds for Coefficients

β_1	β_2	β_3	β_4	β_5	$\Sigma_{i=1}^{5}\beta_i$	β_6	β_7	β_8	β_9	β_{10}	$\Sigma_{i=6}^{10}\beta_i$
Prior 1											
1.15	2.22	.80	1.55	.89	1.65	.81	.69	.92	.63	1.10	−.49
−.74	.02	−1.42	−.71	−.93	1.2	−1.53	−1.52	−.90	−1.66	−1.17	−2.15
Prior 2											
1.04	2.09	.66	1.41	.78	1.61	−.13	.44	.45	−.25	.37	−1.00
−.62	.156	−1.28	−.57	−.82	1.24	−1.22	−.63	−.31	−.33	−.14	−2.00
Prior 3											
.27	2.12	−.62	.68	−.03	1.46	.12	.04	.39	−.04	.43	−.87
−.08	1.70	−1.05	.21	−.38	1.35	−.78	−.81	−.31	−.92	−.44	−1.45

Table 5.3

Ideal Points for Prior 1 (Rotation Invariant Average Regressions)

β_1	β_2	β_3	β_4	β_5	β_6	β_7	β_8	β_9	β_{10}	β_{11}
.28	.28	.28	.28	.28	−.31	−.31	−.31	−.31	−.31	−.15
.50	.45	.27	.14	.07	−.28	−.25	−.26	−.29	−.30	−.15
.61	.59	.23	.55	−.02	−.41	−.24	−.19	−.23	−.19	−.16
.65	.72	.16	−.00	−.05	−.55	−.22	−.12	−.21	−.12	−.16
.65	.86	.06	−.04	−.04	−.68	−.18	−.06	−.22	−.07	−.16
.60	1.03	−.06	−.04	−.03	−.77	−.14	−.02	−.27	−.03	−.16
.50	1.23	−.21	.01	−.04	−.80	−.12	.01	−.34	.01	−.16
.36	1.50	−.45	.15	−.10	−.74	−.18	.09	−.44	.06	−.16
.13	1.96	−.91	.56	−.33	−.42	−.53	.33	−.72	.23	−.15

possible constraints is still too large. It includes the higher order polynomial constraint $(\beta_1-\beta_2)=(\beta_2-\beta_3)$, but it also includes the unlikely constraint $(\beta_1-\beta_2)=-(\beta_2-\beta_3)$. Ideally, we would require the matrix $(\mathbf{MM}')^{-1}$ to be positive, but this too creates computational burdens.

The rotation invariant average regressions for prior 1 are given in Table 5.3. The first point is constrained least squares given all the constraints implicit in the prior. The last point is constrained least squares given *none* of the constraints; that is, it is just the unconstrained least-squares estimate. The intermediate points are weighted averages of constrained least-squares points. The next to last point is a weighted average of all regressions that involve *one* constraint. The next point uses constraints two at a time....

Any point on the contract curve is a weighted average of these average regressions points. A trace of the contract curve risks missing important features of highly variable curves, but the rotation invariant average regressions cannot.

Selected points on the contract curve implied by prior 1 are reported in Table 5.4. There are several observations that can now be made. First of all, observe that the long-run coefficients (the sums) are relatively insensitive to this form of prior information. It simply makes little difference how confident you are that the differences of the coefficients are small. Individual coefficients are in contrast quite sensitive to the precision of the prior. Next, observe that whereas both end points of the contract curve are peculiar, intermediate points have attractively smooth coefficient patterns. Thus although neither extreme form of prior information—diffuse prior or zero difference prior—implies acceptable estimates, "partial" imposition of the prior constraints does yield sensible estimates. Another thing to notice is that the income coefficients are smoothed easily, but the price coefficients resist smoothing. There is pretty clear evidence that the re-

Table 5.4

Selected Points on the Contract Curve, Prior 1

Rel. Like.[a]	β_1	β_2	β_3	β_4	β_5	$\sum_{i=1}^{5} \beta_i$	β_6	β_7	β_8	β_9	β_{10}	$\sum_{i=6}^{10} \beta_i$
.94	.28	.28	.28	.28	.28	1.4	−.31	−.31	−.31	−.31	−.31	−1.55
.96	.50	.48	.25	.14	.08	1.45	−.35	−.26	−.24	−.28	−.26	−1.39
.97	.60	.65	.18	.04	.00	1.47	−.48	−.23	−.16	−.23	−.17	−1.27
.99	.49	1.24	−.24	.05	−.07	1.47	−.74	−.17	.03	−.35	.02	−1.21
1.0	.13	1.95	−.91	.56	−.33	1.4	−.42	−.53	.33	−.72	.23	−1.11

[a]Relative likelihood of the reported point to the maximum likelihood value, computed with σ^2 set to s^2.

sponse to the income stimulus is more rapid than the response to the price stimulus. In fact, the data seem to suggest that other lagged price variables might be added to the equation. Finally, observe that whereas the maximum of the first coefficient reported in Table 5.4 is .60, the maximum ideal point in Table 5.3 is .65. Although the value .65 is not attainable, numbers above .60 are attainable, and to some extent Table 5.4 is misleading.

A comparison of Table 5.2 with Tables 5.3 and 5.4 reveals the importance of the choice of "metric." Tables 5.3 and 5.4 make use of the assumption that $\boldsymbol{\beta}'\mathbf{R}'\mathbf{R}\boldsymbol{\beta}$ is small, whereas Table 5.2 uses only the assumption that $\boldsymbol{\beta}'\mathbf{R}'\mathbf{H}^*\mathbf{R}\boldsymbol{\beta}$ is small where $\mathbf{H}^*$ may be any symmetric positive definite matrix. With a suitable choice of $\mathbf{H}^*$, β_1 may be as large as 1.15 or as small as $-.74$. But Tables 5.3 and 5.4 reveal that if you are willing to restrict $\mathbf{H}^*$ to be proportional to the identity matrix, then β_1 cannot exceed .65 nor fall short of .13.

The next step in the analysis is to select a particular point or set of points on the contract curve. Formally, this can be done by specifying completely the prior distribution which has to this point been defined only in terms of the surfaces on which the density is constant. This is a step I tend to resist, since I have a very difficult time finding sensible questions that would reveal with any accuracy my opinions about the density values. I do think it makes sense to examine the contract curve in several ways. In this case I note that, with a relative likelihood deterioration only to .97, I can get a pattern of coefficients that makes me happy. Incidentally, you may infer from this last sentence that there are features of my prior I have not formally used, namely, that the coefficients should not change sign and should decay in absolute value. For this reason, too, I resist formal methods for selecting points on the contract curve.

CHAPTER 6

SIMPLIFICATION SEARCHES

In the two previous chapters we have considered specification searches that are intended to introduce into a data analysis uncertain prior information. Hypothesis-testing searches arise when more than one model or hypothesis receive positive a priori probability. Interpretive searches involve prior density functions that, although allocating zero probability to all but one hypothesis, do concentrate the prior probability in certain regions of the parameter space. In the case of hypothesis-testing searches the statistical testing selects among a set of hypotheses with no presumption that in a large sample one of the hypotheses will be favored. In contrast, interpretive searches recognize that in a sufficiently large sample the most general hypothesis will necessarily be favored. The intent is not to select among legitimately competing models but rather to "improve" the estimate of the parameters by using an a priori estimate when the data evidence is too weak to yield a reliable sample estimate.

In this chapter we discuss a variety of search that has yet another motivation: simplification. The most general models appropriate for inference with nonexperimental data are usually so cluttered with variables of an incidental nature that they are nearly impossible to comprehend directly. It is thus incumbent on the researcher to find vehicles for communication of his results. He might, for example, focus his discussion on a particular parameter of special interest or perhaps on a linear combination of parameters. Alternatively, the researcher might seek from the data an indication of the "important" variables. We call this a simplification search.

Thus the function of simplification search is not to ask if a restricted specification is true, nor to ask if a restricted specification might lead to better parameter estimates, but rather to ask if a restricted specification that is undeniably simpler and more easily understood is not also "significantly" inferior to the more general model for some hypothetical or real decisions. If it is, we reject the hypothesis that the benefits of the restriction outweigh the costs.

Formal analysis of simplification problems requires a precise definition of the costs and benefits of simplicity. The costs of simplicity may be assessed in the context of some hypothetical decision problems, but the benefits are likely to elude precise definition. Consequently we concentrate our formal attention on the cost side, but we first comment informally on the likely benefits from simplification.

Justifications for simplicity can usefully be divided into two categories. The first makes a "metaphysical" reference to the inherent simplicity of Nature, or at least to man's belief in such. The second category of justifications accepts a complex Nature but rests simplicity on the finiteness and fallibility of Man's perceptive and reasoning faculties. Briefly, simplicity is preferred because "Nature is simple" or because "Man is simple."

The "Nature is simple" hypothesis has, I think, little support among philosophers and statisticians. Jeffreys' is a widely cited exception. He writes [1961, p. 4] "It is asserted, for instance, that the choice of the simplest law is purely a matter of economy of description or thought, and has nothing to do with any reason for believing the law...I say, on the contrary, that the simplest law is chosen because it is the most likely to give correct predictions; that the choice is based on a reasonable degree of belief;..."

Jeffreys is asserting not only that constrained hypotheses should be assigned positive probability but also that they ought to be assigned greater prior probability than any alternative, more complex hypotheses. Such a preference for simple models might be inductively derived. Simple hypotheses could usually yield better predictions. But it is not enough to observe merely that people act as if simple models had a greater degree of believability. Any observed preference for simple models may derive not from the inherent superior believability of parsimonious models but rather from the undeniable difficulties encountered in working with complex descriptions of reality. Nor do I know of any proper empirical evidence to support the assertion that simpler models generally yield better predictions. There is the oft-told story of overfitting in which a naive researcher fits a polynomial of degree $T-1$ given T pairs of observations (y_i, x_i). This undoubtably does yield inferior predictions relative to a polynomial of fixed lower degree. But that can be fully remedied by assigning a proper prior distribution to the parameters of the higher-degree polynomial. I interpret this example as an illustration of the illogic of using a prior that is built to be dominated by the data evidence when the data evidence simply is too weak to do it. It is hardly evidence in favor of simpler models.

I have indicated in the chapter on hypothesis testing that I know of few cases in which I would assign positive probability to a restricted (simple) model. Even then I can find nothing that compels me to favor the simpler model in the assignment of probability. It is the other set of reasons for simplicity that I find persuasive: Simplicity is desirable because it is conducive to the transmission and accumulation of knowledge. It greatly facilitates communication between and among observers and theorists. A complex, novel theory that might take years to filter accurately to other researchers can transmit rapidly (but inaccurately) if it is simplified. Possessors of what they regard to be superior knowledge for their own personal gain are likely to engage in this kind of marketing activity. Many who buy the product may never realize that there is more to the theory than the catchy slogans used to advertise it.

Philosophers have argued in various ways that simplicity encourages progress. Popper (1972) favors simpler models because they are more easily contradicted, which might at first glance seem to hasten the rejection of inferior models. This would be true if the simple model were assigned positive probability, but if such a model is derived from a more complex system of belief, apparently falsifying evidence can be taken to mean only that the simple version does not work under all conditions. In that situation simplicity protects a system of belief from falsification and thereby apparently impedes progress. On the other hand, protection of a system of belief from potential falsification is an essential feature of normal science, according to Kuhn (1962). Filling in the details of a theory and working out all its implications requires a vast amount of tedious work. Such labor would hardly be performed by doubters or even agnostics who would imagine the value of their efforts overnight crashing to zero.

Neither the "Nature is simple" nor the "Man is simple" hypothesis implies any unambiguous definitions or methods of measuring the benefits of simplicity. The number of uncertain parameters is a possible mechanical measure of simplicity, but it cannot be generally satisfactory. If we take (as I do) simplicity as a consequence of man's and society's shortcomings, the definition of simplicity necessarily changes from social millieu to social millieu. It is thus impossible and even undesirable to define simplicity precisely, and we instead must content ourselves with the satisfaction that the participants in any social information process can know themselves what simplicity is and what it is not.

The prototypical example of this is the construction of a map (Polanyi, 1964). We may take as a theory of the world an enormously detailed globe which identifies every object down to the smallest grain of sand. The complexity of this theory effectively prevents us from using it for any purpose whatsoever. Instead, we simplify it in the form of a set of maps. I use one map to find my way to the subway station, another to select the station at which to depart. The pilot of the airplane uses yet another to navigate from Boston to Washington. Each map is a greatly simplified version of the theory of the world; each is designed for some class of decisions and works relatively poorly for others.

The construction of a language is another good example of a simplification problem. The number of aurally and visually discernible words and word patterns is absolutely enormous, perhaps limitless. With as few characters as are in our alphabet we could form $26^5 > 10^7$ distinct five-letter words. Such a vocabulary would be beyond the reach of even the most verbally talented, and the mistaken use of words used infrequently would greatly distort intended communications. A highly limited vocabulary likewise distorts communications by not distinguishing one complex communication from another, for example, the American overuse of the word "nice" to describe a wide variety of generally pleasing responses to environmental stimuli. An optimal vocabulary ideally solves the tradeoff between miscommunications from too few words and miscommunications from too many.

Incidentally, there is a great danger that a simple language is not only a vehicle for communication but that it also creates an impoverished reality of its own. The art of communication forces an awareness of reality, and the more subtle is the language, the more practice one obtains in distinguishing subtleties. Conversely, a coarse language creates no situations for exercising one's capacities to distinguish subtleties, and those faculties may atrophy like any unused muscle. We may, in fact, be unable anymore to distinguish the great variety of sensations we refer to as "nice." This may also be the case in the communication of scientific theories. We may come erroneously to believe in the simplicity of Nature because that is the way scientific theories are communicated.

I do not think it is possible to define simplicity, which is to say in the language of decision theory that it is difficult to compute precise benefits or precise costs from any simplification. In this chapter the cost of simplification is measured in the context of several simple decision problems, but the benefits are not quantified at all. We hope that what we learn can have implications for more complex and more realistic decisions.

One thing that is important to understand is that simplification is a decision problem which uses as an *input* the current information about the parameters. When a current sample is available, simplification logically follows inference and is confused with the inferential process, at great peril to the coherence of a statistical analysis. I would recommend making as clear a distinction as possible between inference and decision, by discussing in separate sections of a research report first the inferential question of how various prior distributions are influenced by the data and second the decision problem of how given various posterior distributions the model can be simplified. Incidentally, since a data set is taken as given, any probability moments reported in this chapter are necessarily conditional on that data. It is thus notationally convenient to suppress the data when writing conditional probability statements, and it is hoped that this will not cause confusion. For example, the statement $E(\boldsymbol{\beta})=(\mathbf{X}'\mathbf{X})^{-1}\mathbf{X}'\mathbf{Y}$ implies that the conditional mean of $\boldsymbol{\beta}$, $E(\boldsymbol{\beta}|\mathbf{Y},\mathbf{X})$, is equal to the least-squares estimate $(\mathbf{X}'\mathbf{X})^{-1}\mathbf{X}'\mathbf{Y}$.

A point that merits repeating is that a simplified model that might perform adequately for some decision-making circumstances will be unambiguously unacceptable in others. It is, therefore, essential to identify precisely the problem that is considered. Three examples suggest the potentially great diversity of decision problems.

Example 1. Aggregate consumption of apples C_a and aggregate consumption of bananas C_b depend on aggregate GNP Y through the functions $C_a=\alpha_a+\beta_a Y$, and $C_b=\alpha_b+\beta_b Y$. If we wish to predict future levels of consumption of apples and bananas, may we without great detriment to the prediction constrain the marginal propensities to consume to equal each other $\beta_a=\beta_b$ and therefore "remember" only the marginal propensity to consume fruit rather than separate propensities for each fruit?

Example 2. GNP Y is thought to depend on the government deficit G and the money stock M, $Y=\alpha+\beta G+\gamma M$. If we wish GNP to attain some target Y^*, may we effectively assure that goal by selecting an appropriate level of the government deficit G^* while treating money M as if it had no effect ($\gamma=0$), or conversely, might we better control money M and act as if G had no effect?

Example 3. A constant-elasticity-of-substitution production function expresses output as a function of capital and labor inputs. If the elasticity of substitution is equal to one, the investment function assuming profit-maximizing behavior is a function of one explanatory variable rather than two. Given the information generated by observing the production process, may we make inferences about the investment process acting as if certain of its parameters took on special values?

These three examples illustrate, respectively, a prediction problem, a control problem, and an inference problem. Relative to a model of the form $y = z\gamma + w\delta + u$ they ask if we may act *as if* δ were zero (1) if we wanted to predict y, (2) if we wanted to control y, or (3) if we wanted to make inferences about γ. The inference problem is distinguished from the others only in that more data is to be gathered before decisions are to be made. The actual, ultimate decision may, in fact, be either a prediction or a control problem. This is called a presimplification problem, referring to the fact that simplification occurs prior to observation. We make much use of the presimplification notion in Chapter 9, when we discuss postdata model construction.

It is easy to demonstrate the inappropriateness of classical hypothesis testing at a fixed level of significance for the simplification problem. Suppose the prior distribution were diffuse. The only information conveyed by the fact that the hypothesis $\gamma = 0$ is or is not rejected at the 5% level of significance is the information that the posterior 95% credible interval includes or does not include the point $\gamma = 0$. Thus you may reject the hypothesis $\gamma = 0$ even though with near certainty γ is infinitesimal. (Figure 6.1a) And you may accept the hypothesis even though with high probability γ is enormous. (Figure 6.1b) It is thus important to distinguish the words "statistically significant" from the words "economically signifi-

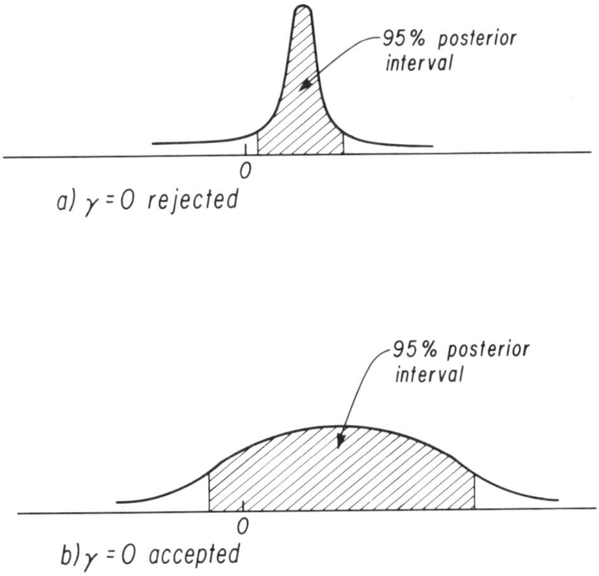

Fig. 6.1 Posterior distributions.

cant." The former measures the amount of information in the data; the latter measures the size of the coefficient in the context of some decision problem.

A more subtle point is that classical tests have built into them rather strong and often unwarranted assumptions about the behavior of the explanatory variables. Consider again the model $y = z\gamma + w\delta + u$ with δ and γ assumed known exactly, and with the explanatory variables z and w satisfying the auxiliary relationship $w = rz + \varepsilon$, where u and ε are independent random variables and r is known. The hypothesis $\delta = 0$ can be used to simplify the model, yielding either $H_0 : y = z\gamma$ or $H_0' : y = z(\gamma + r\delta)$, where the H_0' hypothesis allows the included variable to play partly the role of the excluded variable.

If prediction were the only goal, the hypothesis H_0' is unambiguously superior, since it yields a lower expected loss. But for other reasons H_0 may be a better simplification. A simplification is intended to facilitate communication, and H_0' may be difficult to communicate, since it seems to say that the marginal effect of z on y is $(\gamma + r\delta)$ when, in fact, it is only γ. It seems desirable *at least* to distinguish the hypotheses "we may act as if δ were zero" or "we may act as if w_T were zero" from the hypothesis "we may compensate for not observing w_T," the former pair implying the simplification H_0 and the latter implying H_0'. Classical hypothesis testing makes use of the second-form H_0' with r implicitly estimated in a special way to be discussed subsequently. The other form of simplification is discussed in Section 6.2.

The remainder of this chapter consists of three sections and a conclusion. In the first section we report Lindley's (1968) formal decision-theoretic solution to a prediction-simplification problem. Among the lessons to be learned is the great importance of assumptions about the process that generates the explanatory variables. In fact, the simplification problem depends as much if not more on the process that generates these variables than it does on the regression process linking the dependent variable to the explanatory variables. That observation is used in Section 6.2 to argue in favor of the kind of simplification that makes fewer demands on our knowledge of the explanatory variable process and that also communicates relatively clearly. The third section emphasizes the dependence of the simplification process on the decision problem under consideration by reporting Lindley's (1968) analysis of a control-simplification problem and by contrasting that solution with the prediction-simplification problem.

6.1 Simplification for Conditional Prediction

As an example of a conditional-prediction problem, consider the two-variable linear regression model

$$y_t = \alpha + z_t\gamma + w_t\delta + u_t \tag{6.1}$$

where α, δ and γ are unobservable scalar parameters, u_t is an unobservable error, and y_t, z_t, and w_t are observable variables. Suppose that u_t ($t = 0,\ldots,T$) is a sequence of independent normal random variables with zero means and known variance σ^2. Let a set of T previous observations of the process be $(\mathbf{Y},\mathbf{z},\mathbf{w})$, which together with a multivariate prior distribution for the parameters (α,γ,δ) imply a multivariate posterior distribution with mean

$$E([\alpha,\gamma,\delta]|\mathbf{Y},\mathbf{z},\mathbf{w}) = [\bar{\alpha},\bar{\gamma},\bar{\delta}].$$

In making a conditional prediction of the next outcome, say, Y_T, we assume that both the explanatory variables z_T and w_T are potentially observable prior to the announcement of the prediction, hence the adjective "conditional" modifying prediction. It is perhaps obvious, but it is demonstrated here that if the penalty for prediction error is quadratic, the optimal prediction given both z_T and w_T is

$$\hat{y}_T = \bar{\alpha} + z_T\bar{\gamma} + w_T\bar{\delta} \tag{6.2}$$

where $\bar{\gamma}$ and $\bar{\delta}$ are the posterior means of γ and δ. There will, of course, be prediction errors, partly because of the residual error process u_t and partly because the actual values of the parameters α, γ, and δ are not known.

Suppose, now, that we wished to determine if it is worth the expense to observe the second variable w_T. If w_T is not observed, we must estimate it, by say, $\hat{w}_T$, and predict y_T as a function of z_T only as

$$\hat{y}_T^* = \bar{\alpha} + z_T\bar{\gamma} + \hat{w}_T\bar{\delta}. \tag{6.3}$$

The squared discrepancy between Equations (6.2) and (6.3) is a measure of the error induced by not observing w_T:

$$(\hat{y}_T - \hat{y}_T^*)^2 = (w_T - \hat{w}_T)^2\bar{\delta}^2 \tag{6.4}$$

Note especially that this error depends on the mean of $\bar{\delta}$ but not on its variance. Note also that in testing the hypothesis $\delta = 0$ in the sense of this chapter, that is, by computing numbers like (6.4), we are partly asking the question "is δ small?" but more importantly we are asking also "how well can we forecast w_T?" To answer the latter question, we must model the process that generates w_T and z_T—there is no way the simplification question can be answered without such a model.

One model (that should, I think, be of little interest to economists operating with time-series data) is the multivariate random model, in which the explanatory variables are treated as if they were drawn randomly from a population with fixed mean vector and covariance matrix. In particular, assume that (z_t, w_t) come from a normal population with mean $\boldsymbol{\mu}' = (\mu_z, \mu_w)$

and covariance matrix

$$\mathbf{V}=\begin{bmatrix} v_{11} & v_{12} \\ v_{21} & v_{22} \end{bmatrix}.$$

If we knew the parameters of this distribution, a prediction of w_T would be generated by the conditional regression

$$\hat{w}_T=E(w_T|z_T)=\mu_w+v_{21}v_{11}^{-1}(z_T-\mu_z)$$

with conditional variance

$$E\left[(w_T-\hat{w}_T)^2|z_T\right]=v_{22}-v_{21}v_{11}^{-1}v_{12}.$$

Sample counterparts of these unknown parameters can be used if the prior for $\boldsymbol{\mu}$ and $\mathbf{V}$ is diffuse and if the number of observations is large.[1] We would then have

$$\hat{w}_T=E(w_T|z_T,\mathbf{z},\mathbf{w})=\bar{w}+(\mathbf{w}'\mathbf{M}\mathbf{z})(\mathbf{z}'\mathbf{M}\mathbf{z})^{-1}(z_T-\bar{z}) \tag{6.5}$$

$$E\left[(w_T-\hat{w}_T)^2|z_T,\mathbf{z},\mathbf{w}\right]=\frac{\mathbf{w}'\mathbf{M}\mathbf{w}-(\mathbf{w}'\mathbf{M}\mathbf{z})(\mathbf{z}'\mathbf{M}\mathbf{z})^{-1}(\mathbf{z}'\mathbf{M}\mathbf{w})}{T} \tag{6.6}$$

where $\bar{w}$ and $\bar{z}$ are the sample means of $\mathbf{w}$ and $\mathbf{z}$ and $\mathbf{M}$ is the matrix that removes means $\mathbf{M}=\mathbf{I}-\mathbf{1}_T T^{-1}\mathbf{1}_T'$. The predicting equations (6.3) and expected squared error (6.4) thus become

$$\hat{y}_T^*=\bar{\alpha}+\left(\bar{w}-(\mathbf{w}'\mathbf{M}\mathbf{z})(\mathbf{z}'\mathbf{M}\mathbf{z})^{-1}\bar{z}\right)\bar{\delta}+z_T\left(\bar{\gamma}+(\mathbf{w}'\mathbf{M}\mathbf{z})(\mathbf{z}'\mathbf{M}\mathbf{z})^{-1}\bar{\delta}\right) \tag{6.7}$$

$$E(\hat{y}_T-\hat{y}_T^*)^2=E(w_T-\hat{w}_T)^2\bar{\delta}^2=\frac{\left[\mathbf{w}'\mathbf{M}\mathbf{w}-\mathbf{w}'\mathbf{M}\mathbf{z}(\mathbf{z}'\mathbf{M}\mathbf{z})^{-1}\mathbf{z}'\mathbf{M}\mathbf{w}\right]\bar{\delta}^2}{T} \tag{6.8}$$

Two observations may now be made. If the prior for γ and δ were diffuse, the posterior means $\bar{\alpha}$, $\bar{\delta}$, and $\bar{\gamma}$ would be just the least-squares estimates, say, b_0, b_w and b_z. The coefficient of z_T in Equation (6.7) would then be $b_z+(\mathbf{w}'\mathbf{M}\mathbf{z})(\mathbf{z}'\mathbf{M}\mathbf{z})^{-1}b_w$ which is just the estimated coefficient of a regression of $\mathbf{Y}$ on $\mathbf{z}$ alone. Furthermore, the penalty (6.8) can be written as $\chi^2\sigma^2/T$ where χ^2 is the χ-square value for testing the restriction $\delta=0$,

$$\chi^2=\frac{b_w^2\left[\mathbf{w}'\mathbf{M}\mathbf{w}-\mathbf{w}'\mathbf{M}\mathbf{z}(\mathbf{z}'\mathbf{M}\mathbf{z})^{-1}\mathbf{z}'\mathbf{M}\mathbf{w}\right]}{\sigma^2}.$$

Thus the procedure just described measures the increase in the expected prediction error when w_T is not observed in terms of the usual χ^2 variable for testing $\delta=0$. As is discussed in detail subsequently, it differs from classical hypothesis testing in implicitly defining the significance level as a

[1] Using the material from Section 3.4, and the diffuse prior assumption with $T^*=0$, $\mathbf{S}^*=\mathbf{0}$, and $\nu^*=0$, the variance of w_T given z_T is not (6.6) but rather (6.6) times the adjustment $(T+1)/(T-1)$.

decreasing function of the sample size. What is perhaps more important is the fact that the decision theory logic makes unambiguous the otherwise implicit assumptions about the process that generates the explanatory variables. In particular, classical tests are appropriate only if the explanatory variable vectors are independently drawn from the same population.

Let us now repeat this logic for a general model and for general linear restrictions. Write the linear regression process as

$$\begin{bmatrix} \mathbf{Y} \\ y_T \end{bmatrix} = \begin{bmatrix} \mathbf{X} \\ \mathbf{x}_T' \end{bmatrix} \boldsymbol{\beta} + \begin{bmatrix} \mathbf{u} \\ u_T \end{bmatrix}$$

where $\mathbf{Y}$ and $\mathbf{X}$ are $(T \times 1)$ and $(T \times k)$ matrices and are already observed, where y_T is a future outcome of the process and $\mathbf{x}_T$ is a $k \times 1$ vector of future explanatory variables, and where $[\mathbf{u}', u_T]$ is a $(1 \times (T+1))$ vector of errors with mean zero and covariance $\boldsymbol{\Sigma}$. We are asked to predict y_T given $\mathbf{Y}$, $\mathbf{X}$, and $\mathbf{x}_T$ and in particular to minimize squared prediction error $[y_T - \hat{y}(\mathbf{Y}, \mathbf{X}, \mathbf{x}_T)]^2$ with prediction $\hat{y}$. The expected prediction error can be written as

$$E[y_T - \hat{y}(\mathbf{Y}, \mathbf{X}, \mathbf{x}_T)]^2 = E\left(E([y_T - \hat{y}(\mathbf{Y}, \mathbf{X}, \mathbf{x}_T)]^2 | \mathbf{Y}, \mathbf{X}, \mathbf{x}_T)\right),$$

where the expression in the internal brackets is straightforwardly minimized for every value of $(\mathbf{Y}, \mathbf{X}, \mathbf{x}_T)$ by setting

$$\begin{aligned} \hat{y}(\mathbf{Y}, \mathbf{X}, \mathbf{x}_T) &= E(y_T | \mathbf{Y}, \mathbf{X}, \mathbf{x}_T) = \mathbf{x}_T' E(\boldsymbol{\beta} | \mathbf{Y}, \mathbf{X}, \mathbf{x}_T) + E(u_T | \mathbf{Y}, \mathbf{X}, \mathbf{x}_T) \\ &= \mathbf{x}_T' E(\boldsymbol{\beta} | \mathbf{Y}, \mathbf{X}) + E(u_T | \mathbf{Y}, \mathbf{X}) \end{aligned}$$

which is a linear function of $\mathbf{x}_T$. If some part of $\mathbf{x}_T$ is not observed, we assume that the complete vector is predicted as a linear function of that which is observed. That is, letting $\mathbf{x}_T' = (\mathbf{x}_T^{I\prime}, \mathbf{x}_T^{J\prime})$, we assume that $E(\mathbf{x}_T | \mathbf{Y}, \mathbf{X}, \mathbf{x}_T^I) = \mathbf{A}\mathbf{x}_T^I$, and thus the optimal predicting equation becomes

$$\hat{y}(\mathbf{Y}, \mathbf{X}, \mathbf{x}_T^I) = \mathbf{x}_T^{I\prime} \mathbf{A}' E(\boldsymbol{\beta} | \mathbf{Y}, \mathbf{X}) + E(u_T | \mathbf{Y}, \mathbf{X}).$$

Or, to make a long story short, we wish to restrict our attention to predictions linear in $\mathbf{x}_T$

$$\hat{y}(\mathbf{Y}, \mathbf{X}, \mathbf{x}_T) = \mathbf{x}_T' \boldsymbol{\theta}(\mathbf{Y}, \mathbf{X}) \tag{6.9}$$

where the function $\boldsymbol{\theta}$ may be completely free, in which case it is just the posterior mean of $\boldsymbol{\beta}$, or it may be constrained to have certain elements zero to reflect the fact that certain elements of the vector $\mathbf{x}_T$ are not observed prior to the prediction of y_T. Incidentally, Equation 6.9 implicitly includes the $E(u_T | \mathbf{Y}, \mathbf{X})$ term, since $\mathbf{x}_T'$ is assumed to have one element equal to one.

For ease of notation we write the conditional expected value operator $E(\boldsymbol{\theta} | \mathbf{Y}, \mathbf{X})$ henceforth as just $E(\ )$. If $\mathbf{Y}$ and $\mathbf{X}$ are given, $\boldsymbol{\theta}$ is just a vector of constants, and the expected loss can be written as

$$\begin{aligned}
E(y_T-\hat{y}_T)^2 &= E(\boldsymbol{\beta}'\mathbf{x}_T+u_T-\boldsymbol{\theta}'\mathbf{x}_T)^2\\
&= Eu_T^2+E\mathbf{x}_T'(\boldsymbol{\beta}-\boldsymbol{\theta})(\boldsymbol{\beta}'-\boldsymbol{\theta}')\mathbf{x}_T\\
&= Eu_T^2+E\mathbf{x}_T'(\boldsymbol{\beta}-E\boldsymbol{\beta}+E\boldsymbol{\beta}-\boldsymbol{\theta})(\boldsymbol{\beta}-E\boldsymbol{\beta}+E\boldsymbol{\beta}-\boldsymbol{\theta})'\mathbf{x}_T\\
&= Eu_T^2+E\left[\mathbf{x}_T'(\boldsymbol{\beta}-E\boldsymbol{\beta})(\boldsymbol{\beta}-E\boldsymbol{\beta})'\mathbf{x}_T+\mathbf{x}_T'(E\boldsymbol{\beta}-\boldsymbol{\theta})(E\boldsymbol{\beta}-\boldsymbol{\theta})'\mathbf{x}_T\right]\\
&= Eu_T^2+\operatorname{tr} S(\mathbf{x}_T)V(\boldsymbol{\beta})+(E\boldsymbol{\beta}-\boldsymbol{\theta})'S(\mathbf{x}_T)(E\boldsymbol{\beta}-\boldsymbol{\theta}) \qquad (6.10)
\end{aligned}$$

where we have written $S(\mathbf{x}_T)=E\mathbf{x}_T\mathbf{x}_T'$. The three terms in the last line of this expression are the irreducible mean-square error Eu_T^2, a penalty for uncertainty in $\boldsymbol{\beta}$, and an additional penalty for $\boldsymbol{\theta}\neq E\boldsymbol{\beta}$, this last term wholly independent of the uncertainty in $\boldsymbol{\beta}$.

The minimal expected loss if $\boldsymbol{\theta}$ lies in the linear subspace $\mathbf{R}\boldsymbol{\theta}=\mathbf{r}$ is simply the expected posterior loss (6.10) minimized over that linear subspace. This minimization is a simple Lagrangian problem requiring the derivatives of

$$f=(E\boldsymbol{\beta}-\boldsymbol{\theta})'S(\mathbf{x}_T)(E\boldsymbol{\beta}-\boldsymbol{\theta})+2\boldsymbol{\lambda}'(\mathbf{R}\boldsymbol{\theta}-\mathbf{r})$$

to be set to zero. That is,

$$\frac{\partial f}{\partial \boldsymbol{\lambda}}=\mathbf{R}\boldsymbol{\theta}-\mathbf{r}=\mathbf{0} \qquad (6.11)$$

$$\frac{\partial f}{\partial \boldsymbol{\theta}}=-S(\mathbf{x}_T)(E\boldsymbol{\beta}-\boldsymbol{\theta})+\mathbf{R}'\boldsymbol{\lambda}=\mathbf{0}. \qquad (6.12)$$

These can be solved by premultiplying (6.12) by $\mathbf{R}S^{-1}(\mathbf{x}_T)$ and calculating

$$\boldsymbol{\lambda}=\left(\mathbf{R}S^{-1}(\mathbf{x}_T)\mathbf{R}'\right)^{-1}(\mathbf{R}E(\boldsymbol{\beta})-\mathbf{r})$$

which can be inserted into (6.12) to obtain

$$\boldsymbol{\theta}=E(\boldsymbol{\beta})-S^{-1}(\mathbf{x}_T)\mathbf{R}'\left(\mathbf{R}S^{-1}(\mathbf{x}_T)\mathbf{R}'\right)^{-1}(\mathbf{R}E(\boldsymbol{\beta})-\mathbf{r}). \qquad (6.13)$$

The third term in the mean-square error (6.10) becomes

$$(E\boldsymbol{\beta}-\boldsymbol{\theta})'S(\mathbf{x}_T)(E\boldsymbol{\beta}-\boldsymbol{\theta})=(\mathbf{R}E(\boldsymbol{\beta})-\mathbf{r})'\left(\mathbf{R}S^{-1}(\mathbf{x}_T)\mathbf{R}'\right)^{-1}(\mathbf{R}E(\boldsymbol{\beta})-\mathbf{r}). \qquad (6.14)$$

It is obvious from the positive definiteness of the third term in (6.10) that minimal expected posterior loss requires $\boldsymbol{\theta}=E\boldsymbol{\beta}$, or by (6.14) that a restriction increases expected loss unless $\mathbf{R}E\boldsymbol{\beta}=\mathbf{r}$. A simplification thus necessarily decreases expected prediction accuracy. We assume that a simplification has benefits also, and in the absence of any clear quantitative statement of those benefits, a reasonable number to report is the percentage increase in the expected posterior loss due to the restriction $\mathbf{R}\boldsymbol{\theta}=\mathbf{r}$:

$$L^2(\mathbf{R},\mathbf{r})=\frac{(\mathbf{R}E(\boldsymbol{\beta})-\mathbf{r})'(\mathbf{R}S^{-1}(\mathbf{x}_T)\mathbf{R}')^{-1}(\mathbf{R}E(\boldsymbol{\beta})-\mathbf{r})}{Eu_T^2+\operatorname{tr}S(\mathbf{x}_T)V(\boldsymbol{\beta})}. \tag{6.15}$$

With suitable definitions of prior vagueness we have simply the least-squares results (remembering that the expected value operator is conditional on $\mathbf{X}$ and $\mathbf{Y}$)

$$E(\boldsymbol{\beta}|\mathbf{Y},\mathbf{X})=(\mathbf{X}'\mathbf{X})^{-1}\mathbf{X}'\mathbf{Y}$$

$$V(\boldsymbol{\beta}|\mathbf{Y},\mathbf{X})=\sigma^2(\mathbf{X}'\mathbf{X})^{-1}.$$

Further, if the explanatory variables are independent observations from a multivariate process, we would have the $\mathbf{x}_T$ moment matrix be approximately (see Section 3.4)

$$S(\mathbf{x}_T)=E(\mathbf{x}_T\mathbf{x}_T')=\frac{\mathbf{X}'\mathbf{X}}{T}.$$

Using these in (6.13), $\boldsymbol{\theta}$ is seen to be simply the constrained least-squares estimate subject to $\mathbf{R}\boldsymbol{\beta}=\mathbf{r}$. Inserting them into (6.14), we obtain the increase in the posterior expected loss to be T^{-1} times a factor that is well known to be the increase in the error-sum squares due to the restriction. The summary L^2 becomes

$$L^2(\mathbf{R},\mathbf{r})=\frac{T^{-1}\Delta ESS}{\sigma^2\left(1+\frac{k}{T}\right)} \tag{6.16}$$

where ΔESS is the increase in the error sum of squares, k is the number of coefficients, and T is the number of observations. This contrasts with the classical summary statistic $\Delta ESS/\sigma^2$, which is comparedwith $\chi_p^2(\alpha)$ where p is the rank of $\mathbf{R}$ and α is the significance level. Thus the classical counterpart of (6.16) is the ratio $\Delta ESS/\sigma^2\chi_p^2(\alpha)$. In addition to the nonoccurrence of the factor T^{-1} (which for large T necessitates a "significant" finding), the classical summary differs from the subjectivist summary in depending on p, the number of restrictions. The measure (6.16), incidentally, is just the difference in the multiple correlation coefficients of the two models times a factor that tends to a constant as sample size grows, $L^2(\mathbf{R},\mathbf{r})=(R^2-R_0^2)(\mathbf{Y}'\mathbf{M}\mathbf{Y}/T\sigma^2)/(1+kT^{-1})$. Thus if a restriction does not greatly affect the R^2 of an equation, it will not greatly increase the expected squared prediction error.

This rough coincidence of approaches usefully highlights the assumptions that are implicit in the use of classical tests to simplify models for prediction. Of course, there is the diffuse prior assumption. But more importantly, the vectors of explanatory variables are assumed to be $T+1$ independent replications of a multivariate process. Autocorrelation and trends in particular are assumed away. Few economists would find that acceptable. It is also worth stating explicitly that the variance in the denominator of the t statistic does not measure the uncertainty in the coefficient but rather the inverse of the conditional variance of the explanatory variable. From (6.10), it is seen that uncertainty in the coefficients $V(\boldsymbol{\beta})$ does not influence choice of restrictions $\boldsymbol{\theta} \neq E(\boldsymbol{\beta})$.

6.2 Causally Constrained Conditional Predictions

An important aspect of the solution discussed in the previous section is that observed variables are used to forecast correlated unobserved variables under the assumption that the correlation structure is maintained. The prediction effect (the coefficient) of an observed variable thus includes not only its own estimated coefficient but also a part due to the effect of unobserved variables assumed to be correlated with it. Interpreted in terms of hypothesis testing, the change in the R^2 due to a restriction is calculated relative to a restricted equation with a reestimated set of coefficients. Whereas this may make good sense if we intend the test to determine the truth or falsity of the restriction, it makes less sense for the simplification problem. Do we really mean to say that an effect of an explanatory variable is negligible when it can be predicted well from observation of another explanatory variable? This is the question implicit in a classical t test, for example. A direct application of Webster suggests that a variable can be considered negligible if we can neglect it without substantial loss. Neglecting it means not bothering to predict it or otherwise to make adjustment for not observing it. As will be shown, this is the question implicit in classical beta coefficients and variants thereof.

Turning now from semantics to metaphysics, we can find another version of this same argument. To the extent that the full unconstrained model summarizes our beliefs about the causal nature of the world, the recalculation of the coefficients implicit in hypothesis testing constitutes a potential distortion of that causality. That is, since included variables play in part the role of dropped variables, the constrained equation is causally misleading unless the included variables do, in fact, cause the excluded variables. If they do not, the resulting equation is causally inaccurate. An agnostic attitude toward the causality within the explanatory variable set is reflected by reporting the original estimates of the coefficients of the included variables calculated in the context of the unconstrained equation. These may be described as the direct effects of the included variables on the dependent variable. Indirect effects depend on other unspecified causal linkages.

An example is in order to make clear these relatively obscure notions. Suppose the equation of motion of a body falling from rest is

$$\frac{d^2y}{(dt)^2} = g(1-\beta zt)$$

where z measures the wind resistance and t the time since departure. The parameter g is acceleration in a vacuum, and terminal velocity is reached at time $t = 1/\beta z$. Suppose, further, that observations on a set of falling bodies are used to estimate the equation of location

$$y = \frac{\hat{g}t^2}{2} - \frac{\hat{\beta}\hat{g}zt^3}{6}, \qquad R^2 = .98$$

where the circumflexes indicate estimated parameters. For this particular sample of falling bodies (including feathers and bowling balls) the following auxiliary regression is also calculated

$$(zt^3) = \hat{r}t^2 - \hat{\alpha}.$$

The model may be simplified to exclude the wind resistance variable. Two alternative simpler models are

$$y = \hat{g}t^2/2 \qquad R^2 = .70 \qquad (6.17)$$

$$y = \frac{(\hat{g} - \hat{\beta}\hat{g}\hat{r}/3)t^2}{2} + \frac{\hat{\alpha}\hat{\beta}\hat{g}}{6} \qquad R^2 = .95 \qquad (6.18)$$

It is my contention that the first of these equations is the one that should be used to discuss simplification. It asserts that in a vacuum the estimated rate of acceleration is $\hat{g}$ and that for the class of bodies and for the time periods considered, we ought not to think of the experiment as if it were conducted in a vacuum, since one's ability to predict the location of the falling bodies is seriously affected by that assumption. (The R^2 drops from .98 to .7.) Contrast that perfectly clear statement with the statement appropriate for the second equation. "Wind resistance is 'negligible' since by adjusting the rate of acceleration to $\hat{g} - \hat{\beta}\hat{g}\hat{r}/3$ and by acting as if the initial location of the body were $\hat{\alpha}\hat{\beta}\hat{g}/6$ rather than zero, we can track the position of this class of falling bodies almost as well as we would if we actually observed the wind resistance."

In fact, wind resistance is not negligible; rather, it can be compensated for. At the very least we ought to make clear the distinction between these two statements. For reasons I have explained, I think simplification is more appropriately interpreted as the problem of neglecting variables, rather than the problem of compensating for their effects. There is first the semantic argument that if simplification were intended to compensate for rather than to neglect certain secondary influences, we might expect practitioners to use a more appropriate adjective than "negligible." Second, in compensating for a secondary influence, the theory may be fundamentally and nonsensically distorted. Consider the gravity example. If we neglect wind resistance we assert what is completely true: a body falling from rest in a vacuum accelerates at the constant rate $\hat{g}$. Contrast that with the "theory" that results when wind resistance is compensated for: a body falling from rest at the time of departure instantaneously falls to a height $\hat{\alpha}\hat{\beta}\hat{g}/6$ below its initial position, attaining thereby absolutely no velocity, and thereafter falls, accelerating at the constant rate $\hat{g}-\hat{\beta}\hat{g}\hat{r}/3$. The distortion of Newtonian mechanics is obvious and absurd.

Simplification tests with unrecomputed coefficients can be calculated using the same formulas as tests with recomputed coefficients provided that we choose the constraint matrices $\mathbf{R}$ and $\mathbf{r}$ appropriately. If we write the model as $y_T=\alpha+\mathbf{z}_T'\gamma+\mathbf{w}_T'\boldsymbol{\delta}+u_T$, a simplification hypothesis is $\boldsymbol{\delta}=0$. We may prevent recomputation of the coefficients on the $\mathbf{z}$ variables by imposing also the constraint that the coefficients must equal their posterior means, $E(\gamma)$. Thus a causally constrained simplification is implied by the constraint matrices

$$\mathbf{R}=\begin{bmatrix}\mathbf{0} & \mathbf{I} & \mathbf{0}\\ \mathbf{0} & \mathbf{0} & \mathbf{I}\end{bmatrix} \qquad \mathbf{r}=\begin{bmatrix}E(\gamma)\\ \mathbf{0}\end{bmatrix}$$

where the first column of $\mathbf{R}$ is a vector of zeroes multiplying the constant α in the equation.

With these restriction matrices the mean-square–error penalty (6.14) becomes

$$\begin{aligned}&(\mathbf{R}E(\boldsymbol{\beta})-\mathbf{r})'(\mathbf{R}S^{-1}(\mathbf{x}_T)\mathbf{R}')^{-1}(\mathbf{R}E(\boldsymbol{\beta})-\mathbf{r})\\ &=[E(\boldsymbol{\delta})]'V(\mathbf{w}_T)[E(\boldsymbol{\delta})].\end{aligned} \tag{6.19}$$

Dropping variables without the causal constraint requires constraint matrices

$$\mathbf{R}=[\mathbf{0} \quad \mathbf{0} \quad \mathbf{I}], \mathbf{r}=[\mathbf{0}],$$

and the mean-square–error penalty (6.14) becomes

$$[E(\boldsymbol{\delta})]'V(\mathbf{w}_T|\mathbf{z}_T)[E(\boldsymbol{\delta})] \tag{6.20}$$

where $V(\mathbf{w}_T|\mathbf{z}_T)$ is the conditional variance of $\mathbf{w}_T$, given $\mathbf{z}_T$. Penalty (6.20) is smaller than penalty (6.19) depending on the correlation between the included and excluded variables, because the included variables are used to forecast excluded variables.

Given the assumption of diffuse priors, and supposing that δ is a scalar, the criterion (6.19) is just the square of the least-squares coefficient times the sample variance of the variable. If this were divided by the square of the sample variance of the dependent variable, the resulting number would be just the beta coefficient, which can be computed as least squares with variables standardized to have unit variance. Although standardized coefficients are used in other disciplines, in the econometrics literature they are rarely even mentioned. Goldberger (1964, pp. 197–198) is an exception.

To conclude, criterion (6.20), which is equivalent to (6.16) under diffuseness assumptions, ranks variables considered individually for discarding in the same way as traditional t tests. Criterion (6.19), however, provides a ranking identical to the ranking implied by classical beta coefficients. It seems to me, therefore, that the rarely used beta coefficients could be usefully resurrected as indicators of significance when models are being simplified, although the variance of the explanatory variables ought at a minimum be trend and autocorrelated adjusted.

6.3 Simplification for Control

A point that may be obvious is that simplification is problem specific, and, for example, simplification for prediction may be quite different from simplification for control. The one-period control problem of Lindley (1968) illustrates this fact. Suppose a scalar variable y_T is determined by the linear-regression process

$$y_T = \alpha + \gamma' \mathbf{z}_T + \boldsymbol{\delta}' \mathbf{w}_T + u_T \tag{6.21}$$

where $\boldsymbol{\gamma}$ and $\boldsymbol{\delta}$ are vector parameters, α is a scalar parameter, u_T is a residual error with mean zero and variance σ^2, and $\mathbf{z}_T$ and $\mathbf{w}_T$ are vectors of explanatory variables. The control problem is to select $\mathbf{z}_T$ and $\mathbf{w}_T$ in such a way that y_T is likely to be close to some target t. In particular, let us choose the explanatory variables to minimize expected loss where loss is quadratic

$$L(y_T, t) = (y_T - t)^2.$$

Writing the regression process as

$$y_T = \alpha + \boldsymbol{\beta}' \mathbf{x}_T + u_T \tag{6.22}$$

where $\boldsymbol{\beta}' = [\gamma', \boldsymbol{\delta}']$ and $\mathbf{x}_T = [\mathbf{z}'_T, \mathbf{w}'_{cT}]$, the expected loss can be written as a function of $\mathbf{x}_T$ as

$$E(L(y_T, t)|\mathbf{x}_T) = E\left(\left[\alpha + \boldsymbol{\beta}' \mathbf{x}_T + u_T - t)^2\right]|\mathbf{x}_T\right).$$

Setting the derivatives of this expression to zero to obtain the minimizing value of $\mathbf{x}_T$ yields

$$\mathbf{0}=E\left[2\boldsymbol{\beta}\boldsymbol{\beta}'\mathbf{x}_T+2\boldsymbol{\beta}(\alpha-t)\right]$$

which solves to[2]

$$\mathbf{x}_T=(E\boldsymbol{\beta}\boldsymbol{\beta}')^{-1}E\boldsymbol{\beta}(t-\alpha).$$

Substituting this value of $\mathbf{x}_T$ into the expected loss, we obtain the minimum expected loss as

$$\begin{aligned} L_1 &= \min_{\mathbf{x}_T} E\left[L(y_T,t)|\mathbf{x}_T\right] \\ &= \sigma^2+E(\alpha-t)^2-E(t-\alpha)\boldsymbol{\beta}'(E\boldsymbol{\beta}\boldsymbol{\beta}')^{-1}E\boldsymbol{\beta}(t-\alpha). \qquad (6.23) \end{aligned}$$

This expression for the expected loss simplifies nicely in the case when our knowledge of α and $\boldsymbol{\beta}$ derives only from observation of the regression process previously. Letting $\mathbf{Y}$ be the T-dimensional vector of previous observations of the process and $\mathbf{X}$ be the matrix of observations of the explanatory variables, the posterior moments are

$$E(\alpha)=\bar{Y}-\bar{\mathbf{X}}'(\mathbf{X}'\mathbf{M}\mathbf{X})^{-1}\mathbf{X}'\mathbf{M}\mathbf{Y}=b_0$$

$$E(\boldsymbol{\beta})=(\mathbf{X}'\mathbf{M}\mathbf{X})^{-1}\mathbf{X}'\mathbf{M}\mathbf{Y}=\mathbf{b}$$

where $\mathbf{1}$ is a T-dimensional vector of ones and $\mathbf{M}=\mathbf{I}-\mathbf{1}(\mathbf{1}'\mathbf{1})^{-1}\mathbf{1}'$, $\bar{\mathbf{X}}=\mathbf{X}'\mathbf{1}/T$, $\bar{Y}=\mathbf{1}'\mathbf{Y}/T$. Also, the variance matrix can be written as

$$\begin{aligned} V\begin{bmatrix}\alpha\\ \boldsymbol{\beta}\end{bmatrix} &= \sigma^2\begin{bmatrix}\mathbf{1}'\mathbf{1} & \mathbf{1}'\mathbf{X}\\ \mathbf{X}'\mathbf{1} & \mathbf{X}'\mathbf{X}\end{bmatrix}^{-1} \\ &= \sigma^2\begin{bmatrix}T^{-1}(1+\mathbf{1}'\mathbf{X}(\mathbf{X}'\mathbf{M}\mathbf{X})^{-1}\bar{\mathbf{X}}) & -\bar{\mathbf{X}}'(\mathbf{X}'\mathbf{M}\mathbf{X})^{-1}\\ -(\mathbf{X}'\mathbf{M}\mathbf{X})^{-1}\bar{\mathbf{X}} & (\mathbf{X}'\mathbf{M}\mathbf{X})^{-1}\end{bmatrix}. \end{aligned}$$

Using the identity $\bar{Y}=b_0+\bar{\mathbf{X}}'\mathbf{b}$ we may write the regression process as

$$\begin{aligned} y_T-\bar{Y} &= \alpha+\boldsymbol{\beta}'\mathbf{x}_T-b_0-\mathbf{b}'\bar{\mathbf{X}}+u_T \\ &= (\alpha-b_0)+(\boldsymbol{\beta}-\mathbf{b})'\bar{\mathbf{X}}+\boldsymbol{\beta}'\left(\mathbf{x}_T-\bar{\mathbf{X}}\right)+u_T \\ &\equiv \alpha^*+\boldsymbol{\beta}'\mathbf{x}_T^*+u_T \qquad (6.24) \end{aligned}$$

[2]If $\mathbf{x}_T$ were a scalar and if α and β were known to equal $E\alpha$ and $E\beta$, then the instrument x_T^* becomes $x_T^c=(t-E\alpha)/E\beta$, which is called the certainty equivalence control rule. Assuming α and β independent, the optimal rule can be written in terms of the certainty equivalence rule as $x_T^*=(V(\beta)+E^2(\beta))^{-1}E(\beta)(t-E(\alpha))=(1+t_\beta^{-2})^{-1}x_T^c$, where $t_\beta^2=E^2(\beta)/V(\beta)$. Thus the optimal rule is more conservative than the certainty equivalence rule in the sense that the control variable is not turned on as far. The shrinkage factor $(1+t_\beta^{-2})^{-1}$ is a function of the uncertainty in β as measured by t_β^2.

where

$$\alpha^* = (\alpha - b_0) + (\boldsymbol{\beta} - \mathbf{b})'\overline{\mathbf{X}}$$

$$\mathbf{x}_T^* = \mathbf{x}_T - \overline{\mathbf{X}}.$$

Controlling y_T at t is equivalent to controlling $y_T^* = y_T - \overline{Y}$ at $t^* = t - \overline{Y}$, where y_T^* is generated by the process described in (6.24). The expected loss (6.23) attains a simple form since $E(\boldsymbol{\beta}\alpha^*) = \mathbf{0}$

$$\begin{aligned} L_1 &= E(\alpha^* - t^*)^2 - t^{*2}\mathbf{b}'\left(\mathbf{bb}' + \sigma^2(\mathbf{X}'\mathbf{MX})^{-1}\right)^{-1}\mathbf{b} + \sigma^2 \\ &= E\alpha^{*2} + t^{*2}\left(1 - \mathbf{b}'\left(\mathbf{bb}' + \sigma^2(\mathbf{X}'\mathbf{MX})^{-1}\right)^{-1}\mathbf{b}\right) + \sigma^2 \\ &= \frac{\sigma^2}{T} + \sigma^2 + \frac{t^{*2}}{1 + \mathbf{b}'\mathbf{X}'\mathbf{MXb}/\sigma^2} \\ &= \sigma^2(1 + T^{-1}) + \frac{t^{*2}}{1 + \chi^2} \end{aligned} \tag{6.25}$$

where we have used the inverse formula $(\mathbf{xx}' + \mathbf{A})^{-1} = \mathbf{A}^{-1} - \mathbf{A}^{-1}\mathbf{x}(1 + \mathbf{x}'\mathbf{A}^{-1}\mathbf{x})^{-1}\mathbf{x}'\mathbf{A}^{-1}$.

Thus the minimum expected loss is a quadratic function of the deviation of the target from the historical level of the process, $(t - \overline{Y})^2 = t^{*2}$. The coefficient multiplying this term is $(1 + \chi^2)^{-1}$ where χ^2 is the value of the chi-square statistic for testing $\boldsymbol{\beta} = \mathbf{0}$. A large χ^2 statistic thus implies that y_T can be pushed from its historical mean without incurring great expected loss. The part of the expected loss independent of the target is just the variance of y_T assuming that $\mathbf{x}_T$ is set to its historical level $\overline{\mathbf{X}}$,

$$V\left(y_T | \mathbf{x}_T = \overline{\mathbf{X}}\right) = E\alpha^{*2} + \sigma^2 = \sigma^2(1 + T^{-1}).$$

Next consider the possibility that none of the variables is controlled. To compute expected control error it is then necessary to "guess" what the explanatory variables will be. This means modeling the process that generates the explanatory variables. For our purposes it is enough to know the first two moments of $\mathbf{x}_T$, since the expected loss can be written as

$$\begin{aligned} E(y_T - t)^2 &= E(\alpha + \boldsymbol{\beta}'\mathbf{x}_T + u_T - t)^2 \\ &= \sigma^2 + E(\alpha - t)^2 + 2E(\alpha - t)\boldsymbol{\beta}'\mathbf{x}_T + E\boldsymbol{\beta}'\mathbf{x}_T\mathbf{x}_T'\boldsymbol{\beta}. \end{aligned}$$

Taking as we did in the previous section the assumption of an independent multivariate process for the explanatory variables, we have approximately $E\mathbf{x}_T = \overline{\mathbf{X}} = \mathbf{X}'\mathbf{1}/T$, $V\mathbf{x}_T = \mathbf{X}'\mathbf{MX}/T$, where $\mathbf{M} = \mathbf{I} - \mathbf{1}(\mathbf{1}'\mathbf{1})^{-1}\mathbf{1}'$. These together with the least-squares moments for α and $\boldsymbol{\beta}$ imply in the absence of any control

$$\begin{aligned}
E(y_T-t)^2 &= E\left(y_T-\bar{Y}+\bar{Y}-t\right)^2 \\
&= E\left(y_T-\bar{Y}\right)^2+(\bar{Y}-t)^2 \\
&= \sigma^2+E(\alpha^*+\boldsymbol{\beta}'\mathbf{x}_T^*)^2+t^{*2} \\
&= \sigma^2+E\alpha^{*2}+E\operatorname{tr}(\mathbf{x}_T^*\mathbf{x}_T^{*\prime}\boldsymbol{\beta}\boldsymbol{\beta}')+t^{*2} \\
&= \sigma^2+\frac{\sigma^2}{T}+\frac{k\sigma^2}{T}+\frac{\chi^2\sigma^2}{T}+t^{*2} \\
&= \sigma^2\left(1+\frac{k+1+\chi^2}{T}\right)+t^{*2}
\end{aligned} \tag{6.26}$$

where k is the dimensionality of $\boldsymbol{\beta}$ and χ^2 is the chi-square value for testing $\boldsymbol{\beta}=\mathbf{0}$, $\chi^2=\mathbf{b}'\mathbf{X}'\mathbf{M}\mathbf{X}\mathbf{b}/\sigma^2$.

Equation (6.26), the expected loss with no control, is to be contrasted with Equation (6.25), the expected loss with optimal control. Their difference,

$$\frac{\sigma^2(k+\chi^2)}{T}+\frac{t^{*2}\chi^2}{1+\chi^2}, \tag{6.27}$$

measures the incentive to use what is known about the determinants of y_T in a control exercise. If it is desired to assure that y_T attains its historical level $t=\bar{Y}$, the second term drops out ($t^*=0$). The percentage increase in expected losses due to decontrolling $\mathbf{x}_T$ is then

$$\frac{[k+\chi^2]/T}{1+T^{-1}}=\frac{k+\chi^2}{T+1}, \tag{6.28}$$

which attains its minimum of $k/(T+1)$ when $\chi^2=0$. We are thus led to compare χ^2+k with $T+1$ to determine if decontrolling $\mathbf{x}_T$ could be expected to increase expected losses substantially.

If, on the other hand, it is desired to control y_T at some value far from its historical mean, the second term in (6.27) dominates the expected loss. The percentage increase in expected loss would then be just χ^2, and we would want to compare χ^2 with the number one to decide if controlling $\mathbf{x}_T$ is worthwhile.

We have now examined the extreme cases in which either all or none of the elements of the vector $\mathbf{x}_T'=(\mathbf{z}_T',\mathbf{w}_T')$ is under control. The intermediate case when direct control affects only $\mathbf{z}_T$ is more difficult, since it requires a model describing how $\mathbf{z}_T$ affects the distribution of $\mathbf{w}_T$ or, more accurately, how $\mathbf{z}_T$ affects the conditional distribution $f(y_T|\mathbf{z}_T)$. Both the prediction problem of Section 6.1 and the control problem of this section are most elegantly solved by identifying the following minimal assumptions about the conditional moments of y_T:

$$E(y_T|\mathbf{z}_T)=Ey_T+\mathbf{g}'(\mathbf{z}_T-E\mathbf{z}_T)$$
$$V(y_T|\mathbf{z}_T)=a+(\mathbf{z}_T-E\mathbf{z}_T)'\mathbf{A}(\mathbf{z}-E\mathbf{z}_T). \tag{6.29}$$

These assumptions—that the mean is a linear function and that the variance is a quadratic function of $\mathbf{z}_T$—are implicit in the foregoing discussion. The prediction problem of minimizing $E(y_T-\hat{y})^2$ where $\hat{y}$ is a function of $\mathbf{z}_T$ is straightforwardly solved by letting $\hat{y}=E(y_T|\mathbf{z}_T)$ with resultant expected loss $E(y_T-\hat{y})^2=E[y_T-E(y_T|\mathbf{z}_T)]^2=EV(y_T|\mathbf{z}_T)=a+\operatorname{tr}\mathbf{A}V(\mathbf{z}_T)$.

The control problem is equally trivial. We wish to choose $\mathbf{z}_T$ to minimize

$$\min_{\mathbf{z}_T} E\left[(y_T-t)^2|\mathbf{z}_T\right]=\min_{\mathbf{z}_T} E\left(\left[y_T-E(y_T|\mathbf{z}_T)^2\right]|\mathbf{z}_T\right)+\left[t-E(y_T|\mathbf{z}_T)\right]^2$$
$$=\min_{\mathbf{z}_T} V(y_T|\mathbf{z}_T)+\left[t-E(y_T|\mathbf{z}_T)\right]^2.$$

With the foregoing moments the derivatives of this expression with respect to $\mathbf{z}_T$ are

$$2\mathbf{A}(\mathbf{z}_T-E\mathbf{z}_T)-2\mathbf{g}(t-Ey_T-\mathbf{g}'[\mathbf{z}_T-E\mathbf{z}_T])$$

which when set to zero yields the optimizing value of $\mathbf{z}_T$

$$\mathbf{z}_T^*=E\mathbf{z}_T+(\mathbf{A}+\mathbf{g}\mathbf{g}')^{-1}\mathbf{g}(t-Ey_T).$$

The resulting expected loss is

$$E\left[(y_T-t)^2|\mathbf{z}_T=\mathbf{z}_T^*\right]=a+(t-Ey_T)\mathbf{g}'(\mathbf{A}+\mathbf{g}\mathbf{g}')^{-1}\mathbf{A}(\mathbf{A}+\mathbf{g}\mathbf{g}')^{-1}\mathbf{g}(t-Ey_T)$$
$$+\left[t-Ey_T-\mathbf{g}'(\mathbf{A}+\mathbf{g}\mathbf{g}')^{-1}\mathbf{g}(t-Ey_T)\right]^2$$
$$=a+(t-Ey_T)^2\left[1-\mathbf{g}'(\mathbf{A}+\mathbf{g}\mathbf{g}')^{-1}\mathbf{g}\right]$$
$$=a+\frac{(t-Ey_T)^2}{1+\mathbf{g}'\mathbf{A}^{-1}\mathbf{g}} \tag{6.30}$$

Note that this is a quadratic function of $(t-Ey_T)$, the discrepancy between the target and the expected value of y_T.

To be specific, let us again work with the diffuse prior assumption. After some minor manipulation, we may obtain for the constants in the moments (6.29) the following

$$E(y_T)=\overline{Y},\qquad E(\mathbf{z}_T)=\overline{\mathbf{Z}}$$
$$\mathbf{g}=(\mathbf{Z}'\mathbf{M}\mathbf{Z})^{-1}(\mathbf{Z}'\mathbf{M}\mathbf{Y})\qquad\text{(the regression of }\mathbf{Z}\text{ on }\mathbf{Y})$$
$$\mathbf{A}=\sigma^2(\mathbf{Z}'\mathbf{M}\mathbf{Z})^{-1}$$
$$a=\sigma^2+\sigma^2(1+k_w+\chi_\delta^2)T^{-1}$$

where $\chi_\delta^2 = \mathbf{b}_w'(\mathbf{W'MW} - \mathbf{W'MZ})(\mathbf{Z'MZ})^{-1}\mathbf{Z'MW})\mathbf{b}_w/\sigma^2$. Notice that $\mathbf{g}'\mathbf{A}^{-1}\mathbf{g} = \chi^2_{\gamma|\delta=0}$, the chi-square statistic for testing $\gamma = \mathbf{0}$, given that $\boldsymbol{\delta} = \mathbf{0}$. The relevant expected loss is then

$$\min_{\mathbf{z}_T} E\left[(y_T - t)^2 | \mathbf{z}_T\right]$$

$$= \sigma^2(1 + T^{-1}) + \frac{\chi_\delta^2\sigma^2}{T} + \frac{k_w\sigma^2}{T} + \frac{t^{*2}}{1 + \chi^2_{\gamma|\delta=0}}.$$

For control around the historical mean, $t^{*2} = 0$, the percentage increase in the expected loss if $\mathbf{w}_T$ is not controlled is thus

$$\frac{T^{-1}(\chi_\delta^2 + k_w)}{1 + T^{-1}} = \frac{\chi_\delta^2 + k_w}{1 + T}$$

and we are led to compare $\chi_\delta^2 + k_w$ with $(1 + T)$ to determine if $\mathbf{w}_T$ can be decontrolled with little increase in expected error.

For control far from the historical mean the percentage increase in expected losses due to decontrolling $\mathbf{w}_T$ is

$$\frac{\left(1 + \chi^2_{\gamma|\delta=0}\right)^{-1} - \left(1 + \chi_\beta^2\right)^{-1}}{\left(1 + \chi_\beta^2\right)^{-1}}$$

$$= \frac{\chi_\beta^2 - \chi^2_{\gamma|\delta=0}}{1 + \chi^2_{\gamma|\delta=0}}$$

$$= \frac{\chi_\delta^2}{1 + \chi^2_{\gamma|\delta=0}}$$

and we are led to compare χ_δ^2 with $1 + \chi^2_{\gamma|\delta=0}$.

It need not be repeated that these results involve the unlikely assumption that in controlling $\mathbf{z}_T$ we do not alter the process that generates the explanatory variables (in the sense that the conditional distribution $f(\mathbf{w}_T|\mathbf{z}_T)$ is preserved). The assumption of known σ^2 can be altered by inserting its posterior mean where relevant. Mathematically more appropriately, we may treat the vector $(\mathbf{y}_T, \mathbf{x}_T')$ as coming from a multivariate normal distribution with unknown mean and unknown variance matrix. A conjugate prior for the uncertain parameters implies that the marginal distribution of $(\mathbf{y}_T, \mathbf{x}_T')$ is a multivariate Student distribution with means and variances satisfying (6.28) and (6.29). We leave to the tenacious reader the details of that calculation.

6.4 Conclusion

To conclude we may restate, first, the more important formal results of this chapter and then reiterate the more important informal lessons to be learned.

The results of this chapter listed in Table 6.2 make use of the assumptions listed in Table 6.1. If a variable y is generated by a linear regression process with explanatory variables **w** and **z**, if **w** and **z** themselves come from a multivariate normal process, and if priors for the various parameters are appropriately diffuse, then: (1) for a conditional prediction problem, we need not observe w, if the χ^2 statistic for testing whether **w** can be omitted (χ^2_δ) is small relative to $(T+k)$ where T is the number of observations and k is the dimension of $\mathbf{x}'=(\mathbf{w}',\mathbf{z}')$; (2) for control with a target equal to the historical mean of y, **w** may be decontrolled if $\chi^2_\delta + k_w$ is small relative to $(1+T)$, where k_w is the dimension of **w**; (3) for control far from the historical mean, **w** may be decontrolled if χ^2_δ is small relative to $1+\chi^{2}{}_{\gamma|\delta=0}$, one plus the χ^2 value for testing if **z** belongs in the equation given that **w** does not.

The principal caveat that has been repeated ad nauseam is that these results involve a very specific and often unwarranted assumption about the

Table 6.1

Assumptions for Simplification Analysis

Model	
	$y_t=\alpha+\mathbf{z}_t'\gamma+\mathbf{w}_t'\delta+u_t$
	$\equiv\alpha+\mathbf{x}_t'\beta+u_t, \qquad t=0,1,\ldots,T$
	$u_t\sim N(0,\sigma^2), \qquad \sigma^2$ known
	$\mathbf{x}_t\sim N(\mu,\Sigma)$
	μ,Σ,α,β have diffuse priors
Observations	
	$\mathbf{Y}(T\times 1),\mathbf{Z}(T\times k_z),\mathbf{W}(T\times k_w)$,
	$\mathbf{X}=(\mathbf{Z},\mathbf{W})(T\times k_x)$
Statistics	
	$\mathbf{b}=(\mathbf{X}'\mathbf{M}\mathbf{X})^{-1}\mathbf{X}'\mathbf{M}\mathbf{Y}, \qquad \mathbf{M}=\mathbf{I}-\mathbf{1}T^{-1}\mathbf{1}'$
	$\bar{Y}=\mathbf{1}'\mathbf{Y}/T,\bar{\mathbf{X}}=\mathbf{1}'\mathbf{X}/T$
	$b_0=\bar{\mathbf{Y}}-\bar{\mathbf{X}}\mathbf{b}$
	$\mathbf{g}=(\mathbf{Z}'\mathbf{M}\mathbf{Z})^{-1}\mathbf{Z}'\mathbf{M}\mathbf{Y}, \qquad \tilde{b}_0=\bar{Y}-\bar{Z}'\mathbf{g}$
	$\chi^2_\delta=\mathbf{b}_w'[\mathbf{W}'\mathbf{M}\mathbf{W}-\mathbf{W}'\mathbf{M}\mathbf{Z}(\mathbf{Z}'\mathbf{M}\mathbf{Z})^{-1}\mathbf{Z}'\mathbf{M}\mathbf{W}]\mathbf{b}_w/\sigma^2$
	$\chi^2_\beta=\mathbf{b}'\mathbf{X}'\mathbf{M}\mathbf{X}\mathbf{b}/\sigma^2$
	$\chi^2_{\gamma\mid\delta}=\mathbf{g}'\mathbf{Z}'\mathbf{M}\mathbf{Z}\mathbf{g}/\sigma^2$

Table 6.2

Simplification Analysis

	Decision Rules		Expected Losses	
	D_1: unconstrained	D_0: constrained	L_1	$L_0 - L_1$
Prediction	$\hat{y}_T = b_0 + \mathbf{z}'_T\mathbf{b}_z + \mathbf{w}'_T\mathbf{b}_w$	$\hat{y}_T = \tilde{b}_0 + \mathbf{z}'_T\mathbf{g}$	$\sigma^2(1 + T^{-1}k_x)$	$T^{-1}\sigma^2\chi^2_\delta$
Causally Constrained Prediction	same as above	$\hat{y}_T = b_0 + \overline{\mathbf{W}}'\mathbf{b}_w + \mathbf{z}'_T\mathbf{b}_z$	same as above	$T^{-1}\mathbf{b}'_w\mathbf{W}'\mathbf{MWb}_w$
Control $t^* = t - \ddagger\bar{y}$	$\mathbf{x}_T = \overline{\mathbf{X}} - (\mathbf{bb}' + \sigma^2[\mathbf{X}'\mathbf{MX}]^{-1})^{-1}\mathbf{b}t^*$	$\mathbf{z}_T = \overline{\mathbf{Z}} + (\mathbf{gg}' + \sigma^2[\mathbf{Z}'\mathbf{MZ}]^{-1})^{-1}\mathbf{g}t^*$	$\sigma^2(1 + T^{-1})$ $+ t^{*2}(1 + \chi^2_\beta)^{-1}$	$\sigma^2(\chi^2_\delta + k_w)T^{-1}$ $+ t^{*2}([1 + \chi^2_{\gamma\mid\delta}]^{-1} - [1 + \chi^2_\beta]^{-1})$.
Control $t = \overline{Y}$	$\mathbf{x}_T = \overline{\mathbf{X}}$	$\mathbf{z}_T = \overline{\mathbf{Z}}$	$\sigma^2(1 + T^{-1})$	$\sigma^2(\chi^2_\delta T^{-1} + k_w)$
Control (approx.) t^* very large	same as two lines above	same as two lines above	$t^{*2}(1 + \chi^2_\beta)^{-1}$	$t^{*2}([1 + \chi^2_{\gamma\mid\delta}]^{-1} - [(1 + \chi^2_\beta]^{-1})$

process that generates the explanatory variables. No simplification decisions can be made without either an implicit or explicit study of the behavior of the explanatory variables, and we hardly need say that it seems clear that an explicit study of their behavior is highly desirable.

For both prediction and control problems the effects of the excluded variables have been compensated for by adjustment of the included variables, and we have argued at length that it may be desirable not to adjust in this way. Semantically, adjustment is undesirable, because rather than asking if a variable can be neglected, in fact, we ask if it can be compensated for. Metaphysically, adjustment is undesirable, since it implies a causal link between the included and excluded variables. Statistically, the predictions and control that result may be quite inferior if anything happens to change the historical correlations between the variables. Control, especially, is likely to alter those correlations.

CHAPTER 7

PROXY SEARCHES

Variables that are used in theoretical statements often are not directly observable. When a researcher wishes to discriminate empirically among a set of theories, he must describe precisely the observable differences in the theories. In particular, hypothetical variables must be linked at least probabilistically to observable phenomena. In this situation there is a tendency among empirical workers to dismiss the apparent failure of a theory as merely a breakdown in the link between a hypothetical variable and an observed variable. One might report that "the low R^2 can be interpreted to mean that we have yet to find the appropriate proxy variable." If a theory is thereby completely protected from falsification, we might naturally ask if it is completely protected from verification as well. The goal of this chapter is to answer this question.

We would like to determine the extent to which it is possible to make inferences about theoretical parameters when the hypothetical variables are measured with error. As an extreme possibility, the theoretical parameters may be taken as known and observations used only to determine the accuracy of measurement. A less extreme procedure is to identify several possible ways of measuring the hypothetical variable and to select the proxy that yields the "best" results. This procedure is called a "proxy-variable search." At least when the number of proxy variables is finite, this method apparently spends part of the data evidence to pick a proxy variable but leaves part of the evidence to make inferences about the theoretical parameters.

To give an example of a proxy search in economics, the lifetime budget constraint makes it almost tautological to say that "permanent" consumption depends on "permanent" income. A great deal of empirical work has sought to determine the best way to measure these hypothetical constructs without directly questioning the underlying theory.

The basic statistical model we use is summarized by the equations

$$Y_t = \alpha + \chi_t \beta + z_t \gamma + u_t, \tag{7.1}$$

$$\mathbf{x}_t = \boldsymbol{\theta} + \boldsymbol{\delta} \chi_t + \boldsymbol{\varepsilon}_t, \tag{7.2}$$

$$\chi_t = \eta + \omega z_t + e_t. \tag{7.3}$$

Equation (7.1) describes the theoretical dependence of an observable variable Y_t on an observable variable z_t and an unobservable variable χ_t. Equation (7.2) describes the process that yields the vector of measurements $\mathbf{x}_t$ of the unobservable χ_t, and Equation (7.3) indicates the relevant part of the joint distribution of χ_t and z_t.

We could have treated both Y_t and z_t as unobservable as well. Measurement error in Y_t of the sort described by (7.2) has obvious consequences implied by rewriting (7.1) to allow for the measurement error

$$Y_t = \theta_Y + \delta_Y (\alpha + \chi_t \beta + z_t \gamma + u_t) + \varepsilon_{Yt},$$

where θ_Y, δ_Y, and ε_{Yt} describe the measurement error in Y_t. In such an equation, even if χ_t were observable, we could only determine the coefficients (β, γ) up to the scale factor δ_Y. If the measurement error amplification δ_Y is known, we can estimate β and γ by a regression of Y_t on $\delta_Y \chi_t$ and $\delta_Y z_t$. Conversely, if one of the regression coefficients, β or γ, is known, we may use the same estimates to solve for the other two parameters. Thus we have the choice between spending the evidence to estimate the theoretical coefficients or spending the evidence to estimate the measurement error. Intermediate cases would be implied by assigning proper prior distributions to δ_Y, β, and γ, which though conceptually straightforward seems to be mathematically intractable. A possibility not discussed here is multiple methods of measuring Y_t, each with different inherent biases. The multivariate process that results has proportionality restrictions across equations. For a discussion of maximum likelihood and other estimates see Jöreskog and Goldberger (1975).

Measurement error in the explanatory variables presents problems that are *not* conceptually straightforward. Except in Section 7.6 we have chosen to deal with the case in which one variable is subject to measurement error and the other is not. This is intended to approximate either the situation in which one variable is known to be measured with relative accuracy or the situation in which one is interested in the process conditional on the measurable variable z_t rather than its theoretical counterpart.

This raises the question of why we should be interested in the parameters of the theoretical process defined conditional on the unobservables $\mathbf{\chi}_t$ instead of the obviously estimable parameters of the observable process defined conditional only on the observables $\mathbf{x}_t$. For example, a conditional prediction problem in which Y_t is predicted as a function of other observable variables surely requires only the latter parameters. I think the answer to this question has to do with the problem of pooling information from different sources. "Pure" prior information may apply to the theoretical parameters, and even if interest centers on the other parameters it is necessary to know their relationships in order to make use of prior information. When prior information comes from a different experiment with a different measurement device, the two sources of information can be pooled only by identifying what they have in common—the theoretical parameters. Even if pooling is not the immediate goal, it would be terribly unwieldy to have a hundred sets of parameters, all corresponding to a different measuring device, and we thus hypothesize a single set of parameters implied by perfect measurement.

This chapter is designed to proceed step by step toward a discussion of the proxy variable model and its extensions involving many proxies. We first consider simple normal sampling models in which there are inadequate numbers of observations to estimate the unknown parameters. In the Bayesian framework this means that the likelihood function is not integrable and some form of prior density is necessary to compute a proper posterior distribution. The value of analyzing these simple models is that we are able to define concepts and explore peculiarities characteristic of the proxy-variable model in problems that have clearer intuitive resolutions. For example, we learn that maximum likelihood estimates are not defined for some of these models in the sense that at the apparent maximum likelihood point the likelihood function is peculiarly behaved.

In the second section we review the simple errors-in-variables model with $\gamma=0$, $\theta=0$, $\delta=1$, $\omega=0$. No statistical model in the econometric literature has led to so many confusing and erroneous statements as this simple errors-in-variables model. Textbooks tend to suggest that inferences are precluded by the lack of identification; yet intuitively, the observed correlation between $\mathbf{Y}$ and $\mathbf{x}$ seems to contain information about β. In fact, β may be bracketed on one side by the direct least-squares estimate and on the other by the reverse regression estimate equal to the inverse of the regression of $\mathbf{x}$ on $\mathbf{Y}$. The likelihood function attains its maximum along a line corresponding to these values of β and suitably chosen values of the other parameters.

A comprehensive summary of the literature on the errors-in-variables model is provided by Moran (1971). I have selected from that literature only the material that I regard to be most useful. For example, exact prior information about various parameters has been suggested to break the identification log jam, but since such precise prior information is unlikely to be available, I have not included a discussion of it.

The rest of this chapter deals with natural extensions of the simple errors-in-variables model. In the third section we explore the single-proxy-variable model implied by Equations (7.1) to (7.3), and in the fifth section the multiple-proxy-variable model. For each of these, it is possible to compute bounds for γ analogous to the errors-in-variables bound, actually computed by direct and reverse regressions. Without reference to prior information, it is not possible to say anything about β, however. An instrumental variables model is discussed in Section 7.4. This model is a hybrid of the errors-in-variables model, in that there is both a measurement of the unobservable variable and also an independent proxy.

In describing the likelihood function of all these models we must decide first whether Equation (7.3) is part of the "model" or merely a description of one's prior belief about χ_t. Such a distinction to a Bayesian is, of course, meaningless—the "model" is itself merely a description of one's prior belief. But Equation (7.3) with a normally distributed error e_t may be such a special and unlikely "prior" that it may be better to analyze how the posterior distribution is influenced by the model as defined by Equation (7.1) and (7.2) alone. As will be shown, this is not an easy task.

A confusing terminology has been developed to distinguish the model that makes use of (7.3) from the model that does not. The unobservables χ_t that affect the distributions of specific observations are called *incidental parameters*, and the others are called *structural parameters*. The *structural form* of the model makes use of (7.3) to integrate out the incidental parameters and only structural parameters remain. By default, the model consisting only of Equations (7.1) and (7.2) is called the *functional form*. In place of functional and structural form, I would suggest the words conditional and marginal.

One other bit of terminology is used here. Let the joint likelihood function of two parameters be $L(\theta_1,\theta_2)$. A *marginal likelihood* function makes use of a probability distribution for θ_2 to integrate θ_2 from the function: $L^m(\theta_1)=\int L(\theta_1,\theta_2)\, f(\theta_2)\,d\theta_2$. A *concentrated likelihood* function maximizes the joint likelihood for each value of θ_1: $L^c(\theta_1)=\max_{\theta_2} L(\theta_1,\theta_2)$. [Note that the value of θ_2 depends on θ_1, $\hat{\theta}_2=\hat{\theta}_2(\theta_1)$, and $L^c(\theta_1)= L(\theta_1,\hat{\theta}_2(\theta_1))$.]

The principal conclusion of this chapter is that "reverse" regression should be a part of standard operating procedure. The choice of "dependent" variable for least-squares regression has nothing to do with metaphysical notions of causality. The "left-hand side" variable should be the one measured most inaccurately.

Two important mathematically related questions remain unanswered: What confidence intervals should be used for these models? How do priors effect the decision whether to allocate the evidence to inference about theoretical parameters versus inference about measurement error parameters?

7.1 Inferences with Inadequate Observations

In this section we consider the inferential puzzles that arise in several simple models whose common feature is an excess of uncertain parameters relative to the number of observations. This discussion is intended to provide insights into the more complicated proxy variable problems to be discussed in later sections.

MODEL 1. $x \sim N(\chi, \sigma^2)$. Suppose that a single measurement x is made of some unknown quantity χ with a measurement device that generates normally distributed measurement errors with mean zero and variance σ^2. A single observation from a normal distribution does yield the point estimate x, but without resort to other information, it does not seem possible to say anything about how close x is likely to be to χ. Somewhat surprisingly, this proposition is not transparently obvious on examination of the likelihood function.

The likelihood function given the data x may be written

$$L(\chi, \sigma^2; x) \propto (\sigma^2)^{-1/2} \exp\left[-\frac{1}{2\sigma^2}(x-\chi)^2\right]. \tag{7.4}$$

We may attempt to maximize this function by setting the logarithmic derivatives to zero

$$0 = \frac{\partial \log L(\chi, \sigma^2; x)}{\partial \chi} = \frac{-(x-\chi)}{\sigma^2},$$

$$0 = \frac{\partial \log L(\chi, \sigma^2; x)}{\partial \sigma^2} = -\frac{1}{2\sigma^2} + \frac{(x-\chi)^2}{2\sigma^4}.$$

The apparent solution to these equations is

$$\chi = x, \qquad \sigma^2 = (\chi - x)^2 = 0,$$

which suggests, contrary to intuition, that the datum favors $\sigma^2 = 0$.

However, since the exponential term in the function (7.4) involves the ratio of two zeroes at the apparent maximum, a more careful examination of the function is in order. In fact, the function is not properly defined at $(x,0)$ since within any neighborhood of the point, the function takes on any positive value whatsoever. This can be demonstrated by identifying the lines on which the likelihood function is constant. Setting the logarithm of (7.4) to a constant c we obtain

$$-\frac{1}{2}\log\sigma^2-\frac{1}{2}\frac{(x-\chi)^2}{\sigma^2}=c$$

which can be rewritten as

$$(x-\chi)^2=-\sigma^2\log\sigma^2-2c\sigma^2.$$

As σ^2 goes to zero, the right-hand side of this expression goes to zero regardless of the value of c. Thus *every* line of constant likelihood goes through the point $(x,0)$. Such a point may be called an *essential singularity*.

Although the behavior of the likelihood function in the neighborhood of $(x,0)$ is peculiar, it remains to be demonstrated that this point is inferentially uninteresting, since there are points close to $(x,0)$ that are more "likely" than most other points in the parameter space. But, in treating a likelihood function like a probability distribution (a posterior with a diffuse prior), we are interested in the behavior of the function only to the extent that it generates volume under it. Thus, for example, a likelihood function that is uniform between zero and one is inferentially equivalent to a likelihood function that is the same except at the point .5, where it is enormous. Unless the prior allocates positive probability to the point .5, both functions imply the same posterior probabilities.

In an analogous fashion, unless the prior allocates positive probability to the line $\chi=x$, the point $\sigma^2=0$ will not be an "unusually" interesting value. Let us take as our prior for χ a normal distribution with mean m^* and finite variance σ_χ^2; then the density of x conditional on σ^2 but marginal with respect to χ is normal with mean m^* and variance $\sigma^2+\sigma_\chi^2$, and the (marginal) likelihood of σ^2 is

$$L^m(\sigma^2;x)\propto\left(\sigma^2+\sigma_\chi^2\right)^{-1/2}\exp\left[-\frac{(x-m^*)^2}{2\left(\sigma^2+\sigma_\chi^2\right)}\right] \tag{7.5}$$

The mode of this function occurs at σ^2 satisfying $\sigma^2+\sigma_\chi^2=(x-m^*)^2$, provided that $(x-m^*)>\sigma_\chi^2$. Otherwise, the mode is at the origin, and the marginal likelihood at $\sigma^2=0$ takes on the bounded value $(\sigma_\chi^2)^{-1/2}\exp[-\frac{1}{2}(x-m^*)^2/\sigma_\chi^2]$. In words, if the datum x is far from the prior mean m^* in units of the prior standard error, then the datum favors some value of σ^2 greater than zero. If x and m^* are close, the datum favors $\sigma^2=0$; but $\sigma^2=0$ is never a singular point of the marginal likelihood function, even though it is a singular point of the concentrated likelihood function $\max_\chi L(\chi,\sigma^2;x)$.

Similarly, unless the prior allocates positive probability to the line $\sigma^2=0$, the value $\chi=x$ will not be an "unusually" interesting value. Let us take as our prior for σ^{-2} a gamma distribution with location and scale parameters s^{*2} and ν^*. Then, by referring to properties of a normal-gamma distribution, the marginal likelihood becomes the Student function

$$L^m(\chi;x)\propto f_S(\chi|x,s^*,\nu^*)$$

$$\propto\left[\nu^*+\frac{(\chi-x)^2}{s^{*2}}\right]^{-(\nu^*+1)/2} \tag{7.6}$$

Although this is a function that has a maximum at $\chi=x$, for no value of $\nu^*>0$ is the function unbounded at that point.

The conclusion that is appropriate from this discussion is that the likelihood function is difficult to interpret by itself. We have suggested the slogan "the mapping is the message" to indicate that the evidential content of the data is a mapping from priors into posteriors. Sometimes that mapping is obvious from examination of the likelihood function alone. In this case it is not, and overt reference to prior information is required to determine the values of σ^2 that are favored by the datum.

In analogy with the errors-in-variable terminology, the statement that x is distributed normally with mean χ and variance σ^2 could be called the "functional" form, and the statement that x is distributed normally with mean zero and variance $\sigma^2+\sigma_\chi^2$ could be called the "structural" form. The preceding paragraph then concludes that inferences about σ^2 ought to be made in the context of the structural form of the model since the functional form may lead to erroneous conclusions.[1]

[1]It is interesting also to consider the consequences of diffuse prior distributions. The usual degenerate prior for σ^2 is implied by $\nu^*=0$, and the marginal likelihood (7.6) becomes $|\chi-x|^{-1}$, which has a nonintegrable singularity at $\chi=x$ and is, furthermore, nonintegrable in the tails of the distribution. The usual uniform prior for χ would imply a uniform marginal likelihood for σ^2 (integrate (7.4) with respect to χ) which is nonintegrable in the tail. Thus if you desire a proper posterior distribution for χ, you need a proper prior for σ^2, and if you desire a posterior distribution for σ^2 that is different from the prior, you need a proper prior for χ.

This brings up the question of whether the parameter σ^2 is identified or not. It is true that no two sets of parameters imply the same data density. For this reason, given a proper prior, it cannot be the case that the prior and posterior probabilities of any measurable subset will necessarily coincide. Nonetheless, if the prior for χ is uniform, the posterior and prior on σ^2 will necessarily coincide. It does seem intuitively clear that without some knowledge of χ, the datum contains no interpretable information about σ^2. We may wish to enlarge the definition of identification to include this circumstance. Kadane (1975) provides further discussion.

MODEL 2. $x_i \sim N(\chi, \sigma_i^2)$, $i=1,2$. A single observation from a normal distribution yields an estimate but no meaningful measure of the reliability of the estimate. Suppose next that two independent measurements of χ are made with different measuring devices that may have different variances. In this case, some information about the variances may be derived from the difference between the measurements.

The likelihood function for this model is

$$L(\chi, \sigma_1^2, \sigma_2^2; x_1, x_2) \propto (\sigma_1 \sigma_2)^{-1} \exp\left[-\frac{(\chi - x_1)^2}{2\sigma_1^2} - \frac{(\chi - x_2)^2}{2\sigma_2^2} \right]$$

The same pathology as above occurs on the lines

$$(\chi, \sigma_1^2, \sigma_2^2) = (x_1, 0, \sigma_2^2) \qquad \text{and} \qquad (\chi, \sigma_1^2, \sigma_2^2) = (x_2, \sigma_1^2, 0).$$

It is instructive in this case to concentrate the likelihood function by selecting the value $\chi(\sigma_1^2, \sigma_2^2)$ that maximizes the function for a given σ_1^2 and σ_2^2:

$$\chi(\sigma_1^2, \sigma_2^2) = (\sigma_1^{-2} + \sigma_2^{-2})^{-1}(\sigma_1^{-2} x_1 + \sigma_2^{-2} x_2).$$

The concentrated likelihood function is then

$$L(\chi(\sigma_1^2, \sigma_2^2), \sigma_1^2, \sigma_2^2; x_1, x_2) \propto \frac{1}{\sigma_1 \sigma_2} \exp\left[-\frac{(x_2 - x_1)^2}{2(\sigma_1^2 + \sigma_2^2)} \right].$$

In terms of the ratio $r^2 = \sigma_1^2 / \sigma_2^2$ and the sum $d^2 = \sigma_1^2 + \sigma_2^2$, this function can be written

$$L^c(r^2, d^2; x_1, x_2) \propto \frac{r^2 + 1}{r} d^{-2} \exp\left[-\frac{1}{2d^2} (x_2 - x_1)^2 \right].$$

On any ray out of the origin (r fixed) this function attains its maximum at $d^2 = (x_2 - x_1)^2$ independent of r. Holding d fixed the function attains a minimum at $r^2 = 1$ and is unbounded at $r^2 = 0$ and $r^2 = \infty$. The point $r^2 = 1$, $d^2 = (x_2 - x_1)^2$, $\chi = (x_1 + x_2)/2$ is thus a *saddle point* of the likelihood function.

There is one parameter that seems to be unambiguously "estimable" from these data; it is $\sigma_1^2 + \sigma_2^2$ with "estimate" $(x_1 - x_2)^2$. This is a reasonable estimate, since $x_1 - x_2$ is normal with mean zero and variance $\sigma_1^2 + \sigma_2^2$. The value of knowledge of $d^2 = \sigma_1^2 + \sigma_2^2$ is that it implies the constraints $\sigma_1^2 < d^2$, $\sigma_2^2 < d^2$, which in turn may be useful in constraining confidence intervals. To make this clear write the likelihood function in terms of $d^2 = \sigma_1^2 + \sigma_2^2$

and $r^2 = \sigma_1^2/\sigma_2^2$:

$$L(\chi, d^2, r^2; x_1 x_2) \propto f_N(\chi | m(r^2), v(r^2, d^2))$$
$$(d^2)^{-1/2} \exp\left[-\frac{1}{2d^2}(x_1 - x_2)^2\right] \tag{7.7}$$

where

$$m(r^2) = \frac{x_1 + r^2 x_2}{1 + r^2}$$

$$v(r^2, d^2) = \frac{d^2 r^2}{(1 + r^2)^2}.$$

In words, the likelihood function is the product of a conditional normal distribution on χ times a function independent of r^2. The second factor is just the likelihood function of d^2 given the observation of $x_1 - x_2$ distributed normally with mean zero and variance d^2.

If we use a diffuse prior for d^2, $f(d^2) \propto d^{-2}$, we may integrate this likelihood function to obtain a Student distribution on χ conditional on r^2,

$$L^m(\chi, r^2; x_2, x_2) \propto f_S\left(\chi | m(r^2), r^2(x_1 - x_2)^2/(1 + r^2)^2, 1\right).$$

Although a Student function with one degree of freedom has no moments, it is possible to compute shortest size-α confidence intervals, which will be located at $m(r^2)$ and have length proportional to the square root of $(x_1 - x_2)^2 r^2/(1 + r^2)^2$.

A marginal posterior distribution on χ requires us to integrate this function with respect to a prior distribution on r^2. Personally and/or publicly acceptable priors for r^2 are unlikely to be available. An alternative is to describe the mapping of one-point priors into posteriors, that is, to compute posterior intervals for χ conditional on r^2 for all values of r^2. Referring to the formulas above, as we vary r^2 from zero to infinity, we vary the location of the interval from x_1 to x_2, and we vary the length of the interval from zero (see Fig. 7.1) to a maximum at $r^2 = 1$ and back to zero. Thus although it is impossible to compute precise posterior credible intervals, it is possible to give a very reasonable *class* of posterior intervals. Incidentally, the union of these intervals contains χ with probability in excess of α regardless of the prior for r^2.

MODEL 3. $x_{it} \sim N(\chi_t, \sigma^2)$; $t = 1, \ldots, T$; $i = 1, 2$. Whenever possible, it is desirable actually to compute the marginal posterior distribution of the parameter of interest. A model due to Neyman and Scott (1951) provides a dramatic demonstration of the need to marginalize a likelihood function

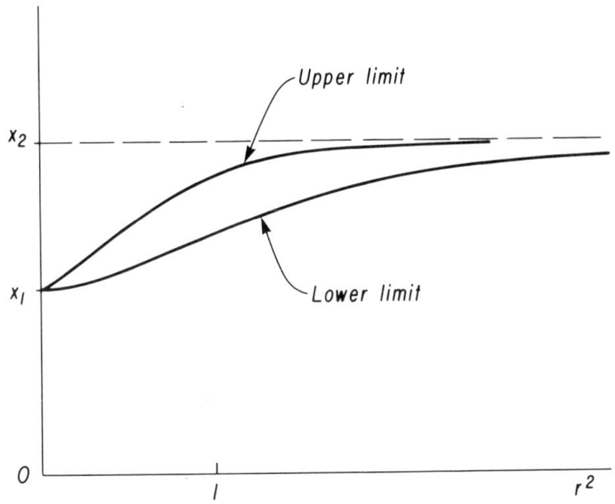

Fig. 7.1 Confidence intervals for χ; $x_1 \sim N(\chi, \sigma_1^2)$, $x_2 \sim N(\chi, \sigma_2^2)$, $r^2 = \sigma_1^2/\sigma_2^2$.

and/or the posterior distribution. Suppose for each of T quantities, $\chi_t (t = 1, \ldots, T)$, we obtain two measurements x_{1t} and x_{2t} distributed independently with mean χ_t and variance σ^2. The likelihood function may then be written as

$$L(\chi_1, \ldots, \chi_t, \sigma^2; \mathbf{X}) \propto (\sigma^{-2})^T \exp\left[-\frac{1}{2\sigma^2} \sum_{t=1}^{T} \left([x_{1t} - \chi_t]^2 + [x_{2t} - \chi_t]^2 \right) \right]$$

$$\propto (\sigma^{-2})^T \exp\left[-\frac{1}{\sigma^2} \sum_{t=1}^{T} \left\{ (\chi_t - \bar{x}_t)^2 + \frac{1}{4}(x_{1t} - x_{2t})^2 \right\} \right] \tag{7.8}$$

where $\mathbf{X}$ is the $t \times 2$ matrix of observations and $\bar{x}_t = (x_{1t} + x_{2t})/2$. Maximizing this function with respect to the parameters yields the estimates $\chi_t = \bar{x}_t$ and

$$\hat{\sigma}^2 = \sum \frac{(x_{1t} - x_{2t})^2}{4T}.$$

Curiously enough, however, the expected value of $(x_{1t} - x_{2t})^2$ is $2\sigma^2$, and it is easy to show that the estimate just reported converges in probability to $2\sigma^2/4 = \sigma^2/2$ as $T \to \infty$.

A marginal likelihood computed by integrating (7.8) with respect to a diffuse distribution for $\chi_1, \chi_2, \ldots, \chi_T$ is easily found to be

$$L^m(\sigma^2; x_1, x_2, \ldots, x_T) \propto (\sigma^{-2})^{T/2} \exp\left[-\frac{1}{4\sigma^2} \sum_t (x_{1t} - x_{2t})^2 \right]$$

which has a mode at $\Sigma(x_{1t}-x_{2t})^2/2T$ which does converge in probability to σ^2.

This peculiar example is not easily made sense of. The following is an attempt. The likelihood function implied by two observations from a normal distribution is

$$L(\chi,\sigma^2|x_1,x_2)\propto(\sigma^2)^{-1}\exp\left[-\frac{1}{\sigma^2}(\chi-\bar{x})^2-\frac{1}{4\sigma^2}(x_1-x_2)^2\right]$$

with a mode at $(\chi,\sigma^2)=(\bar{x},(x_1-x_2)^2/2)$. The contours are as depicted in Figure 7.2. Although the function attains its maximum at $\sigma^2=(x_1-x_2)^2/2$, most of the mass is located at values above the maximum. In fact, the usual degrees-of-freedom adjustment would imply the estimate $(x_1-x_2)^2$, thereby implicitly allowing for the relative "thinness" of the likelihood hill at the maximum. If the number of observations is increased, holding χ fixed, this peculiar shape of the likelihood hill corrects itself, and the joint maximum appropriately indicates the point favored by the data. If, as in the example being discussed here, the mean changes, one never gets to the large-sample situation, and the likelihood hill, in fact, becomes increasingly thinner at the maximum. Thus, loosely speaking, in the limit the function approximates its maximum on a set of measure zero.

By the way, this example has had a significant impact on this author's thinking. I used to be relatively uninterested in the difference between

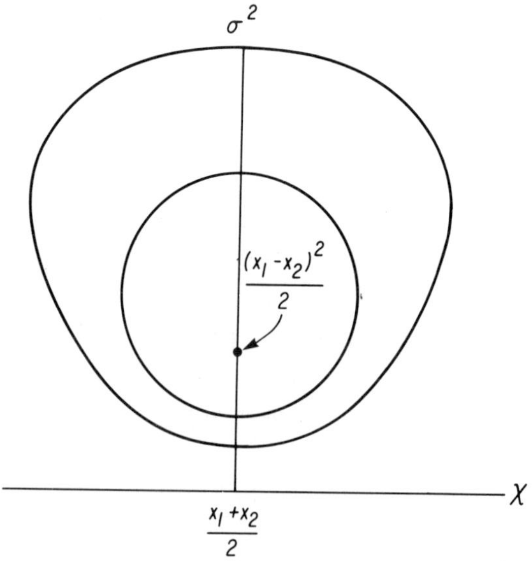

Fig. 7.2 Likelihood contours: $x_i \sim N(\chi,\sigma^2)$, $i=1,2$.

marginal and joint modes, and I would have flippantly asked, "Who chose the (prior) probability distribution that was used to marginalize the likelihood function?" Although that question remains, it is now clear that at least for some problems, marginalization seems essential.

MODEL 4. $x_{it} \sim N(\theta_i + \delta_i \chi_t, \sigma_i^2)$; $t=1,\ldots,T$; $i=1,\ldots,N$. Certain measurement instruments may generate systematic biases. Is it possible by observation of the measurements to determine the extent of the bias? Probably not. In vector notation, a model with biased measurement is

$$\mathbf{x}_t = \boldsymbol{\theta} + \boldsymbol{\delta}\chi_t + \boldsymbol{\varepsilon}_t, \qquad t=1,\ldots,T,$$

where $\mathbf{x}_t$ is a vector of N measurements of the quantity χ_t, $\boldsymbol{\theta}$ and $\boldsymbol{\delta}$ are $N \times 1$ vectors describing the bias, and $\boldsymbol{\varepsilon}_t$ has a multivariate normal distribution with mean vector zero and variance matrix $\mathbf{D} = \operatorname{diag}\{\sigma_1^2, \sigma_2^2, \ldots, \sigma_N^2\}$. We observe in passing that this is the factor analysis model, with a single factor χ_t. (See Harmon, 1960; Jöreskog, 1963; and Lawley and Maxwell, 1964.) As usual, the joint likelihood function has essential singularities, implied by the restrictions that the ith measurement (any i) is errorless, $\sigma_i^2 = 0$ and $x_{it} = \theta_i + \delta_i \chi_t$. Maximizing the rest of the likelihood function subject to this restriction requires us merely to regress all of the other measurements on $\mathbf{x}_i$, thereby estimating the equations $x_{jt} = \theta_j + \delta_j \chi_t = \theta_j + \delta_j (x_{it} - \theta_i)/\delta_i$.

There is also an identification problem, since by altering the scale of the unobservable χ_t with an offsetting change in the scale of $\boldsymbol{\delta}$ we do not alter the distribution of the observables $\mathbf{x}$. Thus $\boldsymbol{\delta}$ can be determined only up to a scale factor: equivalently, we may only conclude that some instruments give relatively high readings and others relatively low readings, but it is not possible to say which is right, if any.

The structural form of this model assumes that the quantities χ_t come from a normal distribution with mean $\bar{\chi}$ and variance σ_χ^2. The vector $\mathbf{x}_t$ is thereby assumed to be drawn from a multivariate normal distribution with mean $\boldsymbol{\theta} + \boldsymbol{\delta}\bar{\chi}$ and variance $\boldsymbol{\delta}\sigma_\chi^2\boldsymbol{\delta}' + \mathbf{D}$. As in the structural model $\boldsymbol{\delta}$ can be multiplied by any constant, and with a suitable rescaling of σ_χ^2 and $\bar{\chi}$, the distribution of the observables is unchanged. We might as well proceed with the assumption that $\sigma_\chi^2 = 1$ and $\bar{\chi} = 0$, keeping in mind that $\boldsymbol{\delta}$ is determinable only up to a scale factor. Maximum likelihood estimation then requires maximization of

$$L(\boldsymbol{\delta}, \mathbf{D}, \boldsymbol{\theta}; \mathbf{x}_1, \ldots, \mathbf{x}_T) \propto |\boldsymbol{\delta}\boldsymbol{\delta}' + \mathbf{D}|^{-T/2}$$

$$\times \exp\left[-\tfrac{1}{2} \sum_t (\mathbf{x}_t - \boldsymbol{\theta})'(\boldsymbol{\delta}\boldsymbol{\delta}' + \mathbf{D})^{-1}(\mathbf{x}_t - \boldsymbol{\theta}) \right].$$

Algorithms for the maximization of such a function are discussed in the factor-analysis literature. The work of Jöreskog (1966) especially should be mentioned here.

7.2 The Errors-in-Variables Problem

The errors-in-variables problem has the feature of all the models just discussed: the ratio of observations to parameters is unhappily small. We must accordingly be alert to the possibility that direct examination of the likelihood surface may be misleading. We nonetheless attempt to explore the likelihood surface directly, a task which may be easier now that we are armed with the knowledge of its potential pitfalls. In this section, and in the sections to follow, we first explore the joint likelihood function of the functional form of the model in which the incidental parameters are treated like the other parameters. The salient feature of this function is its essential singularities. We then explore the (marginal) likelihood function of the structural form of the model in which the incidental parameters are integrated out of the function with respect to a probability function with possibly uncertain hyperparameters.

The errors-in-variables model is mathematically equivalent to Model 4 just discussed with $N=2$, but with a known normalization. In its simplest form we may write the model as

$$Y_t = \beta \chi_t + u_t \tag{7.9}$$

$$x_t = \chi_t + \varepsilon_t \tag{7.10}$$

to indicate that an observation Y_t is linked by a linear process to an unobservable χ_t, which is measured by x_t subject to measurement error ε_t. In effect, we have two measurements of χ_t, one unbiased and the other subject to amplification β.

The least-squares estimate of β suffers from the errors-in-variables "attenuation"—it is biased toward the origin. The bias does not disappear as sample size increases. This sampling property has its counterpart in the likelihood function which in vector notation is

$$L(\chi, \beta, \sigma_u^2, \sigma_\varepsilon^2; \mathbf{Y}, \mathbf{x}) \propto (\sigma_u^2)^{-T/2} \exp\left[-\frac{1}{2\sigma_u^2} (\mathbf{Y} - \chi\beta)'(\mathbf{Y} - \chi\beta) \right]$$

$$(\sigma_\varepsilon^2)^{-T/2} \exp\left[-\frac{1}{2\sigma_\varepsilon^2} (\chi - \mathbf{x})'(\chi - \mathbf{x}) \right].$$

This function, as shown by Solari (1969), has essential singularities at the points satisfying $\sigma_\varepsilon^2 = 0$ and $\chi = \mathbf{x}$ or $\sigma_u^2 = 0$ and $\mathbf{Y} = \chi\beta$. Minimizing the nonpathological part of the function, subject to these two pairs of constraints, yields, respectively, the "two regressions" $\hat{\beta}^D = (\mathbf{x}'\mathbf{x})^{-1}\mathbf{x}'\mathbf{Y}$ and

$\hat{\beta}^R=[(\mathbf{Y}'\mathbf{Y})^{-1}\mathbf{Y}'\mathbf{x}]^{-1}$, the first being the simple regression of $\mathbf{Y}$ on $\mathbf{x}$ and the second being the inverse of the regression of $\mathbf{x}$ on $\mathbf{Y}$. These are the extreme estimates of β analogous to the estimates x_1 and x_2 of χ with $x_1 \sim N(\chi, \sigma_1^2)$ and $x_2 \sim N(\chi, \sigma_2^2)$. These are not maximum likelihood estimates, since the function is not defined at these points. As mentioned before, there is a tendency for the simple regression $\hat{\beta}^D$ to underestimate β ($\hat{\beta}^D$ is "attenuated"), and to compensate for this the "reverse regression" $\hat{\beta}^R$ yields an estimate larger in absolute value than the direct regression. Using the formula $R^2=(\mathbf{Y}'\mathbf{x})^2/\mathbf{x}'\mathbf{x}\mathbf{Y}'\mathbf{Y}$ we may derive the result

$$R^2\hat{\beta}^R=\hat{\beta}^D \tag{7.11}$$

with $0 \leqslant R^2 \leqslant 1$. Furthermore, it may be shown that the reverse regression estimate tends to overestimate β, and the two regressions therefore consistently bound β.

As in the two-observations-per-mean model (Model 2), the stable point of the likelihood function is a saddle point (Solari, 1969). The estimate of β at the saddle point is a geometric average of the two regressions, and $\boldsymbol{\chi}$ is a simple weighted average of $\mathbf{x}$ and $\mathbf{Y}/\beta$, completely analogous to the saddle point discussed previously, in which the estimate of χ was a simple compromise between x_1 and x_2. Another feature of this point is the curious relationship among the estimates $\hat{\beta}^2=\hat{\sigma}_u^2/\hat{\sigma}_\varepsilon^2$. This phenomenon has generated a great deal of confusion in the literature. It is not especially surprising if we were to write the process like Model 2 as

$$x_{1t}=\frac{Y_t}{\beta}=\chi_t+\frac{u_t}{\beta}$$

$$x_{2t}=x_t=\chi_t+\varepsilon_t,$$

and observe that because of the symmetries, it is not surprising to find a saddle point at $\operatorname{Var} x_1=\operatorname{Var} x_2$, that is, at $\sigma_u^2/\beta^2=\sigma_\varepsilon^2$. In the former problem the constraint $\sigma_1^2=\sigma_2^2$ seems reasonable by an appeal to symmetry. No such symmetry exists between Y_t/β and x_t, and it is unlikely that the saddle point will be a point of special interest between the two extremes.

A (prior) distribution for the unobservables χ_t may allow us to make clearer inferences. In the structural form of this model, we assume that each χ_t was drawn from the same normal population $\chi_t \sim N(\bar{\chi}, \sigma_\chi^2)$. This amounts to assuming that the vector (Y_t, x_t) comes from a bivariate normal distribution with mean $(\beta\bar{\chi}, \bar{\chi})$ and variance matrix

$$\boldsymbol{\Sigma}=\begin{bmatrix} \sigma_u^2 & 0 \\ 0 & \sigma_\varepsilon^2 \end{bmatrix}+\begin{bmatrix} \beta \\ 1 \end{bmatrix}\sigma_\chi^2[\beta, 1]$$

$$=\begin{bmatrix} \sigma_u^2+\beta^2\sigma_\chi^2 & \beta\sigma_\chi^2 \\ \beta\sigma_\chi^2 & \sigma_\chi^2+\sigma_\varepsilon^2 \end{bmatrix}. \tag{7.12a}$$

The fact that the mean of Y_t is $\beta\bar{\chi}$ and the mean of x_t is $\bar{\chi}$ suggests an estimator for β: $\bar{Y}/\bar{x}$, the ratio of the observed means. But if the relationship between Y_t and χ_t included the constant α, $Y_t = \alpha + \beta\chi_t$, the mean vector would be $(\alpha + \beta\bar{\chi}, \bar{\chi})$, which would not imply an estimate of β. In fact, we could only solve for α in terms of β, $\alpha = \bar{Y} - \beta\bar{\chi} = \bar{Y} - \beta\bar{x}$. Except in the unlikely event that we know α or have a proper prior for it, inferences about β depend entirely on the covariance matrix of the process and not on the location vector. Henceforth, we proceed as if there were an uncertain α in the theoretical relationship. The likelihood functions reported below have been "concentrated" by setting $\bar{\chi}$ to x and α to $\bar{Y} - \beta_x$. This has the effect of removing the means of the observations, $x_t - \bar{x}$ and $Y_t - \bar{Y}$.

Observe in passing that the conditional distribution of Y_t given x_t is normal with mean $x_t\beta\sigma_\chi^2/(\sigma_\chi^2 + \sigma_\varepsilon^2)$. This suggests that in regressing $\mathbf{Y}$ on $\mathbf{x}$ we obtain the "shrunken" coefficient: $\beta\sigma_\chi^2/(\sigma_\chi^2 + \sigma_\varepsilon^2)$. Conversely, the regression of $\mathbf{x}$ on $\mathbf{Y}$ yields the coefficient $\beta\sigma_\chi^2/(\sigma_u^2 + \beta^2\sigma_\chi^2) = \beta^{-1}(\beta^2\sigma_\chi^2/(\sigma_u^2 + \beta^2\sigma_\chi^2))$, β^{-1} times a factor less than one. The first shrinkage factor is close to one for σ_ε^2 small relative to σ_χ^2, whereas the second is close to one for σ_u^2 small relative to $\beta^2\sigma_\chi^2$. This should generate some further understanding of the content of the "two regressions."

The likelihood function of the structural form, assuming normality and with variables defined around their means, is

$$L(\beta, \sigma_u^2, \sigma_\chi^2, \sigma_\varepsilon^2; \mathbf{Y}, \mathbf{x}) \propto |\mathbf{\Sigma}|^{-T/2} \exp\left[-\tfrac{1}{2} \sum_t (Y_t, x_t) \mathbf{\Sigma}^{-1} (Y_t, x_t)' \right]$$

$$= |\mathbf{\Sigma}|^{-T/2} \exp\left[-\tfrac{1}{2} \operatorname{tr} \mathbf{\Sigma}^{-1} \mathbf{S} \right] \qquad (7.12b)$$

where

$$\mathbf{S} = \begin{bmatrix} \mathbf{Y}'\mathbf{Y} & \mathbf{Y}'\mathbf{x} \\ \mathbf{x}'\mathbf{Y} & \mathbf{x}'\mathbf{x} \end{bmatrix}.$$

The following result describes the maximum likelihood region.

THEOREM 7.1 (ERRORS-IN-VARIABLES BOUND). *The likelihood function (7.12b) with $\mathbf{\Sigma}$ defined by (7.12a) attains its maximum at any value of β between the direct regression estimate $\hat{\beta}^D = (\mathbf{x}'\mathbf{x})^{-1}\mathbf{x}'\mathbf{Y}$ and the reverse regression estimate $\hat{\beta}^R = (\mathbf{x}'\mathbf{Y})^{-1}\mathbf{Y}'\mathbf{Y}$.*

Proof: If $\mathbf{\Sigma}$ is unconstrained, this function attains its maximum of $|\mathbf{S}/T|^{-T/2}\exp[-T]$ at $\mathbf{\Sigma} = \mathbf{S}/T$. This is a feasible value of $\mathbf{\Sigma}$ if we can find values of σ_u^2, σ_χ^2, σ_ε^2, and β such that $\mathbf{\Sigma}(\sigma_u^2, \sigma_\chi^2, \sigma_\varepsilon^2, \beta) = \mathbf{S}/T$. We can, in fact, do this by selecting a value of β, and solving for the other parameters as

$$\sigma_\chi^2 = \frac{\mathbf{x}'\mathbf{Y}}{T\beta}$$

$$\sigma_u^2 = \frac{\mathbf{Y}'\mathbf{Y}}{T} - \beta^2\sigma_\chi^2 = \frac{\mathbf{Y}'\mathbf{Y} - \mathbf{x}'\mathbf{Y}\beta}{T}$$

$$\sigma_\varepsilon^2 = \frac{\mathbf{x}'\mathbf{x}}{T} - \sigma_\chi^2 = \frac{\mathbf{x}'\mathbf{x} - \mathbf{x}'\mathbf{Y}\beta^{-1}}{T}.$$

The constraints $\sigma_u^2 \geqslant 0$ and $\sigma_\varepsilon^2 \geqslant 0$ imply that not all values of β can be associated with the maximum of the likelihood function. These constraints imply $\mathbf{Y}'\mathbf{Y} \geqslant \mathbf{x}'\mathbf{Y}\beta$ and $\mathbf{x}'\mathbf{x} \geqslant \mathbf{x}'\mathbf{Y}\beta^{-1}$, that is,

$$\frac{\beta}{\mathbf{Y}'\mathbf{Y}(\mathbf{x}'\mathbf{Y})^{-1}} = \frac{\beta}{\hat{\beta}^R} \leqslant 1$$

$$\frac{(\mathbf{x}'\mathbf{x})^{-1}\mathbf{x}'\mathbf{Y}}{\beta} = \frac{\hat{\beta}^D}{\beta} \leqslant 1.$$

In words, any value of β between the direct regression $\hat{\beta}^D$ and the reverse regression $\hat{\beta}^R$ with suitably chosen values of the other parameters imply a likelihood value equal to the maximum.

The general features of the concentrated likelihood function

$$L^c(\beta; \mathbf{Y}, \mathbf{x}) = \max_{\sigma_u^2 \geqslant 0,\, \sigma_\varepsilon^2 \geqslant 0,\, \sigma_\chi^2 \geqslant 0} L(\beta, \sigma_u^2, \sigma_\chi^2, \sigma_\varepsilon^2; \mathbf{Y}, \mathbf{x})$$

are difficult to compute. From the preceding discussion we know it has a plateau of height $|\mathbf{S}/T|^{-T/2}\exp[-T]$ between $\hat{\beta}^D$ and $\hat{\beta}^R$. It is also easy to derive

$$L^c(0; \mathbf{Y}, \mathbf{x}) = [\mathbf{x}'\mathbf{x}\mathbf{Y}'\mathbf{Y}/T^2]^{-T/2}\exp[-T]$$

(by observing that given $\beta = 0$, $\mathbf{x}$ and $\mathbf{Y}$ are independent with different variances). Similarly,

$$L^c(\beta; \mathbf{Y}, \mathbf{x}) \geqslant \max_{\sigma_u^2 \geqslant 0,\, \sigma_\varepsilon^2 \geqslant 0,\, \sigma_\chi^2 = 0} L(\beta, \sigma_u^2, \sigma_\chi^2, \sigma_\varepsilon^2; \mathbf{Y}, \mathbf{x})$$
$$= L^c(0; \mathbf{Y}, \mathbf{x}).$$

$$L^c(\beta; Y, \mathbf{x}) \geqslant \max_{\sigma_u^2 \geqslant 0,\, \sigma_\varepsilon^2 = 0,\, \sigma_\chi^2 \geqslant 0} L(\beta, \sigma_u^2, \sigma_\chi^2, \sigma_\varepsilon^2; \mathbf{Y}, \mathbf{x})$$
$$= [(\mathbf{Y} - \mathbf{x}\beta)'(\mathbf{Y} - \mathbf{x}\beta)]^{-T/2}[\mathbf{x}'\mathbf{x}]^{-T/2}T^{T/2}\exp[-T],$$

$$L^c(\beta; \mathbf{Y}, \mathbf{x}) \geqslant \max_{\sigma_u^2 = 0,\, \sigma_\varepsilon^2 \geqslant 0,\, \sigma_\chi^2 \geqslant 0} L(\beta, \sigma_u^2, \sigma_\chi^2, \sigma_\varepsilon^2; \mathbf{Y}, \mathbf{x})$$
$$= [(\mathbf{x} - \mathbf{Y}\beta^{-1})'(\mathbf{x} - \mathbf{Y}\beta^{-1})]^{-T/2}[\mathbf{Y}'\mathbf{Y}]^{-T/2}T^{T/2}\exp[-T].$$

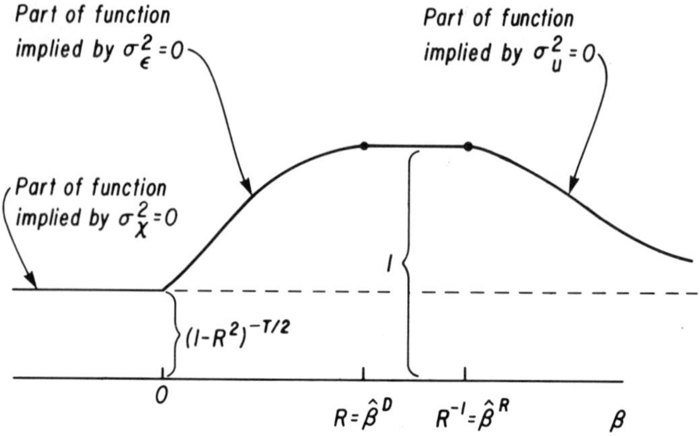

Fig. 7.3 Concentrated likelihood function: the errors-in-variables model.

Note that this last function (whose leading term is associated with the reverse regression) in the limit tends to the value $L^c(0;\mathbf{Y},\mathbf{x})$.

Normalizing so that the likelihood value at the maximum is one and so that $\mathbf{Y}'\mathbf{Y}=\mathbf{x}'\mathbf{x}=1$, and by observing that $|\mathbf{S}|=(\mathbf{x}'\mathbf{x})(\mathbf{Y}'\mathbf{Y})(1-R^2)$, where R^2 is the sample correlation coefficient of $\mathbf{Y}$ and $\mathbf{x}$, we obtain the lower bound for the concentrated likelihood function depicted in Figure 7.3. Note that in addition to having a plateau, this function is uniform for large absolute values of β. To the extent that the concentrated likelihood function is an appropriate one-dimensional description of the evidence about β, we are led to conclude that it is impossible to distinguish one large value of β from another, although values of β with the same sign as $\hat{\beta}^D$ are favored over values with the opposite sign. Decisions that depend on the tails of the distribution thus necessarily are heavily dependent on prior information about β or about the variances.[2]

An approximate posterior distribution for this model has been derived by Lindley and El-Sayyad (1968). They use the ignorance distribution for σ_ε^2, proportional to σ_ε^{-2}, and show that in a large sample β, conditional on the variance ratio $\lambda=\sigma_u^2/\sigma_\varepsilon^2$, is normal with mean β_λ, the root of the quadratic equation (with sign of $\mathbf{x}'\mathbf{Y}$)

$$\beta^2+t\beta-\lambda=0, \qquad t=\frac{\lambda\mathbf{x}'\mathbf{x}-\mathbf{Y}'\mathbf{Y}}{\mathbf{x}'\mathbf{Y}},$$

[2]Note also that the likelihood ratio of $\beta=0$ versus $\beta\neq 0$ is unaffected by the measurement error in χ. It is erroneous to conclude from this fact that a test of the hypotheses $\beta=0$ versus $\beta\neq 0$ is uninfluenced by the measurement error, since, as we have argued in Chapter 4, the relationship between a likelihood ratio and an appropriate hypothesis test is indirect. Nonetheless, it is comforting to know at least that the t value of the regression coefficient has some effect on the measures of dispersion that ought to apply to β.

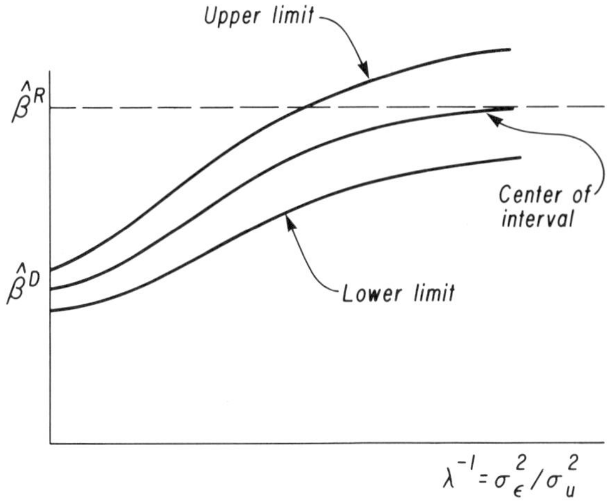

Fig. 7.4 Confidence intervals: the errors-in-variables problem.

and with variance

$$\frac{\beta_\lambda^2\left(\mathbf{x}'\mathbf{x}\mathbf{Y}'\mathbf{Y}-(\mathbf{x}'\mathbf{Y})^2\right)}{T(\mathbf{x}'\mathbf{Y})^2}.$$

There is little sample information about λ, and in the absence of legitimate prior information, it makes sense to consider the class of (approximate) posterior distributions for $\lambda>0$, as we did for Model 2 in the previous section.

The (nonzero) root of the quadratic equation for $\lambda=0$ is $\beta_0=\mathbf{Y}'\mathbf{Y}/\mathbf{x}'\mathbf{Y}$, the reverse regression estimate of β. As λ converges to infinity, the (appropriate) root of the quadratic equation converges to $\beta_\infty=\mathbf{x}'\mathbf{Y}/\mathbf{x}'\mathbf{x}$, the direct regression estimate of β. The standard error is always proportional to β_λ and at $\lambda\to\infty$ the variance takes on the value

$$\left(\frac{\mathbf{x}'\mathbf{Y}}{\mathbf{x}'\mathbf{x}}\right)^2\frac{\mathbf{x}'\mathbf{x}\mathbf{Y}'\mathbf{Y}-[\mathbf{x}'\mathbf{Y}]^2}{T(\mathbf{x}'\mathbf{Y})^2}=\frac{\mathbf{x}'\mathbf{x}\mathbf{Y}'\mathbf{Y}-(\mathbf{x}'\mathbf{Y})^2}{T(\mathbf{x}'\mathbf{x})^2},$$

which happens to be just the usual least-squares estimate of the variance of $\hat{\beta}$. Figure 7.4 indicates this class of intervals which vary from the usual interval located at the direct least-squares $\hat{\beta}^D$ to an interval located at the reverse regression estimate with length expanded by the factor $|\hat{\beta}^R/\hat{\beta}^D|$.

7.3. The Proxy-Variable Problem

A variable x is called a *proxy* for another variable χ if x and χ are positively correlated. Within the normal family we may write $E(x|\chi)=\theta+$

$\delta\chi$, and the assumption of positive correlation is equivalent to $\delta > 0$. In contrast, x is a *measurement* of χ if $E(x|\chi) = \chi$, that is, if $\theta = 0$ and $\delta = 1$.

The simplest proxy variable problem is described by equations analogous to (7.9) and (7.10)

$$Y_t = \beta\chi_t + u_t \tag{7.13}$$

$$x_t = \delta\chi_t + \varepsilon_t \tag{7.14}$$

where, as in the errors-in-variable problem, we take Y_t and x_t to be defined around their means. The parameter of interest may be considered to be β, in which case x_t is a proxy for the explanatory variable χ_t. This model suffers from the identification problem associated with factor analysis (see Model 4): the vector (β, δ) is unique only up to a scale factor, since the distribution of the observables (Y_t, x_t) is unaffected by a scale change in (β, δ) offset by a scale change in χ_t. A regression of $\mathbf{Y}$ and $\mathbf{x}$ can by itself determine *only* the sign of β (given the sign of δ).

The essential singularities of the joint likelihood function implied by (7.13) and (7.14) occur on the lines $(\sigma_u^2, \chi) = (0, \mathbf{Y}/\beta)$ and $(\sigma_\varepsilon^2, \chi) = (0, \mathbf{x})$. Maximizing the nonpathological part of the function subject to these constraints yields the estimates $\hat{\beta}/\hat{\delta} = (\mathbf{x}'\mathbf{x})^{-1}\mathbf{x}'\mathbf{Y}$ and $(\hat{\beta}/\hat{\delta}) = (\mathbf{x}'\mathbf{Y})^{-1}\mathbf{Y}'\mathbf{Y}$, which are just the direct and reverse regression estimates. Note that the consequence of the identification problem is that the location of these modes depends on the ratio β/δ only.

A more interesting model is described by the equations

$$Y_t = \beta\chi_t + \gamma z_t + u_t \tag{7.13'}$$

$$x_t = \delta\chi_t + \varepsilon_t. \tag{7.14'}$$

This differs from the previous model in allowing another variable, z_t, to affect the dependent variable Y_t. As before, the essential singularities are implied by $\sigma_u^2 = 0$ or $\sigma_\varepsilon^2 = 0$. Using these constraints, we may write the model in the first case as $x_t = \delta\chi_t + \varepsilon_t = \delta(Y_t - \gamma z_t)/\beta + \varepsilon_t$ and in the second as $Y_t = (\beta/\delta)x_t + \gamma z_t + u_t$. The first equation implies that γ may be estimated by regressing $\mathbf{x}$ on $\mathbf{Y}$ and $\mathbf{z}$ and by resolving the resulting equation to put $\mathbf{Y}$ on the left-hand side. This may be called the reverse regression estimate of γ and is denoted by $\hat{\gamma}^R$. Referring to the second equation, we see that the essential singularities implied by $\sigma_\varepsilon^2 = 0$ suggest regressing $\mathbf{Y}$ directly on $\mathbf{x}$ and $\mathbf{z}$. The estimate of γ that results is called the direct estimate and is denoted by $\hat{\gamma}^D$.

If the logic of the errors-in-variable model is extended to this proxy-variable problem, we conjecture that in the structural form of this model the concentrated likelihood function has a plateau between $\hat{\gamma}^D$ and $\hat{\gamma}^R$. This is, in fact, the case. The structural form of this model requires a distribution for χ_t. To make the problem interesting, we allow χ_t and z_t to be correlated. In that case, the simple estimate of γ computed by regressing **Y** on **z** alone is potentially misleading, since a variable correlated with **z** has been omitted from the equation. This creates a pressure on the researcher to find a proxy for the unobservable χ_t. Assuming in particular that χ_t is normal with mean ωz_t and variance σ_χ^2

$$\chi_t \sim N(\omega z_t, \sigma_\chi^2)$$

the means and variances of the observables become

$$E[(Y_t, x_t)|z_t] = [(\beta\omega + \gamma)z_t, \delta\omega z_t]$$

$$\Sigma = V[(Y_t, x_t)|z_t] = \begin{bmatrix} \beta^2\sigma_\chi^2 + \sigma_u^2 & \beta\delta\sigma_\chi^2 \\ \beta\delta\sigma_\chi^2 & \delta^2\sigma_\chi^2 + \sigma_\varepsilon^2 \end{bmatrix}.$$

By an argument analogous to the material in the previous section, it may be shown that the likelihood function has a plateau between the direct and reverse regressions. The following result analogous to (7.11) can also be derived.

Let $\hat{\gamma}^S$ be the "simple" regression estimate of γ, $(\mathbf{z}'\mathbf{z})^{-1}\mathbf{z}'\mathbf{Y}$, computed by omitting **x** from the equation. Let $\hat{\gamma}^D$ and $\hat{\gamma}^R$ be the least-squares estimate and the reverse least-squares estimate. Then it is easy to verify that

$$(\hat{\gamma}^D - \hat{\gamma}^S) = R^2(\hat{\gamma}^R - \hat{\gamma}^S), \tag{7.15}$$

where R^2 is the squared multiple correlation coefficient computed when **Y** is regressed on both **x** and **z**. Interpreting this result, we conclude that when an error-ridden proxy variable is included in an equation, the estimates that result are insufficiently far from the estimates when the variable is omitted altogether. The difference should be expanded by a factor no larger than $1/R^2$.

7.4 Instrumental Variables

Suppose next that the variable χ_t is measured with error, but in addition, there is a proxy variable available. The model then consists of the following three equations:

$$Y_t = \beta\chi_t + u_t \tag{7.16}$$

$$x_t = \chi_t + \varepsilon_{1t} \tag{7.17}$$

$$w_t = \delta\chi_t + \varepsilon_{2t}, \tag{7.18}$$

where variables are defined around their means, x_t is a measurement of χ_t, w_t is a proxy variable, and u_t, ε_{1t}, and ε_{2t} are independent normal random variables with zero means and variances σ_u^2, σ_1^2, and σ_2^2, respectively. If the proxy equation were errorless, that is, if σ_2^2 were zero, we could use the relationship $\chi_t = w_t/\delta$ to determine $Y_t = \beta w_t/\delta + u_t$ and $x_{1t} = w_t/\delta + \varepsilon_{1t}$. Thus in regressing **Y** on **w** we would estimate β/δ, and in regressing **x** on **w** we would estimate δ^{-1}. The ratio would then estimate β:

$$\hat{\beta}^{IV} = \frac{\mathbf{Y}'\mathbf{w}}{\mathbf{x}'\mathbf{w}} \tag{7.19}$$

This estimator is known as the "instrumental variables" estimator of β. The variable **w** is said to be an "instrument" for **x**, and the estimator is often derived in the following way. The variable **x** is regressed on **w** to obtain an estimate equal to $(\mathbf{w}'\mathbf{w})^{-1}\mathbf{w}'\mathbf{x}$ and a "predicted" value of **x** equal to $\hat{\mathbf{x}} = \mathbf{w}(\mathbf{w}'\mathbf{w})^{-1}\mathbf{w}'\mathbf{x}$. Then β is estimated by regressing **Y** on the predicted **x**: $(\hat{\mathbf{x}}'\hat{\mathbf{x}})^{-1}\hat{\mathbf{x}}'\mathbf{Y} = \mathbf{Y}'\mathbf{w}/\mathbf{x}'\mathbf{w}$, which is the instrumental variables estimator.

The instrumental variables estimator is due to Reiersol (1945) and to Geary (1949). It is not difficult to demonstrate that $\hat{\beta}^{IV}$ is a consistent estimator of β, for example, see Malinvaud (1970). A careful examination of the likelihood function may help us choose between the consistent estimator $\hat{\beta}^{IV}$ which may have a large small-sample variance and the inconsistent estimators $\hat{\beta}^{D}$ and $\hat{\beta}^{R}$ which may have relatively small variances in small samples.

The likelihood function implied by the functional form of this model has essential singularities on the surfaces $(\sigma^2, \chi) = (0, \mathbf{Y}/\beta)$, $(\sigma_1^2, \chi) = (0, \mathbf{x})$, and $(\sigma_2^2, \chi) = (0, \mathbf{w}/\delta)$. Maximizing the nonpathological part of the likelihood function subject to these constraints yields the estimates $\hat{\beta}^{D} = \mathbf{x}'\mathbf{Y}/\mathbf{x}'\mathbf{x}$, $\hat{\beta}^{R} = \mathbf{Y}'\mathbf{Y}/\mathbf{x}'\mathbf{y}$, and $\hat{\beta}^{IV} = \mathbf{Y}'\mathbf{w}/\mathbf{x}'\mathbf{w}$, respectively. For the models discussed previously, the essential singularities of the functional form are maximum likelihood points of the likelihood function of the structural form. The structural form of the model being discussed here, however, has a single maximum likelihood point, located at one of the three estimates $\hat{\beta}^{D}$, $\hat{\beta}^{R}$, or $\hat{\beta}^{IV}$. It will be shown that if all three estimates have the same sign, then the instrumental variables estimate is maximum likelihood if it lies between $\hat{\beta}^{D}$ and $\hat{\beta}^{R}$, that is, if it satisfies the errors-in-variables bound. Otherwise, the maximum likelihood estimate is one of the end points of the bound, $\hat{\beta}^{D}$ if $\hat{\beta}^{IV}$ is less in absolute value than $\hat{\beta}^{D}$; $\hat{\beta}^{R}$ if $\hat{\beta}^{IV}$ exceeds $\hat{\beta}^{R}$.

The structural form of the model makes use of the assumption that the incidental variables χ_t are independent normal variables with variance σ_χ^2. Equations (7.16), (7.17), and (7.18) then define a trivariate normal process with covariance matrix

$$\mathbf{\Sigma} = \begin{bmatrix} \beta^2\sigma_\chi^2 + \sigma_u^2 & \beta\sigma_\chi^2 & \beta\delta\sigma_\chi^2 \\ \beta\sigma_\chi^2 & \sigma_\chi^2 + \sigma_1^2 & \delta\sigma_\chi^2 \\ \beta\delta\sigma_\chi^2 & \delta\sigma_\chi^2 & \delta^2\sigma_\chi^2 + \sigma_2^2 \end{bmatrix} \tag{7.20}$$

and the likelihood function is

$$L(\boldsymbol{\Sigma};\ \mathbf{Y},\mathbf{x},\mathbf{w}) \propto |\boldsymbol{\Sigma}|^{-T/2}\exp\left[-\frac{1}{2}\operatorname{tr}\boldsymbol{\Sigma}^{-1}\mathbf{S}\right] \tag{7.21}$$

where

$$\mathbf{S}=\begin{bmatrix} \mathbf{Y}'\mathbf{Y} & \mathbf{Y}'\mathbf{x} & \mathbf{Y}'\mathbf{w} \\ \mathbf{x}'\mathbf{Y} & \mathbf{x}'\mathbf{x} & \mathbf{x}'\mathbf{w} \\ \mathbf{w}'\mathbf{Y} & \mathbf{w}'\mathbf{x} & \mathbf{w}'\mathbf{w} \end{bmatrix}.$$

THEOREM 7.2 (INSTRUMENTAL VARIABLES). *Maximization of the likelihood function* (7.21) *subject to the constraint* (7.20) *with the variances* $\sigma_u^2, \sigma_1^2, \sigma_2^2$, *and* σ_χ^2 *nonnegative implies the following estimate of* β:

(*a*) *If* $\hat{\beta}^{IV}$ *and* $\hat{\beta}^{D}$ *are the same sign*

$$\hat{\beta} = \text{median}\left(\hat{\beta}^{D}, \hat{\beta}^{R}, \hat{\beta}^{IV}\right)$$

(*b*) *If* $\hat{\beta}^{IV}$ *and* $\hat{\beta}^{D}$ *are opposite in sign*

if the smallest correlation is

$$\hat{\beta}=\begin{cases} \hat{\beta}^{R} & r_{xw}^2 \\ \hat{\beta}^{D} & r_{yw}^2 . \\ \hat{\beta}^{IV} & r_{xy}^2 \end{cases}$$

Proof: A well-known result is that if $\boldsymbol{\Sigma}$ is unconstrained, maximization of (7.21) with respect to $\boldsymbol{\Sigma}$ yields the estimate

$$\hat{\boldsymbol{\Sigma}} = \frac{\mathbf{S}}{T}.$$

This is the maximum likelihood estimate for the constrained problem if the constraints are not binding, that is, if there exist parameter values such that $\boldsymbol{\Sigma}=\mathbf{S}/T$. The three off-diagonal elements of this matrix determine the following estimates

$$\beta = \frac{\mathbf{Y}'\mathbf{w}}{\mathbf{x}'\mathbf{w}}$$

$$\sigma_\chi^2 = \frac{\mathbf{Y}'\mathbf{x}\mathbf{x}'\mathbf{w}}{T\mathbf{Y}'\mathbf{w}}$$

$$\delta = \frac{\mathbf{Y}'\mathbf{w}}{\mathbf{Y}'\mathbf{x}}.$$

These values together with the diagonal elements of $\boldsymbol{\Sigma}$ imply

$$\sigma_u^2=\frac{\mathbf{Y}'\mathbf{Y}}{T}-\beta^2\sigma_\chi^2=\frac{\mathbf{Y}'\mathbf{Y}-\mathbf{w}'\mathbf{Y}\mathbf{Y}'\mathbf{x}}{T\mathbf{x}'\mathbf{w}}$$

$$\sigma_1^2=\frac{\mathbf{x}'\mathbf{x}}{T}-\sigma_\chi^2=\frac{\mathbf{x}'\mathbf{x}-\mathbf{Y}'\mathbf{x}\mathbf{x}'\mathbf{w}}{T\mathbf{Y}'\mathbf{w}}$$

$$\sigma_2^2=\frac{\mathbf{z}'\mathbf{z}}{T}-\delta^2\sigma_\chi^2=\frac{\mathbf{w}'\mathbf{w}-\mathbf{Y}'\mathbf{w}\mathbf{w}'\mathbf{x}}{T\mathbf{Y}'\mathbf{x}}.$$

If the estimated variances are all positive, these are the maximum likelihood estimates. The constraint $\sigma_\chi^2\geqslant 0$ implies

$$\frac{\mathbf{Y}'\mathbf{x}\mathbf{x}'\mathbf{w}}{\mathbf{Y}'\mathbf{w}}\geqslant 0.$$

This means that $\hat{\beta}^D$ and $\hat{\beta}^{IV}$ must have the same sign. Without loss of generality, we assume that $\mathbf{x}'\mathbf{Y}>0$. Given $\mathbf{x}'\mathbf{Y}>0$, the constraints $\sigma_u^2\geqslant 0, \sigma_1^2\geqslant 0, \sigma_2^2\geqslant 0$ imply, respectively,

$$\hat{\beta}^R=\frac{\mathbf{Y}'\mathbf{Y}}{\mathbf{Y}'\mathbf{x}}\geqslant\frac{\mathbf{w}'\mathbf{Y}}{\mathbf{w}'\mathbf{x}}=\hat{\beta}^{IV} \tag{7.22}$$

$$\hat{\beta}^{IV}=\frac{\mathbf{w}'\mathbf{Y}}{\mathbf{w}'\mathbf{x}}\geqslant\frac{\mathbf{Y}'\mathbf{x}}{\mathbf{x}'\mathbf{x}}=\hat{\beta}^D \tag{7.23}$$

$$\mathbf{Y}'\mathbf{x}-\frac{\mathbf{Y}'\mathbf{w}\mathbf{w}'\mathbf{x}}{\mathbf{w}'\mathbf{w}}\geqslant 0. \tag{7.24}$$

If one or more of the four inequalities is violated, it is not possible to have $\hat{\boldsymbol{\Sigma}}=\mathbf{S}/T$. In that case, the maximum occurs on the boundary of the constraint set. It could occur interior to the constraint set only if the likelihood function had more than one local maximum, which it does not, in fact, have. The constrained maximum can be found by imposing the constraints one at a time and selecting the constraint that yields the highest likelihood value, provided no other constraints are violated once one of the constraints is imposed.

Now consider the maximization of the likelihood function subject to one of the constraints $\sigma_u^2=0$, $\sigma_1^2=0$, or $\sigma_2^2=0$. If $\sigma_u^2=0$, (7.16) becomes $Y_t=\beta\chi_t$ without error and can be substituted into (7.17) and (7.18):

$$x_t=\frac{Y_t}{\beta}+u_{1t}$$

$$w_t=\frac{Y_t\delta}{\beta}+u_{2t}.$$

Maximization of the likelihood function of this normal process leads to the following likelihood value and estimate of β:

$$\sigma_u^2=0:\hat{\beta}=\frac{\mathbf{Y}'\mathbf{Y}}{\mathbf{x}'\mathbf{Y}},\qquad L_1\propto\left[(\mathbf{Y}'\mathbf{Y})(\mathbf{x}'\mathbf{M}_Y\mathbf{x})(\mathbf{w}'\mathbf{M}_Y\mathbf{w})\right]^{-T/2}$$

where $\mathbf{M}_Y=\mathbf{I}-\mathbf{Y}(\mathbf{Y}'\mathbf{Y})^{-1}\mathbf{Y}'$. Similarly, given the other constraints,

$$\sigma_1^2=0:\hat{\beta}=\frac{\mathbf{x}'\mathbf{Y}}{\mathbf{x}'\mathbf{x}}\qquad L_2\propto\left[(\mathbf{x}'\mathbf{x})(\mathbf{Y}'\mathbf{M}_x\mathbf{Y})(\mathbf{w}'\mathbf{M}_x\mathbf{w})\right]^{-T/2}$$

$$\sigma_2^2=0:\hat{\beta}=\frac{\mathbf{w}'\mathbf{Y}}{\mathbf{w}'\mathbf{x}},\qquad L_3\propto\left[(\mathbf{w}'\mathbf{w})(\mathbf{Y}'\mathbf{M}_w\mathbf{Y})(\mathbf{x}'\mathbf{M}_z\mathbf{x})\right]^{-T/2}$$

The constraint $\sigma_X^2=0$ implies the obviously inferior likelihood value $[(\mathbf{Y}'\mathbf{Y})(\mathbf{x}'\mathbf{x})(\mathbf{w}'\mathbf{w})]^{-T/2}$, and that constraint need not be considered further.

If we neglect the exponent and multiply by $(\mathbf{Y}'\mathbf{Y})(\mathbf{x}'\mathbf{x})(\mathbf{w}'\mathbf{w})$, the three likelihood values become, respectively, $[(1-r_{xy}^2)(1-r_{yw}^2)]^{-1}$, $[(1-r_{xy}^2)(1-r_{xw}^2)]^{-1}$and $[(1-r_{yw}^2)(1-r_{xw}^2)]^{-1}$. The first of these three numbers is the largest if r_{xw}^2 is the smallest of the squared correlations, the second if r_{yw}^2 is the smallest, and the third if r_{xy}^2 is the smallest. This establishes part (*b*) of the theorem.

All that remains to be shown to prove part (*a*) is that if $\hat{\beta}^{IV}$ and $\hat{\beta}^D$ have the same sign, then the violation of inequality (7.24) implies that r_{xy}^2 is the smallest correlation; and if $|\hat{\beta}^{IV}|<|\hat{\beta}^D|$, then r_{yw}^2 is the smallest correlation; and if $|\hat{\beta}^R|<|\hat{\beta}^{IV}|$, then r_{xw}^2 is the smallest. For convenience, we continue to assume $r_{xy}>0$. If inequality (7.24) is violated, then in terms of correlations $0<r_{xy}<r_{yw}r_{wx}$. But using $r_{yw}<1$ and $r_{wx}<1$, we obtain $r_{xy}<r_{yw}$ and $r_{xy}<r_{wx}$. Similarly, $|\hat{\beta}^{IV}|<|\hat{\beta}^D|$ implies $0<r_{yw}/r_{xw}<r_{xy}$, or $r_{yw}^2<r_{xy}^2r_{xw}^2$, and r_{yw}^2 is the smallest; $|\hat{\beta}^{\mathrm{R}}|<|\hat{\beta}^{\mathrm{IV}}|$ implies $0<r_{xy}^{-1}<r_{yw}/r_{xw}$ or $r_{xw}^2<r_{yw}^2r_{xy}^2$, and r_{xw}^2 is the smallest.

This theorem may be used to evaluate a statement that is often made about the use of instrumental variables estimators. It is often suggested that $\mathbf{w}$ is a "good" instrument if it is highly correlated with $\mathbf{x}$. On the other hand, if $\mathbf{w}$ and $\mathbf{x}$ are highly correlated, the instrumental variables estimate may not be much different from the direct estimate, and it is then difficult to see how the estimate could be an improvement over direct least squares. The result just proved does shed some light on this puzzle. Normalize the data so that the observed vectors all have length one, and let the three observed correlations be the numbers r_{xy}, r_{xw}, r_{yw}. Since the correlation matrix is positive definite, it must be the case that

$$0\leqslant\det\begin{bmatrix}1 & r_{xy} & r_{yw}\\ r_{xy} & 1 & r_{xw}\\ r_{yw} & r_{xw} & 1\end{bmatrix}=1+2r_{xy}r_{xw}r_{yw}-r_{yw}^2-r_{xw}^2-r_{xy}^2,$$

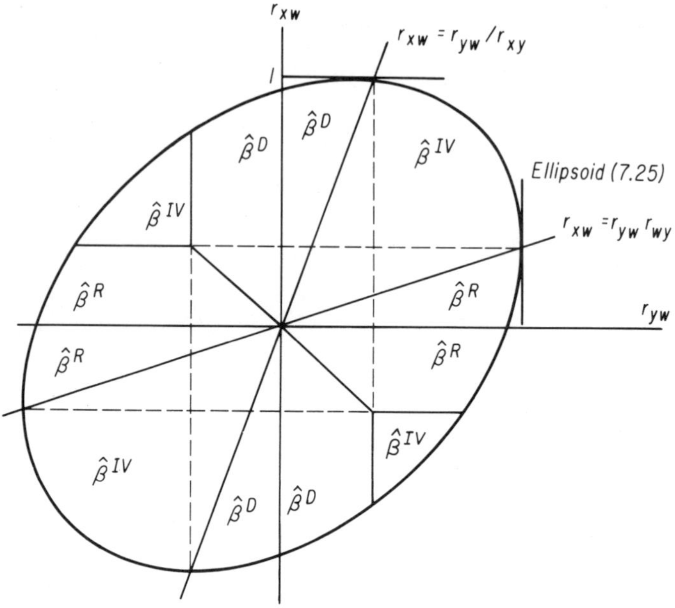

Fig. 7.5 Choice of estimates: the instrumental variables model.

which can be rewritten as

$$\begin{bmatrix} r_{xw} r_{yw} \end{bmatrix} \begin{bmatrix} 1 & -r_{xy} \\ -r_{xy} & 1 \end{bmatrix} \begin{bmatrix} r_{xw} \\ r_{yw} \end{bmatrix} \leqslant 1 - r_{xy}^2. \tag{7.25}$$

Holding fixed r_{xy}, inequality (7.25) implies Figure 7.5. By inspection of this figure, a high value of r_{xw} does not guarantee that $\hat{\beta}^{IV}$ is maximum likelihood. In fact, an increase in r_{xw} may shift the estimate from $\hat{\beta}^{IV}$ to $\hat{\beta}^D$. That is to say, the maximum likelihood estimate may be direct least squares even when r_{xw} is high, if r_{yw} is low. Note, however, that if r_{xw} is one, then the only feasible point is on the ray $r_{xw} = r_{yw}/r_{xy}$, and $\hat{\beta}^{IV} = \hat{\beta}^D$. In that sense, the statement about r_{xw} is valid. (Each ray out of the origin corresponds to a different value for $\hat{\beta}^{IV}$, increasing in the clockwise direction.)

By the way, the maximum likelihood estimate is a peculiar discontinuous function of the data if $\hat{\beta}^{IV}$ and $\hat{\beta}^D$ are opposite in sign. A small change in the data induces a shift from $\hat{\beta}^D$ to $\hat{\beta}^{IV}$, and thus a reversal of the sign of the estimate. Such a property is intuitively nonsensical. May we surmise that measures of uncertainty are in that case enormous, so that the discontinuity is small when measured in units of uncertainty? Or is the mode of the marginal likelihood function the more appropriate point estimate, anyway?

7.5 Multiple Proxy Variables

A generalization of the preceding is the multiple proxy variable model described by the equations

$$Y_t = \beta\chi_t + \gamma z_t + u_t \tag{7.26}$$

$$\mathbf{x}_t = \boldsymbol{\delta}\chi_t + \boldsymbol{\varepsilon}_t \tag{7.27}$$

$$\chi_t = \omega z_t + e_t. \tag{7.28}$$

Equations (7.26) and (7.28) are unchanged from the discussion in Section 7.3. Equation (7.27) is as before, except that it describes the generation of an m-dimensional vector of proxies rather than a single one. We take u_t, $\boldsymbol{\varepsilon}_t$, and e_t to be independent, normal random variables with means zero and variances σ_u^2, Ω, and σ_χ^2, respectively, where Ω is an $m \times m$ covariance matrix.

The essential singularities of the joint likelihood function of the functional form occur when σ_u^2 or when the determinant of Ω is zero. Since I have been unable to connect these points with the maximum likelihood points of the structural form, I do not discuss them further here.

The structural form of the model makes use of (7.28) to integrate out the incidental parameters. In effect, this assumes that $(Y_t, \mathbf{x}_t')$ has a multivariate normal distribution with mean and variance

$$E(Y_t, \mathbf{x}_t') = (\beta\omega z_t + \gamma z_t, \omega z_t \boldsymbol{\delta}')$$

$$V(Y_t, \mathbf{x}_t') = \begin{bmatrix} \beta^2\sigma_\chi^2 + \sigma_u^2 & \beta\sigma_\chi^2\boldsymbol{\delta}' \\ \boldsymbol{\delta}\beta\sigma_\chi^2 & \boldsymbol{\delta}\sigma_\chi^2\boldsymbol{\delta}' + \Omega \end{bmatrix}.$$

As usual, the basic identification problem arises. The distribution of the observables is unaffected by a scale change in $(\beta, \boldsymbol{\delta}')$ offset by a scale change in (ω, σ_χ). Note also that there are proportionality constraints on this $(m+1)$-variate distribution, since both the covariance between Y_t and $\mathbf{x}_t$ $(\beta\sigma_\chi^2\boldsymbol{\delta}')$ and the vector of regression coefficients $(\omega\boldsymbol{\delta}')$ are proportional to $\boldsymbol{\delta}$.

Unconstrained maximum likelihood estimate of the parameters of this process leads to the equations

$$\beta\omega + \gamma = (\mathbf{z}'\mathbf{z})^{-1}\mathbf{z}'\mathbf{Y}$$

$$\omega\boldsymbol{\delta}' = (\mathbf{z}'\mathbf{z})^{-1}\mathbf{z}'\mathbf{X}$$

$$\beta\sigma_\chi^2\boldsymbol{\delta}' = \frac{\mathbf{Y}'\mathbf{M}_z\mathbf{X}}{T}$$

$$\beta^2\sigma_\chi^2 + \sigma_u^2 = \frac{\mathbf{Y}'\mathbf{M}_z\mathbf{Y}}{T}$$

$$\boldsymbol{\delta}\sigma_\chi^2\boldsymbol{\delta}' + \Omega = \frac{\mathbf{X}'\mathbf{M}_z\mathbf{X}}{T}$$

where $M_z = \mathbf{I} - \mathbf{z}(\mathbf{z}'\mathbf{z})^{-1}\mathbf{z}'$ and $\mathbf{X}$ is the $T \times m$ matrix of m proxy variables.

For the sake of producing an analytical result, we assume that the sample moments satisfy the proportionality constraints of the process

$$(\mathbf{z}'\mathbf{z})^{-1}\mathbf{z}'\mathbf{X} \propto \mathbf{Y}'\mathbf{M}_z\mathbf{X}.$$

THEOREM 7.3. *If the sample moments satisfy the proportionality constraints, all "reverse regression" estimates of γ are identical. (A reverse regression estimate is computed by regressing one of the proxies on* $\mathbf{Y}$ *and* $\mathbf{z}$*, and then transforming the estimated equation to write* $\mathbf{Y}$ *as a function of* $\mathbf{z}$ *and the proxy.)*

Proof: A reverse regression estimate of γ can be written

$$\hat{\gamma}_i^R = (\mathbf{z}'\mathbf{z})^{-1}\mathbf{z}'\mathbf{Y} - (\mathbf{z}'\mathbf{z})^{-1}\mathbf{z}'\mathbf{x}_i\hat{\beta}_i^R$$

where $\hat{\beta}_i^R$ is the reverse regression estimate of β:

$$\hat{\beta}_i^R = (\mathbf{Y}'\mathbf{M}_z\mathbf{x}_i)^{-1}\mathbf{Y}'\mathbf{M}_z\mathbf{Y}.$$

The proportionality constraint is

$$\frac{\mathbf{z}'\mathbf{x}_i}{\mathbf{Y}'\mathbf{M}_z\mathbf{x}_i} = \frac{\mathbf{z}'\mathbf{x}_j}{\mathbf{Y}'\mathbf{M}_z\mathbf{x}_j}$$

which implies $\mathbf{z}'\mathbf{x}_i\hat{\beta}_i^R = \mathbf{z}'\mathbf{x}_j\hat{\beta}_j^R$. Using this in the equation for $\hat{\gamma}_i^R$ produces the desired equality $\hat{\gamma}_i^R = \hat{\gamma}_j^R$. This is true, by the way, not only for any one proxy but for any linear combination as well.

THEOREM 7.4 (MULTIPLE PROXY VARIABLES). *If the sample moments satisfy the proportionality constraints, the values of γ corresponding to the maximum of the likelihood function are bounded on one side by the (unique) reverse regression estimate*

$$\hat{\gamma}^R = (\mathbf{z}'\mathbf{z})^{-1}\mathbf{z}'\mathbf{Y} - (\mathbf{z}'\mathbf{z})^{-1}\mathbf{z}'\mathbf{x}_i\hat{\beta}^R$$

where $\hat{\beta}^R = (\mathbf{Y}'\mathbf{M}_z\mathbf{x}_i)^{-1}\mathbf{Y}'\mathbf{M}_z\mathbf{Y}$, *and on the other by*

$$\hat{\gamma}^D = (\mathbf{z}'\mathbf{z})^{-1}\mathbf{z}'\mathbf{Y} - (\mathbf{z}'\mathbf{z})^{-1}\mathbf{z}'\mathbf{x}_i\hat{\beta}^R R^2$$

where R^2 is the multiple correlation coefficient and $\hat{\gamma}^D$ is the estimate computed when $\mathbf{Y}$ *is regressed on* $\mathbf{z}$ *and all the proxy variables*

$$R^2 = \frac{\mathbf{Y}'\mathbf{M}_z\mathbf{X}(\mathbf{X}'\mathbf{M}_z\mathbf{X})^{-1}\mathbf{X}'\mathbf{M}_z\mathbf{Y}}{\mathbf{Y}'\mathbf{M}_z\mathbf{Y}}.$$

Notice also that $\hat{\gamma}^D$ satisfies the equation

$$(\hat{\gamma}^D - \hat{\gamma}^S) = R^2(\hat{\gamma}^R - \hat{\gamma}^S)$$

where $\hat{\gamma}^S = (\mathbf{z}'\mathbf{z})^{-1}\mathbf{z}'\mathbf{Y}$.

Proof: Select any value of β and δ_1 and solve the equations for the other parameters

$$\omega=(\mathbf{z}'\mathbf{z})^{-1}\mathbf{z}'\mathbf{x}_1\delta_1^{-1}$$

$$\gamma=(\mathbf{z}'\mathbf{z})^{-1}\mathbf{z}'\mathbf{Y}-(\mathbf{z}'\mathbf{z})^{-1}\mathbf{z}'\mathbf{x}_1\frac{\beta}{\delta_1}$$

$$\sigma_X^2=\frac{\mathbf{Y}'\mathbf{M}_z\mathbf{x}_1}{T\beta\delta_1}$$

$$\sigma_u^2=\frac{\mathbf{Y}'\mathbf{M}_z\mathbf{Y}}{T}-\frac{\mathbf{Y}'\mathbf{M}_z\mathbf{x}_1}{T}\frac{\beta}{\delta_1}$$

$$\begin{aligned}\boldsymbol{\Omega}&=\frac{\mathbf{X}'\mathbf{M}_z\mathbf{X}}{T}-\boldsymbol{\delta}\sigma_X^2\boldsymbol{\delta}'\\&=\frac{\mathbf{X}'\mathbf{M}_z\mathbf{X}}{T}-\frac{\mathbf{X}'\mathbf{M}_z\mathbf{Y}\sigma_X^2\mathbf{Y}'\mathbf{M}_z\mathbf{X}}{T^2\sigma_X^4\beta^2}\\&=\frac{\mathbf{X}'\mathbf{M}_z\mathbf{X}}{T}-\mathbf{X}'\mathbf{M}_z\mathbf{Y}\left(\frac{\delta_1}{T\beta\mathbf{Y}'\mathbf{M}_z\mathbf{x}_1}\right)\mathbf{Y}'\mathbf{M}_z\mathbf{X}.\end{aligned}$$

The constraint $\sigma_u^2>0$ implies

$$\frac{\beta}{\delta_1}<\frac{\mathbf{Y}'\mathbf{M}_z\mathbf{Y}}{\mathbf{Y}'\mathbf{M}_z\mathbf{x}_1}=\hat{\beta}^R.$$

The constraint $\boldsymbol{\Omega}$ positive definite implies[3]

$$\left(\frac{\mathbf{Y}'\mathbf{M}_z\mathbf{x}_1\beta}{\delta_1}\right)^{-1}<\left(\mathbf{Y}'\mathbf{M}_z\mathbf{X}(\mathbf{X}'\mathbf{M}_z\mathbf{X})^{-1}\mathbf{X}'\mathbf{M}_z\mathbf{Y}\right)^{-1}$$

that is

$$\frac{\beta}{\delta_1}>\frac{\mathbf{Y}'\mathbf{M}_z\mathbf{X}(\mathbf{X}'\mathbf{M}_z\mathbf{X})^{-1}\mathbf{X}'\mathbf{M}_z\mathbf{Y}}{\mathbf{Y}'\mathbf{M}_z\mathbf{x}_1}=\hat{\beta}^R R^2.$$

These two constraints on β/δ_1 determine the hypothesized constraints on γ.

An interesting implication of Theorem 7.4 is that extra proxies are not enough to identify the coefficients, but the bounds for the coefficients are reduced, depending on the multiple R^2 of $\mathbf{Y}$ on all the proxies. If the proxies were independent of each other, that is, if $\boldsymbol{\Omega}$ were diagonal, extra proxies would be more helpful, however.[4]

[3] $\boldsymbol{\Omega}=\mathbf{S}-\boldsymbol{\delta}\sigma^2\boldsymbol{\delta}'$ is positive definite if for any vector ψ, $0\leqslant\psi'\mathbf{S}\psi-\psi'\boldsymbol{\delta}\sigma^2\boldsymbol{\delta}'\psi$, which in turn requires $\sigma^2<\max_\psi\psi'\mathbf{S}\psi/\psi'\boldsymbol{\delta}\boldsymbol{\delta}'\psi=1/\boldsymbol{\delta}'\mathbf{S}^{-1}\boldsymbol{\delta}$.

[4] This does not lend itself to a very tractable analysis and is not discussed here. The constraints that $\boldsymbol{\Omega}$ is proportional to the identity matrix and $\boldsymbol{\delta}$ is the vector of ones does lead to a tractable solution.

7.6 Errors in Many Variables

Another more general model has all k variables measured with error:

$$Y_t = \beta' \chi_t + u_t \tag{7.29}$$

$$\mathbf{x}_t = \chi_t + \varepsilon_t \tag{7.30}$$

where variables are measured around their means, χ_t is a k-dimensional vector of incidental parameters, $\mathbf{x}_t$ is a vector of measurements of χ_t, and ε_t is a vector of measurement errors distributed normally with mean vector zero and diagonal covariance matrix $\mathbf{D}$. The structural form of this model makes use of the assumption that χ_t is normal with mean zero and covariance Ω.

Thus the vector $(Y_t, \mathbf{x}_t)$ has covariance matrix equal to

$$\Sigma = \begin{pmatrix} \sigma_u^2 + \beta'\Omega\beta & \beta'\Omega \\ \Omega\beta & \mathbf{D}+\Omega \end{pmatrix}.$$

The maximum likelihood estimate can be found as before by setting Σ equal to the sample moments

$$\mathbf{S} = \begin{bmatrix} \mathbf{Y'Y} & \mathbf{Y'X} \\ \mathbf{X'Y} & \mathbf{X'X} \end{bmatrix} \cdot T^{-1}$$

where $\mathbf{X}$ is the $T \times k$ matrix of measurements of $\chi_t, t = 1, \ldots, T$. Given the diagonal matrix $\mathbf{D}$, we can solve for the other parameters as

$$\hat{\beta} = \frac{\Omega^{-1}\mathbf{X'Y}}{T} = (\mathbf{X'X} - T\mathbf{D})^{-1}\mathbf{X'Y} \tag{7.31}$$

$$\hat{\Omega} = \frac{\mathbf{X'X}}{T} - \mathbf{D}$$

$$\hat{\sigma}^2 = \frac{\mathbf{Y'Y}}{T} - \beta'\Omega\beta = \frac{\mathbf{Y'Y}}{T} - \frac{\mathbf{Y'X}(\mathbf{X'X}T^{-1} - \mathbf{D})^{-1}\mathbf{X'Y}}{T^2}.$$

Any value of $\hat{\beta}$ corresponding to any nonnegative diagonal matrix $\mathbf{D}$ is, therefore, a maximum likelihood point provided that $\hat{\Omega}$ is positive semi-definite and $\hat{\sigma}_u^2$ is nonnegative.

As far as I know, an exact description of the set of maximum likelihood estimates is not known. It is possible to compute $k-1$ "reverse" regressions in which one of the explanatory variables is used as the dependent variables. Each of these reverse regressions is feasible, given a suitably chosen $\mathbf{D}$, as is the direct regression. It is natural to conjecture that the feasible region is the convex hull of these k regressions. This is correct if all

k regressions are in the same orthant, a result implicitly due to Frisch (1934), proved by Reiersol (1945) and Koopmans (1937), and discussed by Malinvaud (1970, p. 43). When the estimates lie in different orthants, the feasible region is apparently unbounded with sides generated by the k regressions.

It is interesting, in closing, to contrast Equation (7.31) with the matrix-weighted average that is the posterior mean of $\boldsymbol{\beta}$ given a normal prior with mean zero and precision matrix $\mathbf{D}$. For the prior-dependent problem the least-squares estimate is shrunk toward the origin by adding to the $\mathbf{X}'\mathbf{X}$ matrix a positive diagonal matrix. For the errors-in-variables model the least-squares estimate is blown away from the origin by subtracting from the $\mathbf{X}'\mathbf{X}$ matrix a positive diagonal matrix. I have sometimes chided my colleagues who are enamored of ridge regression that the net effect of the two forces may mean that least squares is just right!

7.7 Priors and Proxies

The interesting problems of the relationship between prior information and proxy variables have yet to be discussed. To give an example, when "bad" estimates are implied by a data set, they are sometimes "explained" with reference to the errors-in-variables model. "The coefficient is smaller than it ought to be because of the errors-in-variable attenuation," or "the theory would have worked better if we could have found more appropriate proxy variables." Are these statements and statements like them appropriate? Can it be that there is a form of prior information that allows us to discard or discount especially unreasonable estimates but to ignore the errors-in-variables issues if the coefficients are relatively consistent with the priors? Are more proxies necessarily better than less? What are the inferential consequences of searching for a proxy until one is found that "works"?

Most of these questions refer to measures of dispersion of posterior distribution. To get close to answers, we must generate approximate measures of dispersion. The last time in this chapter that such measures were discussed was in Section 7.2, in which we reported Lindley and El-Sayyad's (1968) analysis of the errors-in-variables model in which diffuse priors were assumed for the regression parameter. Their analysis could be generalized to the more complicated proxy problems, but that work remains to be done. We report here a few simple results that constitute a minor foray into an intriguing and important field of research.

With reference to the simple proxy variable model discussed in Section 7.3, the following proposition can be explored: "a poor proxy is worse than none." With a proper prior distribution this cannot be the case, just as it is *never* better in an *inference* problem to leave a variable out of a regression, unless there is some cost of observation, or in a more general framework, cost of constructing a prior. But it may be the case that the regression computed with the proxy variable left out is a better approximation to the location of the posterior distribution than is the regression with the proxy variable in the equation.

In the material on interpretive searches we began with a prior located at the origin and concluded that a variable may be omitted if its estimated coefficient was not far from zero. A proxy variable may be a candidate for omission (a) if its coefficient is small, (b) if its coefficient is far from the value we would have expected if we had perfect measurement, or (c) if the coefficient on **z** is closer to the value we expected when the proxy is omitted. Although each has been used in practice and each has a certain appeal, none is an obviously valid procedure. Prior information is clearly the foundation of these procedures, and we now turn to a consideration of how priors affect our estimates.

First we observe that prior information about either β or δ can be used to normalize the vector (β,δ), but otherwise, because of the identification problem, it has no effect on the modes of the posterior. It is not uncommon, however, to have prior information about both β and δ. A prior for δ is likely to be located at one, of course, and it may be that the researcher has some more-or-less distinct ideas about β as well. Let us take the extreme case in which β and δ are known exactly. This implies no constraints on the maximum likelihood estimates if no inequalities are violated. In that event, the appropriate estimate of γ is

$$\tilde{\gamma}=(\mathbf{z}'\mathbf{z})^{-1}\mathbf{z}'\mathbf{Y}-(\mathbf{z}'\mathbf{z})^{-1}\mathbf{z}'\mathbf{x}\frac{\beta}{\delta}, \qquad (7.32)$$

which is the least-squares estimate of γ with the constraint that the coefficient on **x** is equal to the ratio β/δ.[5]

In the second best world, in which we can choose between regressions with or without the **x** variable included, we need only determine which of the two resulting estimates is closer to (7.32). This is equivalent to selecting the regression with **x** omitted if the least-squares coefficient on **x** is closer to zero than to β/δ. Incidentally, it is necessary only to know the sign of β/δ to know that it is better to omit **x** if its coefficient has the wrong sign.

We may now consider situations in which there is prior information about γ (possibly zero). We begin with the presumption that if the least-squares estimate of γ is close to the value we expect, we may conclude first that the coefficient on x is likely to be close to β and second that it is desirable to adjust the prior estimate of γ in the direction of the

[5]Note that $\tilde{\gamma}$ is an unbiased estimator of γ.

least-squares estimate. If, on the other hand, the proxy doesn't "work," it produces a coefficient far from the prior estimate. We would be led to conclude that there is little information about β or about γ generated by the experiment.

In fact, it is not enough to have even exact prior information about γ, since the basic identification problem remains; if we multiply the vector (β,δ) by some constant and make an offsetting change in (σ_χ^2,ω), we do not alter the distribution of the observables.[6] The fact that prior information about γ alone is of little value for estimating β is fairly obvious when we consider that $\hat{\gamma}$ may be close to the true value merely if ω is small, that is, if $\boldsymbol{\chi}$ and $\mathbf{z}$ are uncorrelated. It is necessary, but it is not sufficient.

If *both* γ and δ are known, there is no identification problem, and we may solve for the maximum likelihood estimates as

$$\omega = \frac{(\mathbf{z}'\mathbf{z})^{-1}\mathbf{z}'x}{\delta}$$

$$\beta = \frac{(\mathbf{z}'\mathbf{z})^{-1}\mathbf{z}'(\mathbf{Y}-\gamma\mathbf{z})}{\omega}$$

$$= \delta\,(\mathbf{z}'\mathbf{x})^{-1}\mathbf{z}'(\mathbf{Y}-\gamma\mathbf{z})$$

$$\sigma_\chi^2 = \frac{\mathbf{x}'\mathbf{M}_z\mathbf{Y}}{T\beta\delta}$$

$$\sigma_u^2 = \frac{\mathbf{Y}'\mathbf{M}_z\mathbf{Y}}{T} - \beta^2\sigma_\chi^2$$

$$\sigma_\varepsilon^2 = \frac{\mathbf{x}'\mathbf{M}_z\mathbf{x}}{T} - \delta^2\sigma_\chi^2$$

[6]An interesting observation is that these assumptions determine only a subset of the set of all bivariate normal processes. Denoting the means and variances of $(Y_t-\gamma z_t, x_t)$ by

$$(m_1,m_2) = z_t(\beta\omega,\delta\omega)$$

$$\begin{bmatrix} v_{11} & v_{12} \\ v_{12} & v_{22} \end{bmatrix} = \begin{bmatrix} \beta^2\sigma_\chi^2+\sigma_u^2 & \beta\delta\sigma_\chi^2 \\ \beta\delta\sigma_\chi^2 & \delta^2\sigma_\chi^2+\sigma_\varepsilon^2 \end{bmatrix},$$

from the fact that the variances are positive we may derive the constraints

$$\frac{v_{12}}{m_1m_2} \geqslant 0; \quad v_{11} - \frac{m_1}{m_2}v_{12} \geqslant 0; \quad v_{22} - \frac{m_2}{m_1}v_{12} \geqslant 0.$$

It can be shown that if the direct and reverse regressions do not bracket γ, then the data favor a bivariate normal process excluded from this class.

provided the variances are positive. Note that β is an instrumental variables estimator, with $\mathbf{z}$ used as an instrument for $\mathbf{x}$ in the regression of $\mathbf{Y}-\gamma\mathbf{z}$ on $\mathbf{x}$.

An appealing proposition is that if $\mathbf{x}$ is a "good" measurement of χ, then when we regress $\mathbf{Y}$ on $\mathbf{x}$ and $\mathbf{z}$, the coefficient on $\mathbf{z}$ should be close to the known value of γ. If it is not, then it seems unlikely that we could obtain much information about β. Intuition thus suggests that the uncertainty in β may be related to the difference between γ and $\hat{\gamma}$. This, like a long list of other interesting questions, will remain unanswered until some appropriate approximations of the posterior dispersion are available.

CHAPTER 8

DATA-SELECTION SEARCHES

Theoretical models are often vague or have nothing at all to say about the choice of particular observations. Even less frequently do they suggest the circumstances in which two or more observations can be regarded as independent pieces of relevant information. It is thus necessary for the empirical worker both to select a subset of potential observations and to determine the extent to which observations are correlated. To put it another way, the researcher must identify observations or transformations of observations that can be considered to be independent replications of an unchanging "experiment." In practice, this may mean estimating coefficients with different subsets or different transformations of the data set and selecting the result that appears best according to some criteria. We call this a data-selection search.

The fact that this process is data dependent obviously has consequences for the interpretation of the final result of a data-selection search. It seems clear that when the data evidence is partly spent to pick a data set, the regression equation that is finally selected to convey the data evidence at least overstates the precision of the evidence and likely distorts it as well. The function of this chapter is thus to describe the inferences that are appropriate when some of the data are discarded or when the data are transformed by a data-dependent function.

In the case of interpretive searches, a constrained regression can at best approximate the location of a posterior distribution, which is, in fact, a mixture of many constrained regressions. Similarly, a Bayesian will necessarily discard none of the observations but will instead place relatively low weight on some. The extent to which one can approximate his posterior distribution by discarding altogether some observations and assigning equal weight to the remaining ones is not extensively discussed. Nor is the interplay between prior information and data selection extensively discussed. Instead, this chapter reports the likelihood functions implied by statistical models that generate outliers or interdependencies.

Although a data-selection search involves features of interpretive searches, the two may be distinguished in the following way. Let a linear model be written as $Y_t=\alpha_t+\boldsymbol{\beta}_t'\mathbf{x}_t$ where $\mathbf{x}_t$ is a vector of observable explanatory variables, Y_t is the observable dependent variable, and $(\alpha_t,\boldsymbol{\beta}_t)$ is a vector of unobservable parameters applying to the tth observation. Assume that the unobservables $(\alpha_t,\boldsymbol{\beta}_t)$, $t=1,\ldots,T$ have a common mean $(\alpha,\boldsymbol{\beta})$. It is the business of a data-selection search to pick the multivariate distribution of the unobservables around their common mean $(\alpha,\boldsymbol{\beta})$. An interpretive search, on the other hand, is designed to make use of prior information about the means $(\alpha,\boldsymbol{\beta})$.

The usual least-squares logic results from the assumptions that $\boldsymbol{\beta}_t$ does not vary from observation to observation and that α_t is the sum of $\alpha+\varepsilon_t$, an independent normal random variable with constant variance. A first step toward relaxing this assumption is to allow for dependence among the ε_t random variables. This is discussed in Section 8.1 under the heading of "nonspherical disturbances." Alternatively, the assumption of normality may be relaxed. The consequences of fat-tailed nonnormal distributions are described in Section 8.2. The next two sections explore models that let the slope parameters $\boldsymbol{\beta}_t$ vary from observation to observation, but maintain the assumption of normality.

No attempt is made to relax all assumptions simultaneously, and this chapter does not suggest a mechanical approach to the data-selection problem. In that regard, it is like the chapter on interpretive searches, which takes the data distribution as given and analyzes the mapping of prior distributions into posterior distributions. The point is made in that chapter that if the prior could be uniquely determined, there would be a unique interpretation of the data, but ambiguity in the choice of prior implies ambiguity in the posterior distribution. In the case of data-selection searches, if the data distribution could be taken as given, the data would imply a unique likelihood function. But just as it is impossible unambiguously to select a prior, so too is it impossible unambiguously to select a data distribution. Not only must the interpretation of the data evidence thus remain elusive, but also the data evidence itself must be defined imprecisely. A researcher can only report features of the mapping of prior and data distributions into posteriors.

8.1 Nonspherical Disturbances

The conditional distribution of the error vector has heretofore been assumed to be multivariate normal with mean vector zero and covariance matrix $\sigma^2\mathbf{I}$. The further assumption of a gamma distribution for σ^{-2} implies that the error vector is distributed (marginally) multivariate Student with a covariance matrix also proportional to the identity matrix. We now wish to consider the consequences of using a covariance matrix that is not proportional to the identity matrix. We refer to these errors as nonspherical disturbances, thereby implying that the isodensities of the random errors are not spheres, as they are when the assumptions above apply.

Some consideration of nonspherical errors seems absolutely essential with nonexperimental data. Careful elicitation of one's personal opinions about the error process is quite unlikely to lead to zero covariances between the errors, and inferences from observable data may be erroneous if there is no adjustment made for departures from sphericity. To give an example, a time-series data set may be enlarged by a factor of 12 merely by using monthly data instead of annual data. But because of the dependence in the monthly residuals, the number 12 greatly overstates the real gain in information. This has been accurately called "counting your wealth in small change."

The nonspherical regression process may be written as

$$\mathbf{Y}=\mathbf{X}\boldsymbol{\beta}+\mathbf{u} \tag{8.1}$$

where $\mathbf{Y}$ is $T\times 1$, $\mathbf{X}$ is $T\times k$, $\boldsymbol{\beta}$ is $k\times 1$ and $\mathbf{u}$ is a $T\times 1$ normal error vector with mean vector zero and covariance matrix $\sigma^2\boldsymbol{\Sigma}$ and where $\boldsymbol{\Sigma}$ is a known matrix. The likelihood function may then be written as

$$f(\mathbf{Y}|\boldsymbol{\beta},\sigma^2,\boldsymbol{\Sigma})\propto|\sigma^2\boldsymbol{\Sigma}|^{-\frac{1}{2}}\exp\frac{-\frac{1}{2}(\mathbf{Y}-\mathbf{X}\boldsymbol{\beta})'\boldsymbol{\Sigma}^{-1}(\mathbf{Y}-\mathbf{X}\boldsymbol{\beta})}{\sigma^2}.$$

The exponent in this function may be decomposed as

$$(\mathbf{Y}-\mathbf{X}\boldsymbol{\beta})'\boldsymbol{\Sigma}^{-1}(\mathbf{Y}-\mathbf{X}\boldsymbol{\beta})=(\mathbf{Y}-\mathbf{X}\mathbf{b})'\boldsymbol{\Sigma}^{-1}(\mathbf{Y}-\mathbf{X}\mathbf{b})+(\boldsymbol{\beta}-\mathbf{b})'\mathbf{X}'\boldsymbol{\Sigma}^{-1}\mathbf{X}(\boldsymbol{\beta}-\mathbf{b})$$

where $\mathbf{b}$ is a solution to

$$\mathbf{X}'\boldsymbol{\Sigma}^{-1}\mathbf{X}\mathbf{b}=\mathbf{X}'\boldsymbol{\Sigma}^{-1}\mathbf{Y}.$$

When $\mathbf{b}$ is unique, that is, when $\mathbf{X}'\boldsymbol{\Sigma}^{-1}\mathbf{X}$ is invertible, $\mathbf{b}$ is called the generalized least-squares estimate of $\boldsymbol{\beta}$, since it is the value of $\boldsymbol{\beta}$ that minimizes $(\mathbf{Y}-\mathbf{X}\boldsymbol{\beta})'\boldsymbol{\Sigma}^{-1}(\mathbf{Y}-\mathbf{X}\boldsymbol{\beta})$.

By an inspection of these formulas, the consequences of $\boldsymbol{\Sigma}\neq\mathbf{I}$ are first, to alter the location of the likelihood ellipsoid from $(\mathbf{X}'\mathbf{X})^{-1}\mathbf{X}'\mathbf{Y}$ to $(\mathbf{X}'\boldsymbol{\Sigma}^{-1}\mathbf{X})^{-1}\mathbf{X}'\boldsymbol{\Sigma}^{-1}\mathbf{Y}$, second, to alter the shape of the ellipsoids ($\mathbf{X}'\boldsymbol{\Sigma}^{-1}\mathbf{X}$ replaces $\mathbf{X}'\mathbf{X}$), and third, to alter the data estimate of the labeling function (the error sum of squares is computed differently).[1] We know from the discussion of sensitivity analysis in Chapter 5 that each of these may have a significant effect on the posterior distribution. For example, although least squares and generalized least squares may be exactly the same, differences in the precision matrices $\mathbf{X}'\boldsymbol{\Sigma}^{-1}\mathbf{X}$ and $\mathbf{X}'\mathbf{X}$ may mean that the two posterior distributions, corresponding to these two error processes, are quite different.

A more common situation arises when $\boldsymbol{\Sigma}$ is not known with certainty. It is then necessary to select a personal prior distribution for $\boldsymbol{\Sigma}$, which may be done by first writing $\boldsymbol{\Sigma}$ as a deterministic function of some vector of parameters $\boldsymbol{\theta}$, and then by assigning to $\boldsymbol{\theta}$ some hopefully convenient prior. As far as I know, there is no convenient way to analyze such models. The posterior distribution of $\boldsymbol{\beta}$ may be written as

$$f(\boldsymbol{\beta}|\mathbf{Y},\mathbf{X}) \propto \int_{\boldsymbol{\theta}} f(\boldsymbol{\beta}|\mathbf{Y},\mathbf{X},\boldsymbol{\theta}) f(\boldsymbol{\theta}|\mathbf{Y},\mathbf{X}) d\boldsymbol{\theta} \tag{8.2}$$

where $f(\boldsymbol{\beta}|\mathbf{Y},\mathbf{X},\boldsymbol{\theta})$ is a tractable distribution of $\boldsymbol{\beta}$, given some covariance matrix $\boldsymbol{\Sigma}(\boldsymbol{\theta})$, and where $f(\boldsymbol{\theta}|\mathbf{Y},\mathbf{X})$ is proportional to the product of a (tractable) marginal likelihood function $f(\mathbf{Y}|\boldsymbol{\theta},\mathbf{X})$ and a prior for $\boldsymbol{\theta}$. Unfortunately, numerical methods are required to evaluate the integral (8.2)

The most commonly analyzed form of nonspherical disturbances is first-order autocorrelation with

$$\boldsymbol{\Sigma} = \sigma^2 \begin{bmatrix} 1 & \rho & \rho^2 & \cdots & \rho^{T-1} \\ \rho & 1 & \rho & \cdots & \rho^{T-2} \\ \rho^2 & \rho & 1 & \cdots & \rho^{T-3} \\ \vdots & \vdots & \vdots & & \vdots \\ \rho^{T-1} & \rho^{T-2} & \rho^{T-3} & \cdots & 1 \end{bmatrix} \tag{8.3}$$

and

$$\boldsymbol{\Sigma}^{-1} = \sigma^{-2}(1-\rho^2)^{-1} \begin{bmatrix} 1 & -\rho & 0 & \cdots & 0 & 0 \\ -\rho & 1+\rho^2 & -\rho & \cdots & 0 & 0 \\ 0 & -\rho & 1+\rho^2 & \cdots & 0 & 0 \\ \vdots & \vdots & \vdots & & \vdots & \vdots \\ 0 & 0 & 0 & \cdots & 1+\rho^2 & -\rho \\ 0 & 0 & 0 & \cdots & -\rho & 1 \end{bmatrix}.$$

[1] A labeling function assigns to every isolikelihood ellipsoid a particular likelihood value.

The marginal likelihood $f(\mathbf{Y}|\rho)$ can be computed without difficulty, since $\boldsymbol{\Sigma}$ can be diagonalized by the transformation $\boldsymbol{\Sigma}^{-1}=\mathbf{C}'\mathbf{C}\sigma^{-2}$ where

$$\mathbf{C}=(1-\rho^2)^{-\frac{1}{2}}\begin{bmatrix} (1-\rho^2)^{\frac{1}{2}} & 0 & 0 & \cdots & 0 & 0 \\ -\rho & 1 & 0 & \cdots & 0 & 0 \\ 0 & -\rho & 1 & \cdots & 0 & 0 \\ \vdots & \vdots & \vdots & & \vdots & \vdots \\ 0 & 0 & 0 & \cdots & 1 & 0 \\ 0 & 0 & 0 & \cdots & -\rho & 1 \end{bmatrix}.$$

Thus the vector $\mathbf{Y}^*=\mathbf{CY}$ is distributed normally with mean $\mathbf{CX}\boldsymbol{\beta}$ and variance $\mathbf{C\Sigma C}'=[\mathbf{C}'^{-1}\boldsymbol{\Sigma}^{-1}\mathbf{C}^{-1}]^{-1}=\sigma^2[\mathbf{C}'^{-1}\mathbf{C}'\mathbf{C}\mathbf{C}^{-1}]=\sigma^2\mathbf{I}$. Then, for example, if the parameters $(\boldsymbol{\beta},\sigma^2)$ are assigned a conjugate prior, the marginal distribution of $\mathbf{Y}^*=\mathbf{CY}$ is given by Equation (4.13) in Chapter 4, since all the assumptions leading to that equation are satisfied.

If, in particular, we take the diffuse prior assumption with ν_1 and $\mathbf{N}^*$ in that equation set to zero, we obtain the marginal density of $\mathbf{Y}^*$ as

$$f(\mathbf{Y}^*|\rho)\propto|\mathbf{X}'\mathbf{C}'\mathbf{CX}|^{-1/2}(ESS(\rho))^{-T/2}$$

where

$$ESS(\rho)=\left[\mathbf{CY}-\mathbf{CXb}(\rho)\right]'\left[\mathbf{CY}-\mathbf{CXb}(\rho)\right].$$

Of course, $\mathbf{Y}^*$ is not observed, and it is necessary to transform this density into a density of $\mathbf{Y}=\mathbf{C}^{-1}\mathbf{Y}^*$, a transformation with Jacobian $|\mathbf{C}|=(1-\rho^2)^{(1-T)/2}$:

$$f(\mathbf{Y}|\rho)\propto(1-\rho^2)^{(1-T)/2}|\mathbf{X}'\mathbf{C}'\mathbf{CX}|^{-1/2}(ESS(\rho))^{-T/2}.$$

Under the same assumptions,[2] $\boldsymbol{\beta}$ given ρ has a Student distribution with parameters

$$E(\boldsymbol{\beta}|\rho)=\mathbf{b}(\rho)=(\mathbf{X}'\mathbf{C}'\mathbf{CX})^{-1}\mathbf{X}'\mathbf{C}'\mathbf{CY}$$
$$\mathbf{H}^{**}(\rho)=\mathbf{X}'\mathbf{C}'\mathbf{CX}/s^2$$
$$\nu=T$$
$$\nu s^2=ESS(\rho).$$

As a result, the marginal posterior distribution of $\boldsymbol{\beta}$ is a mixture of Student distributions.[3]

[2]This follows directly from the material in Chapter 4. Incidentally, there are certain technical differences between this treatment of autocorrelation and Zellner's (1971, pp. 86–98), having to do with the distribution of the first observation and also the choice of diffuse prior.

[3]See Zellner (1971, pp. 86–98) for an explicit expression for the result of an essentially similar integral.

As an interesting alternative way of looking at these distributions, observe that $\boldsymbol{\Sigma}^{-1}$ can be decomposed as

$$\sigma^2(1-\rho^2)\boldsymbol{\Sigma}^{-1}=(1-2\rho+\rho^2)\mathbf{I}$$

$$+\rho\begin{bmatrix} 1 & -1 & 0 & \cdots & 0 & 0 \\ -1 & 2 & -1 & \cdots & 0 & 0 \\ 0 & -1 & 2 & \cdots & 0 & 0 \\ \vdots & & \vdots & & \vdots & \vdots \\ 0 & 0 & 0 & \cdots & 2 & -1 \\ 0 & 0 & 0 & \cdots & -1 & 1 \end{bmatrix}$$

$$+(\rho-\rho^2)\begin{bmatrix} 1 & 0 & 0 & \cdots & 0 & 0 \\ 0 & 0 & 0 & \cdots & 0 & 0 \\ 0 & 0 & 0 & \cdots & 0 & 0 \\ \vdots & \cdot & \vdots & & \vdots & \vdots \\ 0 & 0 & 0 & & 0 & 0 \\ 0 & 0 & 0 & \cdots & 0 & 1 \end{bmatrix}$$

$$\equiv(1-2\rho+\rho^2)\mathbf{I}+\rho\mathbf{A}+\rho(1-\rho)\mathbf{B}.$$

In the event that prior information is relatively diffuse, $\mathbf{N}^*=\mathbf{0}$, the (generalized least-squares) posterior mean of $\boldsymbol{\beta}$ is

$$E(\boldsymbol{\beta}|\mathbf{Y},\rho)=(\theta_1\mathbf{X}'\mathbf{X}+\theta_2\mathbf{X}'\mathbf{A}\mathbf{X}+\theta_3\mathbf{X}'\mathbf{B}\mathbf{X})^{-1}(\theta_1\mathbf{X}'\mathbf{Y}+\theta_2\mathbf{X}'\mathbf{A}\mathbf{Y}+\theta_3\mathbf{X}'\mathbf{B}\mathbf{Y}),$$

where

$$\theta_1=(1-2\rho+\rho^2)$$
$$\theta_2=\rho$$
$$\theta_3=\rho(1-\rho).$$

Finally, since $\mathbf{B}$ is almost a zero matrix, we may write the conditional posterior mean approximately as

$$E(\boldsymbol{\beta}|\mathbf{Y},\rho)\doteq(\theta_1\mathbf{X}'\mathbf{X}+\theta_2\mathbf{X}'\mathbf{A}\mathbf{X})^{-1}(\theta_1\mathbf{X}'\mathbf{X}\mathbf{b}+\theta_2\mathbf{X}'\mathbf{A}\mathbf{X}\mathbf{b}^{\Delta})$$

where $\mathbf{b}$ is the usual least-squares estimate and $\mathbf{b}^{\Delta}$ is the "first difference" estimate, $\mathbf{b}^{\Delta}=(\mathbf{X}'\mathbf{A}\mathbf{X})^{-1}\mathbf{X}'\mathbf{A}\mathbf{Y}$ computed by least squares with data in first difference form

$$Y_t-Y_{t-1}=\sum_j\beta_j(x_{jt}-x_{jt-1}).$$

In words, the generalized least-squares estimate of $\boldsymbol{\beta}$ is approximately a matrix-weighted average of ordinary least squares and least squares with data in first differences. It is worthwhile recalling here the discussion of matrix-weighted averages in Chapter 5, in particular, the fact that the average may be quite unlike either of its components $\mathbf{b}$ or $\mathbf{b}^{\Delta}$.

The first-order autocorrelation matrix (8.3) is one very special kind of interdependence. For other models the reader is referred to Granger and Newbold (1976) or to Box and Jenkins (1970).

8.2 Outliers and Nonnormal Errors

Researchers who share with computers the task of examining data points almost always want to discard extreme observations. In some cases the resultant apparent inferences are greatly affected by the choice of observations. Were the assumptions implicit in the least-squares estimate actually accepted, there would be no logic to such a procedure. But in the rejection of outliers, a researcher is implicitly rejecting the assumption of normality, in particular, he is opting for a distribution with fatter tails than a normal distribution. Just as it is undesirable to choose a prior or a model after having seen the data, it is also undesirable to choose an implicit data density in this way. First of all, the inferences that are thus gathered are not fully legitimate. Second, the rejection of outliers tends to ignore the fact that it may be quite important to know the probability laws according to which the outliers are generated. In this section we consider formal models that involve "outlier rejection." It must be said at the outset that these models are relatively intractable and that cheaper, "data analytic" methods of discarding outliers may be preferred.[4] In fact, the discussion in this section may be nothing more than an apology for reasonable procedures.

Consider first the following model, due to Box and Tiao (1968). Let the values $Y_1, Y_2, \ldots, Y_T$ be a random sample from a normal population with mean μ and variance σ^2. Assume that $Z_t = Y_t$ is usually observed but that occasionally an outlier occurs and instead $Z_{t'} = Y_{t'} + e_{t'}$ is observed, where $e_{t'}$ is distributed normally with mean 0 and variance $\sigma^2 + \phi^2$. If σ^2 and ϕ^2 as well as the occurrence of the outliers are known, then the mode of the likelihood function of μ is located at the weighted mean of the observations with weight σ^{-2} on the regular observations and weight $(\sigma^2+\phi^2)^{-1}$ on the outliers. If I is the set of n regular observations and $\bar{I}$ is the set of $T-n$ outliers, the likelihood function is

$$f(Y_1, Y_2, \ldots, Y_T | \mu, \sigma^2, \phi^2) \propto (\sigma^2)^{-n/2} \exp\left[-\frac{1}{2\sigma^2} \sum_I (Y_t - \mu)^2 \right]$$

$$(\sigma^2+\phi^2)^{-(T-n)/2} \exp\left[-\frac{1}{2(\sigma^2+\phi^2)} \sum_{\bar{I}} (Y_t - \mu)^2 \right].$$

[4]By the term "data analytic" methods I refer to procedures such as discarding observations more than three standard deviations from the mean. These procedures do not refer explicitly to data densities. For reviews see Hogg (1974) and Huber (1972).

If the set $\bar{I}$ of outliers were known in advance, the analysis is relatively straightforward. The more interesting case with $\bar{I}$ uncertain involves great complications, since there are 2^T different subsets $\bar{I}$. As a result, there are 2^T different "models," each corresponding to a different allocation of outliers, and the posterior distribution is a mixture, $\Sigma_i w_i f_i(\mu)$, of 2^T distributions.

Note that the implicit data density in this example is a mixture of normals

$$f(Z_t|\mu,\sigma^2,\phi^2)=\pi f_N(Z_t|\mu,\sigma^2)+(1-\pi)f_N(Z_t|\mu,\sigma^2+\phi^2)$$

where $1-\pi$ is the probability of an outlier and $f_N(Z|\mu,\sigma^2)$ indicates a normal density with mean μ and variance σ^2. Another fat-tailed distribution that is a continuous mixture of normals is the Student function

$$f_S(Z|\mu,s^2,\nu)=\int f_N(Z|\mu,\sigma^2)f_\gamma(\sigma^{-2}|s^2,\nu)d\sigma^{-2} \tag{8.4}$$

An analysis of Student sampling has been done by Blattberg and Gonedes (1975).

Another class of nonnormal distributions has been analyzed extensively by Box and Tiao (1973, Chapter 3). They explore the class of exponential power distributions

$$f(Y|\mu,\phi,\beta)=k\phi^{-1}\exp\left(-\tfrac{1}{2}\left|\frac{Y-\mu}{\phi}\right|^{2/(1+\beta)}\right),\quad -\infty<Y<\infty, \tag{8.5}$$

where k is a normalizing constant depending on β and the parameters satisfy $\phi>0$, $-\infty<\mu<\infty$, $-1<\beta\leqslant 1$, with $\beta=0$ corresponding to the normal distribution.

8.3 Pooling Disparate Evidence

The inferential models so far discussed have been constructed to answer questions of the form: given that a coin lands heads up, what conclusions may be drawn about the probability of getting a head if the same coin is flipped again? We now turn to questions of the form: given that coin A lands heads up, what conclusions may be drawn about the probability of getting a head if coin B is flipped? No objective distinction should be made between these two inferential problems. *Any* inductive inference depends on the *subjective* link between observed and unobserved events. Although there may be general agreement that two flips of the same coin are more closely linked than two flips of two different coins, there can be no incontrovertible argument to that effect. Thus in this section we consider the kind of joint priors for p_A and p_B that would induce us to use the observed behavior of coin A to draw inferences about p_B, the probability of a head if coin B is flipped. This also compels us in the next section to consider nontrivial joint priors for $(p_1, p_2, \ldots)$ where p_i is the probability of getting a head on the ith flip of some particular coin.

The regression model is written as

$$Y_{it} = \boldsymbol{\beta}_{it}'\mathbf{x}_{it} + \varepsilon_{it} \qquad \begin{array}{l} i=1,\ldots,N \\ t=1,\ldots,T \end{array} \tag{8.6}$$

where $\boldsymbol{\beta}_{it}$ is a $(k \times 1)$ vector of coefficients and $\mathbf{x}_{it}$ is a $(k \times 1)$ vector of observable explanatory variables possibly including a constant. As an example of such a model, Y_{it} may be purchases of oranges by individual i in time period t, and x_{it} may be the income of individual i in period t. The errors ε_{it} and the parameters $\boldsymbol{\beta}_{it}$ are doubly subscripted to reflect the fact that the demand for oranges varies across individuals and across time (as social tastes for oranges change or as the individual grows older, for example).

The question of interest is whether observation of $(Y_{j\tau}, \mathbf{x}_{j\tau}')$ yields information about $(\varepsilon_{it}, \boldsymbol{\beta}_{it})$ for $t \neq \tau$. Any solution to this problem depends on the prior distribution for the $(k+1)NT$ unobservables. One extreme would have the vector $(\varepsilon_{it}, \boldsymbol{\beta}_{it}')$ independent of $(\varepsilon_{j\tau}, \boldsymbol{\beta}_{j\tau}')$ for all $(i,t) \neq (j,\tau)$, and it is obvious, in that case, that there is no information in the observation of the jth process, $Y_{j\tau}$, about the unobservables of the ith process $(\varepsilon_{it}, \boldsymbol{\beta}_{it}')$. Other assumptions imply pooling of evidence in one way or another. Let us for now postpone the issues raised by the variability of parameters over time and impose the conditions

$$\boldsymbol{\beta}_{it} = \boldsymbol{\beta}_i \qquad \text{for all } t.$$

This implies a set of N equations

$$Y_{it} = \boldsymbol{\beta}_i'\mathbf{x}_{it} + \varepsilon_{it}, \qquad i=1,\ldots,N \tag{8.7}$$

each of which describes the generation of T observations. Vectors of observations and errors of the ith process are denoted by

$$\mathbf{Y}_i = \begin{bmatrix} Y_{i1} \\ Y_{i2} \\ \vdots \\ Y_{iT} \end{bmatrix}, \qquad \mathbf{X}_i = \begin{bmatrix} \mathbf{x}_{i1}' \\ \mathbf{x}_{i2}' \\ \vdots \\ \mathbf{x}_{iT}' \end{bmatrix}$$

and the whole set of observations by

$$\mathbf{Y} = \begin{bmatrix} \mathbf{Y}_1 \\ \mathbf{Y}_2 \\ \vdots \\ \mathbf{Y}_N \end{bmatrix}, \quad \mathbf{X} = \begin{bmatrix} \mathbf{X}_1 & \mathbf{0} & \cdots & \mathbf{0} \\ \mathbf{0} & \mathbf{X}_2 & \cdots & \mathbf{0} \\ \vdots & \vdots & & \vdots \\ \mathbf{0} & \mathbf{0} & \cdots & \mathbf{X}_N \end{bmatrix} \quad \boldsymbol{\beta} = \begin{bmatrix} \boldsymbol{\beta}_1 \\ \boldsymbol{\beta}_2 \\ \vdots \\ \boldsymbol{\beta}_N \end{bmatrix} \quad \boldsymbol{\varepsilon} = \begin{bmatrix} \varepsilon_1 \\ \varepsilon_2 \\ \vdots \\ \varepsilon_N \end{bmatrix}$$

This allows us to write the set of regression equations compactly as

$$\mathbf{Y}=\mathbf{X}\boldsymbol{\beta}+\boldsymbol{\varepsilon}. \tag{8.8}$$

The generalized least-squares estimate of $\boldsymbol{\beta}$ is thus

$$\mathbf{b}=(\mathbf{X}'\boldsymbol{\Sigma}^{-1}\mathbf{X})^{-1}\mathbf{X}'\boldsymbol{\Sigma}^{-1}\mathbf{Y}$$

where $\boldsymbol{\Sigma}$ is the $NT\times NT$ variance matrix of $\boldsymbol{\varepsilon}$. Furthermore, assuming a normal prior for $\boldsymbol{\beta}$ with mean $\mathbf{b}^*$ and variance $(\mathbf{H}^*)^{-1}$, the posterior moments of $\boldsymbol{\beta}$ are

$$E(\boldsymbol{\beta}|\mathbf{Y},\mathbf{X})=(\mathbf{H}^*+\mathbf{X}'\boldsymbol{\Sigma}^{-1}\mathbf{X})^{-1}(\mathbf{H}^*\mathbf{b}^*+\mathbf{X}'\boldsymbol{\Sigma}^{-1}\mathbf{X}\mathbf{b}) \tag{8.9}$$

$$V(\boldsymbol{\beta}|\mathbf{Y},\mathbf{X})=(\mathbf{H}^*+\mathbf{X}'\boldsymbol{\Sigma}^{-1}\mathbf{X})^{-1}. \tag{8.10}$$

Several special cases of these formulas are now discussed.

MULTIVARIATE REGRESSIONS: UNCORRELATED COEFFICIENTS

The first model of interest allows the errors ε_{it} to be correlated across equations but constrains to zero the prior covariance between coefficients of different equations. The initial reaction that this is unlikely to lead to pooling of evidence has led Zellner (1971) to call the multivariate regression model a system of "seemingly unrelated regressions."

Using the notation of equation (8.8) let the prior moments of $\boldsymbol{\beta}$ be

$$E(\boldsymbol{\beta})=\mathbf{b}^*,\ V(\boldsymbol{\beta})=\mathbf{V}^*=\mathbf{H}^{*-1}=\begin{bmatrix}\mathbf{H}_1^{*-1} & \mathbf{0} & \cdots & \mathbf{0}\\ \mathbf{0} & \mathbf{H}_2^{*-1} & \cdots & \mathbf{0}\\ \vdots & \vdots & & \vdots\\ \mathbf{0} & \mathbf{0} & \cdots & \mathbf{H}_N^{*-1}\end{bmatrix} \tag{8.11}$$

thereby indicating that there is no a priori correlation between the coefficients in the various processes. The processes are related in the sense that the residual terms are correlated:

$$E(\varepsilon_{it}\varepsilon_{j\tau})=\begin{cases}\sigma_{ij} & \text{if } t=\tau\\ 0 & \text{otherwise}\end{cases}.$$

Thus, letting $\boldsymbol{\Omega}$ be the $N\times N$ matrix of contemporaneous covariances $\boldsymbol{\Omega}=\{\sigma_{ij}\}$, the covariance matrix of $\boldsymbol{\varepsilon}$ becomes

$$\boldsymbol{\Sigma}=V(\boldsymbol{\varepsilon})=\boldsymbol{\Omega}\otimes\mathbf{I}_T$$

$$\equiv\begin{bmatrix}\sigma_{11}\mathbf{I}_T & \sigma_{12}\mathbf{I}_T\cdots & \sigma_{1N}\mathbf{I}_T\\ \sigma_{21}\mathbf{I}_T & \sigma_{22}\mathbf{I}_T\cdots & \sigma_{2N}\mathbf{I}_T\\ \vdots & \vdots & \vdots\\ \sigma_{N1}\mathbf{I}_T & \sigma_{N2}\mathbf{I}_T\cdots & \sigma_{NN}\mathbf{I}_T\end{bmatrix}, \tag{8.12}$$

where $\mathbf{I}_T$ denotes a $T \times T$ identity and where $\otimes$ is the Kronecker product. With (8.11) and (8.12), the posterior moments (8.9) and (8.10) become

$$E(\boldsymbol{\beta}|\mathbf{Y},\mathbf{X},\boldsymbol{\Omega}) = \left(\mathbf{H}^* + \mathbf{X}'[\boldsymbol{\Omega}\otimes\mathbf{I}_T]^{-1}\mathbf{X}\right)^{-1}\left(\mathbf{H}^*\mathbf{b}^* + \mathbf{X}'[\mathbf{I}_T\otimes\boldsymbol{\Omega}]^{-1}\mathbf{Y}\right) \quad (8.13)$$

$$V(\boldsymbol{\beta}|\mathbf{Y},\mathbf{X},\boldsymbol{\Omega}) = \left(\mathbf{H}^* + \mathbf{X}'[\boldsymbol{\Omega}\otimes\mathbf{I}_T]^{-1}\mathbf{X}\right)^{-1} \quad (8.14)$$

with[5]

$$[\boldsymbol{\Omega}\otimes\mathbf{I}_T]^{-1} = \boldsymbol{\Omega}^{-1}\otimes\mathbf{I}_T\,.$$

The further assumption that the explanatory variables are the same in each equation $\mathbf{X}_i = \mathbf{X}_j$ involves no loss in generality, since the list of explanatory variables may be implicitly varied by using prior distributions that concentrate the probability of some parameters in the neighborhood of the origin. In Kronecker notation, let $\mathbf{X} = \mathbf{I}_N \otimes \mathbf{X}_i$, and the formula for the posterior mean becomes

$$E(\boldsymbol{\beta}|\mathbf{Y},\mathbf{X},\boldsymbol{\Omega}) = \left(\mathbf{H}^* + \boldsymbol{\Omega}^{-1}\otimes\mathbf{X}_i'\mathbf{X}_i\right)^{-1}\left(\mathbf{H}^*\mathbf{b}^* + (\boldsymbol{\Omega}^{-1}\otimes\mathbf{X}_i')\mathbf{Y}\right)$$

since

$$(\mathbf{I}_N\otimes\mathbf{X}_i')(\boldsymbol{\Omega}^{-1}\otimes\mathbf{I}_T)(\mathbf{I}_N\otimes\mathbf{X}_i) = (\boldsymbol{\Omega}^{-1}\otimes\mathbf{X}_i')(\mathbf{I}_N\otimes\mathbf{X}_i) = \boldsymbol{\Omega}^{-1}\otimes\mathbf{X}_i'\mathbf{X}_i$$

and

$$(\mathbf{I}_N\otimes\mathbf{X}_i')(\boldsymbol{\Omega}^{-1}\otimes\mathbf{I}_T)\mathbf{Y} = (\boldsymbol{\Omega}^{-1}\otimes\mathbf{X}_i')\mathbf{Y}.$$

Two extreme cases imply no pooling of information across equations. If prior information is relatively diffuse, $\mathbf{H}^* = \mathbf{0}$, the posterior location becomes

$$\begin{aligned} E(\boldsymbol{\beta}|\mathbf{Y},\mathbf{X},\boldsymbol{\Omega}) &= (\boldsymbol{\Omega}^{-1}\otimes\mathbf{X}_i'\mathbf{X}_i)^{-1}(\boldsymbol{\Omega}^{-1}\otimes\mathbf{X}_i')\mathbf{Y} \\ &= \left[\boldsymbol{\Omega}\otimes(\mathbf{X}_i'\mathbf{X}_i)^{-1}\right]\left[\boldsymbol{\Omega}^{-1}\otimes\mathbf{X}_i'\right]\mathbf{Y} \\ &= \left[\mathbf{I}_N\otimes(\mathbf{X}_i'\mathbf{X}_i)^{-1}\mathbf{X}_i\right]\mathbf{Y} \\ &= \begin{bmatrix} (\mathbf{X}_1'\mathbf{X}_1)^{-1}\mathbf{X}_1'\mathbf{Y}_1 \\ (\mathbf{X}_2'\mathbf{X}_2)^{-1}\mathbf{X}_2'\mathbf{Y}_2 \\ \vdots \\ (\mathbf{X}_N'\mathbf{X}_N)^{-1}\mathbf{X}_N'\mathbf{Y}_N \end{bmatrix}, \end{aligned} \quad (8.15)$$

which makes use of the assumptions $\mathbf{X}_i = \mathbf{X}_j$ for all i, j. In words, if prior information is relatively diffuse about the parameters $\boldsymbol{\beta}$, the posterior

[5]See Appendix 1 for the algebra of Kronecker products. Two properties are used here $(\mathbf{A}\otimes\mathbf{B})(\mathbf{C}\otimes\mathbf{D}) = \mathbf{AC}\otimes\mathbf{BD}$ and $(\mathbf{A}\otimes\mathbf{B})^{-1} = (\mathbf{A}^{-1}\otimes\mathbf{B}^{-1})$.

location can be computed by performing least squares, equation by equation. Similarly, if $\mathbf{\Omega}$ is a diagonal matrix, the formulas can be written as

$$E(\boldsymbol{\beta}_i|\mathbf{Y},\mathbf{X},\mathbf{\Omega})=(\mathbf{H}^*{}_i+\sigma_{ii}^{-1}\mathbf{X}_i'\mathbf{X}_i)^{-1}(\mathbf{H}_i^*\mathbf{b}_i^*+\sigma_{ii}^{-1}\mathbf{X}_i'\mathbf{Y}_i), \quad i=1,\ldots,N,$$

and the N equations may be analyzed separately.

Thus when the coefficients in the various equations are a priori independent with covariance matrix (8.11), the desire to pool evidence springs from the coincidence of correlated errors and prior information about the coefficients. Since correlation of the error terms may in many circumstances be significant, the critical bottleneck to pooling across equations is the formation of legitimate prior information about the coefficients. It may seem strange that prior information about the coefficients in one equation induces this kind of pooling. More peculiar still is the fact that if $\boldsymbol{\beta}_i$ is the parameter of interest, until $\mathbf{Y}_i$ is observed, there is no informational value in observations of the other processes $\mathbf{Y}_j$, since the prior and posterior distributions of $\boldsymbol{\beta}_i$ coincide, $f(\boldsymbol{\beta}_i)=f(\boldsymbol{\beta}_i|\mathbf{Y}_j)$ for $j\neq i$.[6] Thus the only effect of $\mathbf{Y}_j$ is to alter the interpretation of $\mathbf{Y}_i$. This may all become clear if it is pointed out that the conditional mean of the residual in the ith equation depends on the true residual in the jth equation, $E(\boldsymbol{\varepsilon}_i|\mathbf{Y}_j,\mathbf{X},\boldsymbol{\beta}_j)=\sigma_{ij}\sigma_{jj}^{-1}(\mathbf{Y}_j-\mathbf{X}_j\boldsymbol{\beta}_j)$. To adjust for this nonzero mean it would be appropriate to regress $\mathbf{Y}_i-\sigma_{ij}\sigma_{jj}^{-1}(\mathbf{Y}_j-\mathbf{X}_j\boldsymbol{\beta}_j)$ on $\mathbf{X}_i$.[7] When the prior for $\boldsymbol{\beta}_j$ is diffuse, the best estimate of $\mathbf{Y}_j-\mathbf{X}_j\boldsymbol{\beta}_j$ is just the vector of residuals in the jth equation. This vector is by construction orthogonal to $\mathbf{X}_j$ and hence to $\mathbf{X}_i$. This variable can, therefore, have no effect on the estimate of $\boldsymbol{\beta}_i$ when it is subtracted from $\mathbf{Y}_i$.

The point I have been leading up to may now be clear. The pooling phenomenon associated with multivariate regression is very subtly based on prior information. "Seemingly unrelated" regression estimates should be used with the same kind of care that we would apply to problems that call for more overt forms of prior information. Mechanical, thoughtless use of such routines is to be discouraged.

MULTIVARIATE REGRESSION: CORRELATED COEFFICIENTS

A more direct reason for pooling evidence across equations is a priori correlation of the coefficients. A time series of observations on many individuals is an example; the individuals are unlikely to be identical, but we do expect them to be "similar." For each individual we would specify a

[6]The reader is asked to verify this. Gary Chamberlain has pointed out to me a similar less confusing situation. Suppose in the usual normal regression model $\mathbf{Y}$ depends on two explanatory variables $\mathbf{Y}=\mathbf{x}_1\beta_1+\mathbf{x}_2\beta_2+\mathbf{u}$. The conditional distribution $f(\beta_1|\mathbf{x}_2)$ is independent of $\mathbf{x}_2$, yet $f(\beta_1|\mathbf{Y},\mathbf{x}_2)$ depends on $\mathbf{x}_2$.

[7]The resultant estimate of $\boldsymbol{\beta}_i$ is $\mathbf{b}_i-\sigma_{ij}\sigma_{jj}^{-1}(\mathbf{b}_j-\boldsymbol{\beta}_j)$ where $\mathbf{b}_i$ is the usual least-squares estimate.

different regression equation, but we would have a prior distribution that summarizes the feeling that the coefficients are likely to be similar.

As an illustration of the intuitively compelling reasons for some kind of pooling of observations across processes, suppose we estimated for eight different individuals a linear consumption function with consumption expenditures as the dependent variable and income as the explanatory variable. The following table of least-squares results summarizes the data information about the slope coefficient (the marginal propensity to consume) in these equations:

individual	1	2	3	4	5	6	7	8
least-squares estimate	.81	.80	.82	.86	.85	.84	.86	.83
standard error	2.1	2.4	1.8	1.2	3.6	4.0	5.1	1.1

Note that the standard errors are very large, and a 95% posterior interval for individual one's coefficient, assuming a relatively diffuse prior, would be approximately $.81 \pm 4.2$. But notice also that the marginal propensities to consume are very close for all individuals. This fact intuitively makes us more confident about the number .81 than is suggested by this interval. Furthermore, we may want to adjust the number .81 upward to make it more representative of the class of estimates.

It goes without saying that it is not always desirable to pool evidence in this way. If equation 1 were a consumption function, equation 2 a production function, equation 3 an investment function, and so forth, the peculiar coincidence of coefficients would be regarded as a statistical artifact, and no pooling would be desirable. But for "similar" processes, pooling is intuitively sensible and, in fact, is necessary to avoid the following "clairvoyant" paradox: In a population of individuals that contains no clairvoyants, you will come to believe with essential certainty that someone is a clairvoyant.

As an example of the clairvoyant problem, suppose N different coins are flipped T times each, in an effort to find a coin that lands heads up with high probability. Let p_i be the probability of a head if coin i is flipped, and take as observations T flips of each of N different coins. Suppose that a prior distribution for these probabilities is selected that would not imply pooling of the evidence across different coins. In particular, let p_i, $i = 1,\ldots,N$ be a set of N independent identically distributed random variables. If the number of coins is large enough, there will almost certainly be at least one coin that yielded all heads, even if the probability of a head is one-half for all coins. If, furthermore, T is large enough, the evidence of T heads in T flips will lead to the conclusion that this coin will almost certainly yield a head again. Thus you will conclude that there is a coin that usually lands heads up. This is a perfectly proper Bayesian procedure, and fault cannot be found with it on logical grounds. If you do not like its implications, it is only because you do not like the prior distribution. The prior implicitly says that the probabilities p_i are drawn independently from an urn that does contain some values of p close to one. Given enough such selections, you should indeed obtain a p_i close to one. An alternative prior distribution might have the variance of this urn be uncertain. The probabilities p_i would then be a priori dependent, $f(p_1,\ldots,p_N)=\int_\theta[\Pi_i f(p_i|\theta)]d\theta$, where θ is a variance parameter. This prior would imply pooling of the evidence across different coins and would not necessarily lead you to conclude that there is a biased (clairvoyant) coin.

Three "everyday" examples may make the point most forcefully.

Example 1. Several days before a United States presidential election television newsmen find a town that has always voted for the winner of the past elections. A preference poll of the town's inhabitants is then used to predict the outcome of the election.

Example 2. One thousand individuals are sent a letter describing a revolutionary new investment advisory service. Half are told that stock A will rise in value, half are told that it will fall in value. If stock A rises in value those 500 who were so informed are sent another letter. Half are told that stock B will rise; half are told it will fall. By this process you will end up with approximately ten individuals who have been given seven accurate stock tips in a row. It is then time to begin charging for the investment advice.

Example 3. At the end of the first month of the baseball season, there are always some hitters with batting percentages above .400. But at the end of the season, it is very rare to have even one hitter with an average above .400. (For an analysis of batting averages, see Efron and Morris, 1975.)

Returning now to the regression problem, the two significant features of the pooling phenomenon above—shrinking estimates toward a common mean and reducing standard errors—can be effected by selecting a prior covariance matrix $(\mathbf{H}^*)^{-1}$ that does not have the block diagonal form. A convenient way to construct such a matrix is to assume that the vectors $\boldsymbol{\beta}_i$ are an exchangeable normal process, that is, to assume that your opinions about the vectors are normal and unaffected by their ordering. In that event you will act as if the coefficients were selected randomly from a fixed normal urn:[8]

$$\boldsymbol{\beta}_i=\boldsymbol{\beta}+\mathbf{u}_i$$

[8]The notion of exchangeability is due to deFinetti (1937) and is skillfully exploited by Lindley and Smith (1972), from which this section is derived.

with $\mathbf{u}_i$ distributed normally with mean vector zero and covariance matrix $V(\mathbf{u}_i)=\mathbf{V}$. A normal prior for the "hyperparameter" $\bar{\boldsymbol{\beta}}$ with mean ξ and variance $V(\bar{\boldsymbol{\beta}})$, implies that $\boldsymbol{\beta}_i$ is the sum of two normal vectors and is itself normal with moments

$$E\left(\begin{bmatrix} \beta_1 \\ \beta_2 \\ \vdots \\ \beta_N \end{bmatrix}\right) = \begin{bmatrix} \xi \\ \xi \\ \vdots \\ \xi \end{bmatrix} = \mathbf{1}_N \otimes \xi$$

$$V\left(\begin{bmatrix} \beta_1 \\ \beta_2 \\ \vdots \\ \beta_N \end{bmatrix}\right) = \begin{bmatrix} \mathbf{V}+V(\bar{\beta}) & V(\bar{\beta}) & \cdots & V(\bar{\beta}) \\ V(\bar{\beta}) & \mathbf{V}+V(\bar{\beta}) & & V(\bar{\beta}) \\ \vdots & \vdots & & \vdots \\ V(\bar{\beta}) & V(\bar{\beta}) & \cdots & \mathbf{V}+V(\bar{\beta}) \end{bmatrix}$$

$$= [\mathbf{I}_N \otimes \mathbf{V}] + [\mathbf{1}_N \otimes \mathbf{I}_k] V(\bar{\beta}) [\mathbf{1}_N \otimes \mathbf{I}_k]'.$$

The variance matrix may be inverted to obtain the precision matrix.[9]

$$\begin{aligned} \mathbf{H}^* = V^{-1}(\beta) &= [\mathbf{I}_N \otimes \mathbf{V}]^{-1} - [\mathbf{I}_N \otimes \mathbf{V}]^{-1}[\mathbf{1}_N \otimes \mathbf{I}_k] \\ &\quad \left[(\mathbf{1}_N \otimes \mathbf{I}_k)'(\mathbf{I}_N \otimes \mathbf{V})^{-1}(\mathbf{1}_N \otimes \mathbf{I}_k) + V^{-1}(\bar{\beta})\right]^{-1} \\ &\quad [\mathbf{1}_N \otimes \mathbf{I}_k]'[\mathbf{I}_N \otimes \mathbf{V}]^{-1} \\ &= [\mathbf{I}_N \otimes \mathbf{V}^{-1}] - [\mathbf{1}_N \otimes \mathbf{V}^{-1}]\left[\mathbf{1}_N'\mathbf{1}_N \otimes \mathbf{V}^{-1} + V^{-1}(\bar{\beta})\right]^{-1}[\mathbf{1}_N' \otimes \mathbf{V}^{-1}]. \end{aligned}$$

Finally, under the further natural assumption that information about the mean $\bar{\beta}$ is relatively weak, $V^{-1}(\bar{\beta})$ may be set to a zero matrix to obtain the (singular) precision matrix:

$$\begin{aligned} V^{-1}(\beta) &= [\mathbf{I}_N \otimes \mathbf{V}^{-1}] - [\mathbf{1}_N \otimes \mathbf{V}^{-1}][N \otimes \mathbf{V}^{-1}]^{-1}[\mathbf{1}_N' \otimes \mathbf{V}^{-1}] \\ &= [\mathbf{I}_N \otimes \mathbf{V}^{-1}] - [\mathbf{1}_N N^{-1} \otimes \mathbf{I}_k][\mathbf{1}_N' \otimes \mathbf{V}^{-1}] \\ &= \mathbf{I}_N \otimes \mathbf{V}^{-1} - \mathbf{1}_N N^{-1}\mathbf{1}_N' \otimes \mathbf{V}^{-1} \\ &= (\mathbf{I}_N - \mathbf{1}_N N^{-1}\mathbf{1}_N') \otimes \mathbf{V}^{-1}. \end{aligned}$$

[9]Using the formula $(\mathbf{A}+\mathbf{BDB}')^{-1} = \mathbf{A}^{-1} - \mathbf{A}^{-1}\mathbf{B}(\mathbf{B}'\mathbf{A}^{-1}\mathbf{B}+\mathbf{D}^{-1})^{-1}\mathbf{B}'\mathbf{A}^{-1}$.

Note that this precision matrix times the prior location is zero, independent of the choice of ξ

$$[(\mathbf{I}_N-\mathbf{1}_N N^{-1}\mathbf{1}'_N)\otimes\mathbf{V}^{-1}][\mathbf{1}_N\otimes\xi]$$
$$=(\mathbf{1}_N-\mathbf{1}_N N^{-1}\mathbf{1}'_N\mathbf{1}_N)\otimes\mathbf{V}^{-1}\xi=\mathbf{1}_N 0\otimes\mathbf{V}^{-1}\xi=\mathbf{0}_{kN}$$

where $\mathbf{0}_{kN}$ is a zero vector of length kN.

The posterior moments may now be written as

$$\mathbf{b}^{**}=E(\boldsymbol{\beta}|\mathbf{Y},\mathbf{X},\Omega)=\left[(\mathbf{I}_N-\mathbf{1}_N N^{-1}\mathbf{1}'_N)\otimes\mathbf{V}^{-1}\right.$$
$$\left.+\mathbf{X}'(\Omega^{-1}\otimes\mathbf{I}_T)\mathbf{X}\right]^{-1}\left[\mathbf{X}'(\Omega^{-1}\otimes\mathbf{I}_T)\mathbf{X}\right]\mathbf{b} \tag{8.16}$$

$$V(\boldsymbol{\beta}|\mathbf{Y},\mathbf{X},\Omega)=\left[(\mathbf{I}_N-\mathbf{1}_N N^{-1}\mathbf{1}'_N)\otimes\mathbf{V}^{-1}+\mathbf{X}'(\Omega^{-1}\otimes\mathbf{I}_T)\mathbf{X}\right]^{-1} \tag{8.17}$$

where $\mathbf{b}$ is a solution to the normal equations

$$\mathbf{X}'(\Omega^{-1}\otimes\mathbf{I}_T)\mathbf{X}\mathbf{b}=\mathbf{X}'(\Omega^{-1}\otimes\mathbf{I}_T)\mathbf{Y}.$$

These formulas can be greatly simplified in the event that Ω is a diagonal matrix.

It is then convenient to write the posterior distribution of $\boldsymbol{\beta}$ and $\bar{\boldsymbol{\beta}}$ as the product of the conditional distribution of $\boldsymbol{\beta}$ given $\bar{\boldsymbol{\beta}}$ times a marginal on $\bar{\boldsymbol{\beta}}$. Conditional on $\bar{\boldsymbol{\beta}}$, $\boldsymbol{\beta}_i$ is independent of $\boldsymbol{\beta}_j$, $i\neq j$, with moments

$$E(\boldsymbol{\beta}_i|\bar{\boldsymbol{\beta}},\mathbf{Y},\mathbf{X},\Omega)=(\mathbf{V}^{-1}+\sigma_{ii}^{-1}\mathbf{X}'_i\mathbf{X}_i)^{-1}(\mathbf{V}^{-1}\bar{\boldsymbol{\beta}}+\sigma_{ii}^{-1}\mathbf{X}'_i\mathbf{X}_i\mathbf{b}_i) \tag{8.18}$$

$$V(\boldsymbol{\beta}_i|\bar{\boldsymbol{\beta}},\mathbf{Y},\mathbf{X},\Omega)=(\mathbf{V}^{-1}+\sigma_{ii}^{-1}\mathbf{X}'_i\mathbf{X}_i)^{-1}. \tag{8.19}$$

These are the usual formulas if the prior were normal with moments $\bar{\boldsymbol{\beta}}$ and $\mathbf{V}$. To compute the distribution of $\bar{\boldsymbol{\beta}}$ given $\mathbf{Y}$ it is necessary to write the distribution of $\mathbf{Y}$ given $\bar{\boldsymbol{\beta}}$ as

$$\mathbf{Y}_i=\mathbf{X}_i\boldsymbol{\beta}_i+\boldsymbol{\varepsilon}_i=\mathbf{X}_i\bar{\boldsymbol{\beta}}+\mathbf{X}_i\mathbf{u}_i+\boldsymbol{\varepsilon}_i,$$

which is the usual regression process with mean $\mathbf{X}_i\bar{\boldsymbol{\beta}}$ and variance $\mathbf{X}_i\mathbf{V}\mathbf{X}'_i+\sigma_{ii}\mathbf{I}_T$. With $\bar{\boldsymbol{\beta}}$ a priori diffuse and given all the vectors $\mathbf{Y}_i$, we have straightforwardly

$$E(\bar{\boldsymbol{\beta}}|\mathbf{Y},\Omega,\mathbf{X})=\left[\sum_i\mathbf{X}'_i(\mathbf{X}_i\mathbf{V}\mathbf{X}'_i+\sigma_{ii}\mathbf{I}_T)^{-1}\mathbf{X}_i\right]^{-1}\left[\sum_i\mathbf{X}'_i(\mathbf{X}_i\mathbf{V}\mathbf{X}'_i+\sigma_{ii}\mathbf{I}_T)^{-1}\mathbf{Y}_i\right],$$

which is the generalized least-squares estimate of $\bar{\boldsymbol{\beta}}$ and

$$V(\bar{\boldsymbol{\beta}}|\mathbf{Y},\Omega,\mathbf{X})=\left[\sum_i\mathbf{X}'_i(\mathbf{X}_i\mathbf{V}\mathbf{X}'_i+\sigma_{ii}\mathbf{I}_T)^{-1}\mathbf{X}_i\right]^{-1}. \tag{8.20}$$

These last two formulas can be further simplified by observing that

$$\begin{aligned}\mathbf{X}_i'(\mathbf{X}_i\mathbf{V}\mathbf{X}_i'+\sigma_{ii}\mathbf{I}_T)^{-1}&=\sigma_{ii}^{-1}\mathbf{X}_i'-\sigma_{ii}^{-1}\mathbf{X}_i'\mathbf{X}_i\left(\sigma_{ii}^{-1}\mathbf{X}_i'\mathbf{X}_i+\mathbf{V}^{-1}\right)^{-1}\mathbf{X}_i'\sigma_{ii}^{-1}\\&=\sigma_{ii}^{-1}\mathbf{X}_i'\mathbf{X}_i\left[(\mathbf{X}_i'\mathbf{X}_i)^{-1}-\sigma_{ii}^{-1}\left(\sigma_{ii}^{-1}\mathbf{X}_i'\mathbf{X}_i+\mathbf{V}^{-1}\right)^{-1}\right]\mathbf{X}_i'\\&=\sigma_i^{-1}\mathbf{X}_i'\mathbf{X}_i\left[\sigma_{ii}^{-1}\mathbf{I}_k+(\mathbf{X}_i'\mathbf{X}_i)^{-1}\mathbf{V}^{-1}-\sigma_{ii}^{-1}\mathbf{I}_k\right]\\&\quad\left(\sigma_{ii}^{-1}\mathbf{X}_i'\mathbf{X}_i+\mathbf{V}^{-1}\right)^{-1}\mathbf{X}_i'\\&=\sigma_{ii}^{-1}\mathbf{V}^{-1}\left(\sigma_{ii}^{-1}\mathbf{X}_i'\mathbf{X}_i+\mathbf{V}^{-1}\right)^{-1}\mathbf{X}_i'.\end{aligned}$$

Thus we may write

$$\begin{aligned}E\left(\bar{\boldsymbol{\beta}}|\mathbf{Y},\boldsymbol{\Omega},\mathbf{X}\right)=&\left[\sum_i\left(\sigma_{ii}^{-1}\mathbf{X}_i'\mathbf{X}_i+\mathbf{V}^{-1}\right)^{-1}\mathbf{X}_i'\mathbf{X}_i\sigma_{ii}^{-1}\right]^{-1}\\&\times\left[\sum_i\left(\sigma_{ii}^{-1}\mathbf{X}_i'\mathbf{X}_i+\mathbf{V}^{-1}\right)^{-1}\mathbf{X}_i'\mathbf{X}_i\mathbf{b}_i\sigma_{ii}^{-1}\right]\end{aligned}\tag{8.21}$$

where $\mathbf{b}_i=(\mathbf{X}_i'\mathbf{X}_i)^{-1}\mathbf{X}_i'\mathbf{Y}_i$.

The pooling of information across processes with this kind of correlation structure and with the processes otherwise independent ($\boldsymbol{\Omega}$ diagonal) is thus summarized by two equations. Equation (8.18) describes the posterior location of $\boldsymbol{\beta}_i$ as a compromise between the least-squares estimate $\mathbf{b}_i$ and the grand mean $\bar{\boldsymbol{\beta}}$, which is itself a matrix-weighted average (8.21) of each of the least-squares points.

ERROR-COMPONENTS MODEL

What is known as the error-components model, introduced into the econometric literature by Balestra and Nerlove (1966), is a special case of the multivariate model discussed in the previous subsection. The model assumes that the slope vectors in the various processes are identical and also constrains the contemporaneous precision matrix $\boldsymbol{\Omega}^{-1}$ to be proportional to a special matrix. The model is written as

$$y_{it}=\sum_{j=1}^{k-1}x_{ijt}\bar{\beta}_j+\bar{\beta}_0+\alpha_i+\gamma_t+\varepsilon_{it}\tag{8.22}$$

to indicate that there are $k-1$ slope parameters ($\bar{\beta}_j, j=1,\ldots,k-1$) common to every process and that the process level or constant includes four additive variables: a constant $\bar{\beta}_0$; a component α_i, common to all observations of the ith process; a component γ_t, common to all observations in the tth period; and an independent normal error ε_{it}, assumed to have mean zero and variance σ_ε^2. Furthermore, α_i and γ_t are assumed to be independently drawn from normal distributions with zero means and variances σ_α^2 and σ_γ^2. Finally, the vector $\bar{\boldsymbol{\beta}}$ is assigned a diffuse prior distribution.

The reader may verify that this model is a special case of the multivariate regression model with correlated errors and correlated parameters. For expository purposes it is most convenient to concentrate on the posterior distribution of $\bar{\boldsymbol{\beta}}'=(\bar{\beta}_0,\bar{\beta}_1,\ldots,\bar{\beta}_{k-1})$ by integrating out the parameters α_i and γ_t. The regression process may then be written as

$$\mathbf{Y}=\begin{bmatrix}\mathbf{X}_1\\ \mathbf{X}_2\\ \vdots\\ \mathbf{X}_N\end{bmatrix}\bar{\boldsymbol{\beta}}+\mathbf{u}\equiv\mathbf{X}\bar{\boldsymbol{\beta}}+\mathbf{u}$$

where $\mathbf{u}$ is the normal random variable

$$\mathbf{u}=\begin{bmatrix}\mathbf{1}_T & 0 & & 0\\ 0 & \mathbf{1}_T & & 0\\ & & \ddots & \\ 0 & 0 & & \mathbf{1}_T\end{bmatrix}\boldsymbol{\alpha}+\begin{bmatrix}\mathbf{I}_T\\ \mathbf{I}_T\\ \vdots\\ \mathbf{I}_T\end{bmatrix}\boldsymbol{\gamma}+\boldsymbol{\varepsilon}$$

$$=(\mathbf{I}_N\otimes\mathbf{1}_T)\boldsymbol{\alpha}+(\mathbf{1}_N\otimes\mathbf{I}_T)\boldsymbol{\gamma}+\boldsymbol{\varepsilon}$$

with mean $\mathbf{0}_{NT}$ and variance

$$\begin{aligned}V(\mathbf{u})&=\sigma_\alpha^2(\mathbf{I}_N\otimes\mathbf{1}_T)(\mathbf{I}_N\otimes\mathbf{1}_T')+\sigma_\gamma^2(\mathbf{1}_N\otimes\mathbf{I}_T)(\mathbf{1}_N'\otimes\mathbf{I}_T)+\sigma^2\mathbf{I}_{NT}\\ &=\sigma_\alpha^2(\mathbf{I}_N\otimes\mathbf{1}_T\mathbf{1}_T')+\sigma_\gamma^2(\mathbf{1}_N\mathbf{1}_N'\otimes\mathbf{I}_T)+\sigma^2\mathbf{I}_{NT}.\end{aligned}$$

The posterior mean of $\bar{\boldsymbol{\beta}}$ given the data $\mathbf{Y}$ and beginning with a diffuse prior is just the generalized least-squares estimate

$$E\left(\bar{\boldsymbol{\beta}}|\mathbf{Y},\sigma_\alpha^2,\sigma_\gamma^2,\sigma^2,\mathbf{X}\right)=\left(\mathbf{X}'V^{-1}(\mathbf{u})\mathbf{X}\right)^{-1}\mathbf{X}'V^{-1}(\mathbf{u})\mathbf{Y}.$$

In the special case in which there are no time effects, $\sigma_\gamma^2=0$, the variance matrix may be inverted to

$$\begin{aligned}V^{-1}(\mathbf{u})&=\left[\sigma^2\mathbf{I}_{NT}+\sigma_\alpha^2(\mathbf{I}_N\otimes\mathbf{1}_T)(\mathbf{I}_N\otimes\mathbf{1}_T')\right]^{-1}\\ &=\sigma^{-2}\mathbf{I}_{NT}-\sigma^2(\mathbf{I}_N\otimes\mathbf{1}_T)\left([\mathbf{I}_N\otimes\mathbf{1}_T'][\mathbf{I}_N\otimes\mathbf{1}_T]\sigma^{-2}+\sigma_\alpha^{-2}\mathbf{I}_N\right)^{-1}\\ &\quad(\mathbf{I}_N\otimes\mathbf{1}_T')\sigma^{-2}\\ &=\sigma^{-2}\mathbf{I}_{NT}-\sigma^{-4}\left(T\sigma^{-2}+\sigma_\alpha^{-2}\right)^{-1}(\mathbf{I}_N\otimes\mathbf{1}_T\mathbf{1}_T')\\ &=\theta_1\mathbf{I}_{NT}+\theta_2\mathbf{B}\end{aligned}$$

where

$$\mathbf{B}=\mathbf{I}_{NT}-(\mathbf{I}_N\otimes\mathbf{1}_T)T^{-1}(\mathbf{I}_N\otimes\mathbf{1}'_T)$$

$$\theta_1=\sigma^{-2}\left(\frac{\sigma^2}{T\sigma_\alpha^2+\sigma^2}\right)$$

$$\theta_2=\sigma^{-2}\left(\frac{T\sigma_\alpha^2}{T\sigma_\alpha^2+\alpha^2}\right).$$

The posterior location then becomes

$$E(\boldsymbol{\beta}|\mathbf{Y},\mathbf{X},\sigma_\alpha^2,\sigma^2)=(\theta_1\mathbf{X}'\mathbf{X}+\theta_2\mathbf{X}'\mathbf{B}\mathbf{X})^{-1}(\theta_1\mathbf{X}'\mathbf{Y}+\theta_2\mathbf{X}'\mathbf{B}\mathbf{Y}),$$

which is a matrix-weighted average of the least-squares estimate $(\mathbf{X}'\mathbf{X})^{-1}\mathbf{X}'\mathbf{Y}$ and the least-squares with dummy variables estimate $(\mathbf{X}'\mathbf{B}\mathbf{X})^{-1}\mathbf{X}'\mathbf{B}\mathbf{Y}$, corresponding, respectively, to the extreme prior variances of $\sigma_\alpha^2=0$ and $\sigma_\alpha^2=\infty$. As discussed extensively in Chapter 5, a matrix-weighted compromise between two estimates may turn out to be apparently unlike either of the estimates, and the usual procedure of looking at only the end points of this contract curve may be highly misleading.[10]

[10]Maddala (1971) writes this matrix-weighted average in a slightly different form. He sets

$$V^{-1}(\mathbf{u})=\theta_1\mathbf{A}+\theta_2\mathbf{B}$$

with

$$\mathbf{B}=\mathbf{I}_{NT}-(\mathbf{I}_N\otimes\mathbf{1}_T)T^{-1}(\mathbf{I}_N\otimes\mathbf{1}'_T)$$

$$\mathbf{A}=(\mathbf{I}_N\otimes\mathbf{1}_T)(\mathbf{I}_N\otimes\mathbf{1}'_T)$$

$$\theta_2=\sigma^{-2}$$

$$\theta_1=T^{-1}(T\sigma_\alpha^2+\sigma^2)^{-1}$$

The posterior location then becomes

$$E(\boldsymbol{\beta}|\mathbf{Y},\sigma_\alpha^2,\sigma^2,\mathbf{X})=(\theta_1\mathbf{X}'\mathbf{A}\mathbf{X}+\theta_2\mathbf{X}'\mathbf{B}\mathbf{X})^{-1}(\theta_1\mathbf{X}'\mathbf{A}\mathbf{Y}+\theta_2\mathbf{X}'\mathbf{B}\mathbf{Y}),$$

which is a matrix-weighted average of least-squares "with dummy variables," $(\mathbf{X}'\mathbf{B}\mathbf{X})^{-1}\mathbf{X}'\mathbf{B}\mathbf{Y}$ corresponding to $\sigma_\alpha^2=\infty$, and least-squares using as observations the average values of the variables for each process (with N observations of the form $\mathbf{1}'\mathbf{Y}_i/T,\mathbf{1}'\mathbf{X}_i/T$):

$$(\mathbf{X}'\mathbf{A}\mathbf{X})^{-1}\mathbf{X}'\mathbf{A}\mathbf{Y}=\left(\sum_i\mathbf{X}'_i\mathbf{1}\mathbf{1}'\mathbf{X}_i\right)^{-1}\left(\sum_i\mathbf{X}'_i\mathbf{1}\mathbf{1}'\mathbf{Y}_i\right)$$

However, $\theta_1/\theta_2=T^{-1}\sigma^2/(\sigma^2+T\sigma_\alpha^2)$ must be less than T^{-1}, and as a result the whole curve décolletage is not relevant. In fact, $\theta_2/\theta_1=T^{-1}$ produces the least-squares estimate $(\mathbf{X}'\mathbf{X})^{-1}(\mathbf{X}'\mathbf{Y})$, and the segment of the curve connecting this point to least-squares with average data should be regarded as a curiosum.

8.4 Time-Varying Parameters

The usual analysis of a regression process implicitly or explicitly rests on the assumption that the parameters that govern the generation of the data are "more-or-less" the same for all data points. Formal analysis requires the much stricter assumption of perfect constancy, although in practice this, like all other assumptions, is thought to hold in some approximate sense. The parameters are thought to be "sufficiently" constant to allow a fruitful analysis based on the constancy assumption. If the parameters do vary, then estimators describe "some kind of weighted" average of the parameter in question. The vagueness in this informal relaxation of the constancy assumption clearly leaves much to be desired, particularly in poorly specified models, in which "parameters" are functions of the time-varying correlations between the included and excluded variables.

An interesting example of conflicting behavior occurs when a data set is arbitrarily selected. Data sets are often truncated because of the possibility of structural shifts, yet the resulting data subset is then analyzed as if structural changes were impossible. For example, pre-1953 data may be excluded from the analysis on the basis of structural changes. Paradoxically, the same researcher who discards pre-1953 data on the basis of structural change proceeds to analyze the remaining data with simple regression methods. It is, of course, most unlikely that the economic world would undergo an important and fundamental change in 1953 and thereafter remain relatively stagnant. In fact, when we decide to ignore pre-1953 data we are likely to feel that the 1954 data point is only marginally relevant as well.

The heteroscedastic model with declining variances is often suggested in such circumstances, since it can be used to discount the importance of the earlier data points. Although this discounting is intellectually appealing, it is based on an unacceptable assumption about the behavior of the error term; specifically, data points are weighted by the precision of the error term which is assumed to be small for the older observations. However, one's desire to discount the older observations is not related to the precision of the process. Rather, as one gathers older and older data, he begins to question the appropriateness of the constancy assumption. He is likely to be interested in the most recent structure, and the more distant the data point, the less related the structure, and the more meaningless the information obtained. Although the heteroscedastic discount is appealing, there is no assurance that it accurately reflects the decay in the informational value associated with the changing structure, since it is based on the decay in informational value associated with a decreasing process precision.[11]

From a Bayesian point of view, this problem is straightforward. Every data point may be assumed to be generated by a unique regression process; one observation is made from each process, and the pooling of evidence across processes as described in Section 8.3 applies. Of course, the prior distribution reflects the fact that the regression parameters are thought to be roughly constant over time. This statistical model is the natural extreme of the Bayesian view of inference. Inasmuch as no two data points are related objectively in any way at all, it is impossible to make objective inferences about the nature of the world. Inferences are possible only if subjective prior information is available, that is, only if you (irrationally?) believe the world is orderly.

The model with time-varying parameters has three important implications. First, the discounting of the evidence in earlier observations is based on structural change. Second, the diffuseness of predictions increases naturally as we attempt to project the current structure farther and farther into the future. That is, the value of sample information decays with time, paralleling the decaying relationship between the sample process and the future process. It is intuitively clear that our ability to predict and/or control economic systems decays with time. Stochastic control systems built around a constant parameter assumption result in solutions that rest on greatly overestimated knowledge of the system's future. This tends to result in reckless current decisions, which ignore important elements of uncertainty in the future. The third implication of this model is that "outliers" are legitimately discarded when they suggest structural change. Extreme data points require a suitable adjustment of the regression coefficients applying to the outlier period, and regression coefficients applying to other periods may be insensitive to the presence of the outlier.

The model being discussed is

$$y_t = \mathbf{x}_t' \boldsymbol{\beta}_t + u_t, \qquad t = 1, \ldots, T, \tag{8.23}$$

[11] It should be pointed out that it is possible to build a formal model of time-varying parameters that does imply heteroscedasticity. Cooley and Prescott (1973a,b) write a model as $y_t = \beta x_t + \alpha_t$, where α_t is the time-varying parameter. The stochastic model they suggest can be described by the equations $\alpha_t = u_t + \varepsilon_t, u_t = u_{t-1} + v_t$, with v_t and ε_t being independent spherical normal random variables. Conditional on u_T, say, the variance of α_t is an increasing function of $|T-t|$, which is a heteroscedastic feature. But the model also implies a special kind of correlation between the residuals.

where $\boldsymbol{\beta}_t$ is a k-dimensional vector of parameters applying in the tth period, $\mathbf{x}_t$ is a $(k\times 1)$ vector of explanatory variables, and u_t is an independent normal random error with mean zero and variance σ^2. This can be written in the form of a multivariate regression as

$$\begin{bmatrix} y_1 \\ y_2 \\ \vdots \\ y_T \end{bmatrix} = \begin{bmatrix} \mathbf{x}_1' & \mathbf{0} & \cdots & \mathbf{0} \\ \mathbf{0} & \mathbf{x}_2' & \cdots & \mathbf{0} \\ \vdots & \vdots & & \\ \mathbf{0} & \mathbf{0} & \cdots & \mathbf{x}_T' \end{bmatrix} \begin{bmatrix} \boldsymbol{\beta}_1 \\ \boldsymbol{\beta}_2 \\ \vdots \\ \boldsymbol{\beta}_T \end{bmatrix} + \begin{bmatrix} u_1 \\ u_2 \\ \vdots \\ u_T \end{bmatrix}.$$

The prior distribution that is commonly used for time-varying parameters is normal with mean $E\boldsymbol{\beta}_i=\mathbf{b}^*$ and variance matrix

$$V\left(\begin{bmatrix} \boldsymbol{\beta}_1 \\ \boldsymbol{\beta}_2 \\ \boldsymbol{\beta}_3 \\ \vdots \\ \boldsymbol{\beta}_T \end{bmatrix}\right) = \begin{bmatrix} \mathbf{V} & \boldsymbol{\phi}\mathbf{V} & \boldsymbol{\phi}^2\mathbf{V} & \cdots & \boldsymbol{\phi}^{T-1}\mathbf{V} \\ \mathbf{V}\boldsymbol{\phi}' & \mathbf{V} & \boldsymbol{\phi}\mathbf{V} & \cdots & \boldsymbol{\phi}^{T-2}\mathbf{V} \\ \mathbf{V}\boldsymbol{\phi}'^2 & \mathbf{V}\boldsymbol{\phi}' & \mathbf{V} & \cdots & \boldsymbol{\phi}^{T-3}\mathbf{V} \\ \vdots & \vdots & \vdots & & \vdots \\ \mathbf{V}\boldsymbol{\phi}'^{T-1} & \mathbf{V}\boldsymbol{\phi}'^{T-2} & \mathbf{V}\boldsymbol{\phi}'^{T-3} & \cdots & \mathbf{V} \end{bmatrix}.$$

The reader may verify that the conditional moments of such a normal process are

$$E(\boldsymbol{\beta}_t|\boldsymbol{\beta}_{t-1},\boldsymbol{\beta}_{t-2},\ldots)=\mathbf{b}^*+\boldsymbol{\phi}(\boldsymbol{\beta}_{t-1}-\mathbf{b}^*) \tag{8.24}$$

$$V(\boldsymbol{\beta}_t|\boldsymbol{\beta}_{t-1},\boldsymbol{\beta}_{t-2},\ldots)=\mathbf{V}-\boldsymbol{\phi}\mathbf{V}^{-1}\boldsymbol{\phi}. \tag{8.25}$$

The important feature of these moments is that they depend only on the most recent value of the parameter vector. This allows us to write the prior as $f(\boldsymbol{\beta}_T|\boldsymbol{\beta}_{T-1})f(\boldsymbol{\beta}_{T-1}|\boldsymbol{\beta}_{T-2})\cdots f(\boldsymbol{\beta}_1)$.

Suppose first that we observe y_1 only. The distribution conditional on y_1 only is proportional to

$$\begin{aligned} f(\boldsymbol{\beta}|y_1) &\propto f(\boldsymbol{\beta}_T|\boldsymbol{\beta}_{T-1})f(\boldsymbol{\beta}_{T-1}|\boldsymbol{\beta}_{T-2})\cdots f(\boldsymbol{\beta}_1)f(y_1|\boldsymbol{\beta}_1) \\ &= f(\boldsymbol{\beta}_T|\boldsymbol{\beta}_{T-1})f(\boldsymbol{\beta}_{T-1}|\boldsymbol{\beta}_{T-2})\cdots f(\boldsymbol{\beta}_2|\boldsymbol{\beta}_1)f(\boldsymbol{\beta}_1|y_1). \end{aligned}$$

In words, the observation of y_1 affects only the marginal distribution of $\boldsymbol{\beta}_1$ and not the conditional distributions. In the usual way the moments are

$$E(\boldsymbol{\beta}_1|y_1)=(\sigma^{-2}\mathbf{x}_1\mathbf{x}_1'+\mathbf{V}^{-1})^{-1}(\sigma^{-2}\mathbf{x}_1y_1+\mathbf{V}^{-1}\mathbf{b}^*)$$

$$V(\boldsymbol{\beta}_1|y_1)=(\sigma^{-2}\mathbf{x}_1\mathbf{x}_1'+\mathbf{V}^{-1})^{-1}.$$

The moments of $\boldsymbol{\beta}_2$ given y_1 can be computed by integrating out $\boldsymbol{\beta}_1$ from the joint distribution $f(\boldsymbol{\beta}_2|\boldsymbol{\beta}_1)f(\boldsymbol{\beta}_1|y_1)$ which we can do simply by using the moments of $\boldsymbol{\beta}_1$ and the formulas (8.24) and (8.25):

$$E(\boldsymbol{\beta}_2|y_1)=\mathbf{b}^*+\boldsymbol{\phi}\left[E(\boldsymbol{\beta}_1|y_1)-\mathbf{b}^*\right] \tag{8.26}$$

$$V(\boldsymbol{\beta}_2|y_1)=\mathbf{V}-\boldsymbol{\phi}\mathbf{V}^{-1}\boldsymbol{\phi}'+\boldsymbol{\phi}\mathbf{V}(\boldsymbol{\beta}_1|y_1)\boldsymbol{\phi}'. \tag{8.27}$$

Next, suppose y_2 is observed. If interest centers on $\boldsymbol{\beta}_2$, the moments just reported can be used as if they were prior moments, since we can write

$$\begin{aligned} f(\boldsymbol{\beta}_2|y_1,y_2) &\propto \left[\int_{\boldsymbol{\beta}_1} f(\boldsymbol{\beta}_2|\boldsymbol{\beta}_1)f(\boldsymbol{\beta}_1)f(y_1|\boldsymbol{\beta}_1)\,d\boldsymbol{\beta}_1\right]f(y_2|\boldsymbol{\beta}_2) \\ &= f(\boldsymbol{\beta}_2|y_1)f(y_2|\boldsymbol{\beta}_2) \end{aligned}$$

where $f(\boldsymbol{\beta}_2|y_1)$ has the moments (8.26) and (8.27). Thus, as usual, we have the moments of $\boldsymbol{\beta}_2$ as

$$\begin{aligned} E(\boldsymbol{\beta}_2|y_1,y_2) &= \left(V^{-1}(\boldsymbol{\beta}_2|y_1)+\sigma^{-2}\mathbf{x}_2\mathbf{x}_2'\right)^{-1} \\ &\quad\times\left(V^{-1}(\boldsymbol{\beta}_2|y_1)E(\boldsymbol{\beta}_2|y_1)+\sigma^{-2}\mathbf{x}_2y_2\right) \\ V(\boldsymbol{\beta}_2|y_1,y_2) &= \left(V^{-1}(\boldsymbol{\beta}_2|y_1)+\sigma^{-2}\mathbf{x}_2'\mathbf{x}_2\right)^{-1}. \end{aligned}$$

Repeated application of this logic leads to the recursive relationships due to Kalman (1960)

$$\begin{aligned} E(\boldsymbol{\beta}_t|y^{t-1}) &= (\mathbf{I}-\boldsymbol{\phi})\mathbf{b}^*+\boldsymbol{\phi}E(\boldsymbol{\beta}_{t-1}|\mathbf{y}^{t-1}) \\ E(\boldsymbol{\beta}_t|\mathbf{y}^t) &= V(\boldsymbol{\beta}_t|\mathbf{y}^t)\left(V^{-1}(\boldsymbol{\beta}_t|\mathbf{y}^{t-1})E(\boldsymbol{\beta}_t|\mathbf{y}^{t-1})+\sigma^{-2}\mathbf{x}_t y_t\right) \end{aligned}$$

where $\mathbf{y}^t=(y_t,y_{t-1},\ldots,y_1)$, and

$$\begin{aligned} V^{-1}(\boldsymbol{\beta}_t|\mathbf{y}^t) &= V^{-1}(\boldsymbol{\beta}_t|\mathbf{y}^{t-1})+\sigma^{-2}\mathbf{x}_t\mathbf{x}_t' \\ V(\boldsymbol{\beta}_t|\mathbf{y}^{t-1}) &= \boldsymbol{\phi}V(\boldsymbol{\beta}_{t-1}|\mathbf{y}^{t-1})\boldsymbol{\phi}'+\mathbf{V}-\boldsymbol{\phi}\mathbf{V}^{-1}\boldsymbol{\phi}'. \end{aligned}$$

8.5 Inferences about the Hyperparameters

The reader should have objected before reaching this point that a large number of parameters or hyperparameters whose values are likely to be relatively uncertain have been treated as if they were known. Conceptually it is straightforward to assign a probability distribution to any unknown parameters and to proceed directly to Bayes' rule—probably by integrating out the parameters of little interest. In most cases this is a most unpleasant task. The purpose of this chapter is not to solve real inference problems but only to illustrate how data sets ought to be massaged in a number of interesting circumstances. Since the massaging concepts and principles seem little affected by the uncertainty in the hyperparameter, the treatment to this point is adequate for the purpose at hand.

Nor do I wish now to deal with the tedious algebra that would be required to treat uncertain hyperparameters. Typically, this involves assigning a hyperparameter some diffuse distribution and either integrating it out of the posterior analytically or writing the equations that would be jointly solved to find the modes of the posterior distribution. In some cases, particularly with the time-varying parameter models, it is still an open question as to which parameters may be assigned diffuse priors and which may not, if a proper posterior is desired.

For treatments of an uncertain autocorrelation coefficient the reader may consult Zellner (1971, Chap. 7), for multivariate regressions with an uncertain covariance matrix, see Zellner (1971, Chap. 8). Lindley and Smith (1972), Geisser (1966), and Box and Tiao (1964) deal with many different multivariate models. Swamy (1971) and Hildreth and Houck (1968) also discuss inference about the parameters of a (prior) distribution. For time-varying parameters there are many papers and references in a special volume of the *Annals of Economic and Social Measurement*, National Bureau of Economic Research (1973).

Another model of time-varying parameters—switching regressions—has been analyzed by Quandt (1958). For a review see Brown et al. (1975).

CHAPTER 9

DATA-INSTIGATED MODELS[1]

The theory of statistical inference takes as given a fixed set of maintained hypotheses. A critical feature of many real learning exercises is, however, the search for *new* hypotheses that explain the given data. An example is a judicial proceeding in which the lawyers for the defense spend their time looking for hypotheses that are plausible given the available facts and that discredit the prosecutor's hypothesis of their client's guilt. Once the proceeding gets to the court, it may concentrate on the statistical inference issue of identifying the data evidence in favor of a set of fairly well-defined hypotheses. But before it gets there, the participants scramble for hypotheses that explain the given evidence. When the search for new hypotheses is successful, the following dilemma must be confronted: how can we say whether the data favor or cast doubt on the new hypothesis, when the new hypothesis was, in fact, constructed to explain the data?

A fictitious example illustrates this dilemma. In a large survey involving many questions it is discovered that coffee drinking and heart disease are correlated, a fact which suggests some control of coffee consumption. The lawyers for the defense, the American Coffee Institute, argue that coffee drinkers tend to fill their tea-

[1]This chapter is taken from Leamer (1974).

pots with the first water out of the tap in the morning, that this water is brackish from sitting in the pipes overnight, and that it is brackish water not coffee that causes heart disease. They recommend plastic pipes. After allowing for the consumption of brackish water by the individuals surveyed, the correlation between coffee consumption and heart disease is greatly reduced, to the point of insignificance. Is coffee guilty or innocent?

A real example of the phenomenon is reported in an interview by Jones (1974), subtitled "Princeton Professor Charles Westoff Finds Twenty Percent Increase in Frequency of Sexual Activity Among Married Americans." (Reprinted from *People Weekly*, © 1974 Time Inc.)

What did you find out?

We started by looking at the relationship between coital frequency and method of contraception in order to determine whether methods with high and low frequency were the same in 1965 and 1970. Then I noticed, almost in passing, that there seemed to be a 20 percent increase in frequency of sexual intercourse between 1965 and 1970.

What was your reaction?

The figures excited my curiosity. I tried to explain the increase at first by looking at obvious reasons, such as the fact that the entire population was younger in 1970 than in 1965, and we know that young people have a high coital frequency and that it declines steadily with age. That explained only a small part of the increase. I then checked our hypothesis that the increased use of the modern birth control methods might explain the increase, but that explained only about a third of it.

Did you then accept the fact of increased sexual activity among married couples?

I was still not sure. It could have been that the apparent increase was not real but rather a reporting phenomenon. That is, because of the more permissive atmosphere surrounding sex, people talk about it much more freely than before and perhaps even feel a pressure to be "with it." The reported increase in sexual activity could be a matter of exaggeration in 1970 and/or under-reporting in 1965.

How did you resolve this "exaggeration factor"?

There was only one test I could think of, and it is hardly definitive. Since the same 20 percent increase in coital frequency showed up more or less among women using different birth control methods, or no method at all, I reasoned that among women who did not practice any contraception, either because they wanted to get pregnant or for other reasons, that a 20 percent increase in sexual intercourse might be reflected in a decrease in the length of time it took such women to become pregnant. So we looked at that and, much to my surprise, saw a substantial change which, quite fortuitously, also showed up as a 20 percent reduction in the time required to conceive in the absence of contraception.

In the end, I was forced to two conclusions. First, that there is a striking relationship between frequency of sexual intercourse and type of contraceptive method, with the greatest frequency associated with the pill, the IUD, and male sterilization. Second, that there has been an increase of about 20 percent in the frequency of sexual intercourse between married couples under the age of 45 between 1965 and 1970.

How do you account for the increase?

I can only speculate about that. I have already mentioned the influence of more effective and more convenient contraceptive methods. The increasing availability of legal abortion has also reduced anxieties among many women about unwanted pregnancies. There has been an increase in openness and permissiveness about sex in our society during this period. Another possible cause results from the fact that divorce rates have been going up, with the consequence that the average duration of marriage was lower in 1970 than in 1965. To exaggerate it, there were more women on their honeymoons in 1970 than in 1965.

This is a delightful example of how research with nonexperimental data frequently proceeds. The observed fact of 20% increase in sexual activity sequentially stimulated the three hypotheses:

1. A younger population.
2. Greater use of birth control devices.
3. Reporting problems.

The first two could not fully account for the increase, and the third was eliminated by the clever use of outside information. This led to two additional hypotheses, not actually examined:

4. Increased permissiveness. (How do we measure it?)
5. Fewer years of marriage.

This is a very clear case of observations in search of hypotheses. If one of the hypotheses turned out to be "successful," can we say that it is favored by the same data? This problem is quite outside the scope of Bayesian statistical theory. The formal Bayesian learning model describes a superbeing who begins his existence with a joint probability function on all uncertain events. Empirical learning amounts to nothing more than the transformation of a marginal to a conditional distribution. In contrast, much of our informal nonnumerical day-to-day learning, and at least some of the more formal statistical-numerical learning, begins without any explicit joint distribution. If the Bayesian learning model is used, it must, therefore, make use of a joint distribution that is constructed given the observed data. This is both philosophically and practically questionable, since it clearly risks double-counting the data evidence.

I like to describe this as Sherlock Holmes inference. Sherlock solves the case by weaving together all the bits of evidence into a plausible story. He would think it indeed preposterous if anyone suggested that he should construct a function indicating the probability of the particular evidence at hand for all possible hypotheses and then assign prior probabilities to the hypotheses. He advises instead, "No data yet.... It is a capital mistake to theorize before you have all the evidence. It biases the judgments."[2]

There is, incidentally, a tendency among social scientists, particularly those most trained in statistical inference, to disparage Sherlock Holmes inference. "Boy, he really went on a fishing expedition that time, didn't he?" The fact that Sherlock Holmes procedures invalidate statistical inference is even sometimes taken to mean that Sherlock Holmes inference is "unscientific." Nothing could be further from the truth. In fact, a strong argument can be made that statistical inference, not Sherlock Holmes inference is unscientific. The nineteenth century French physiologist Bernard (1927, pp. 137–138) writes (quoted by Cornfield, 1975)

> A great surgeon performs operations for stones by a single method; later he makes a statistical summary of deaths and recoveries, and he concludes from these statistics that the mortality law for this operation is two out of five. Well, I say that this ratio means literally nothing scientifically and gives no certainty in performing the next operation. What really should be done, instead of gathering facts empirically, is to study them more accurately, each in its special determinism...by statistics, we get a conjecture of greater or less probability about a given case, but never any certainty, never any absolute determinism...only basing itself on experimental determinism can medicine become a true science....

Of course, this overstates the case, but there can be no doubt that an essential part of the scientific method is a careful examination of the anomalies of the data, with the intent of finding plausible explanations if possible. Kuhn (1969, pp. 9–10) makes this point forcefully in explaining why astronomy is a science and why astrology is not:

> Compare the situations of the astronomer and the astrologer. If an astronomer's prediction failed and his calculations checked, he could hope to set the situation right. Perhaps the data were at fault: old observations could be re-examined and new measurements made, tasks which posed a host of calculational and instrumental puzzles. Or perhaps theory needed adjustment, either by the manipulation of epicycles, eccentrics, equants, etc., or by more fundamental reforms of astronomical technique. For more than a millennium these were the theoretical and mathematical puzzles around which, together with their instrumental counterparts, the astronomical research tradition was constituted. The astrologer, by contrast, had

[2]Doyle (1888).

no such puzzles. The occurrence of failures could be explained, but particular failures did not give rise to research puzzles, for no man, however skilled, could make use of them in a constructive attempt to revise the astrological tradition. There were too many possible sources of difficulty, most of them beyond the astrologer's knowledge, control, or responsibility. Individual failures were correspondingly uninformative, and they did not reflect on the competence of the prognosticator in the eyes of his professional compeers... In short, though astrologers made testable predictions and recognized that the predictions sometimes failed, they did not and could not engage in the sorts of activities that normally characterize all recognized sciences.

An implication of both of these quotations is that Sherlock Holmes procedures are an essential feature of scientific learning. But when models are instigated by the data, the traditional theories of inference are, regrettably, invalidated. It does seem intuitively clear that the data evidence is weaker than it would have been if a complete set of models had been hypothesized before observation commenced. It thus seems desirable to have a method by which evidence can be formally discounted when postdata model construction occurs. This would have the desirable benefit of putting a price on this kind of data mining. Researchers would then be encouraged more carefully to consider the cost of hypothesis specification relative to the costs of data evidence deterioration through Sherlock Holmes procedures.

I propose in this chapter a method of discounting evidence that parallels a formal decision-theoretic analysis of a presimplification problem in which models are simplified before observation to avoid observation or processing costs. During the analysis, relatively inexpensive tests may indicate that the simplification is undesirable, and the full model may be resurrected. Postdata model construction may thus be interpreted as the data-dependent decision that presimplification is undesirable.

For example, given the two-variable linear regression model $\mathbf{Y}=\mathbf{x}\beta+\mathbf{z}\gamma+\mathbf{u}$ and the auxiliary regression $\mathbf{z}=\mathbf{x}r+\varepsilon$, it is not necessary to observe $\mathbf{z}$ in order to make inferences about β, if either γ or r is zero. Even if neither is identically zero, it may be uneconomical to suffer the costs of observing $\mathbf{z}$. However, once $\mathbf{Y}$ and $\mathbf{x}$ are observed, you may change your mind about observing $\mathbf{z}$, possibly because the sample correlation between $\mathbf{Y}$ and $\mathbf{x}$ is too low or the wrong sign.

This formal decision theory proolem requires a supermind, capable of fully specifying an unsimplified model and the relevant prior distributions. But the principal reason most of us use presimplified models is to avoid the (unlimited?) cost of a full probability assessment. Once a full assessment is made, it seems likely that the true ("believed") complete model would be used. Although a simplified model thus cannot usefully result from the formal decision-theory apparatus, we can think of our models as if they were so derived. In fact, the informal construction of a "working hypothesis" parallels closely the formal decision-theory problem. Models are constructed not as reality but rather as simplifications useful for some implicit or explicit decisions.

The reason for adopting this attitude toward models is that it implies constraints on priors for models constructed after the data analysis commences. The implication of these constraints is that data evidence is discounted in an appealing way when it results from a postdata model search. In the two-variable model mentioned previously, the conditional mean of $\mathbf{Y}$ given $\mathbf{x}$ is $\mathbf{x}(\beta+r\gamma)$. The regression of $\mathbf{Y}$ on $\mathbf{x}$ thus yields an estimate of $\beta+r\gamma$. If you interpret this as an estimate of β, then you have revealed that you think $r\gamma$ is small. If you then decide to observe $\mathbf{z}$, and if you do not improperly alter your prior, you will shrink the estimate of γ toward the revealed prior mean of zero. Thus the data evidence will have to be strong enough to overcome this prejudice, and in that sense the evidence is discounted.

This chapter is divided into six sections. The phenomenon of concept formation is further introduced in section 9.1. The implications (or rather the nonimplications) of stopping rules for inference are discussed in Section 9.2. A surprising conclusion of this chapter is that suspicion of postdata model construction should derive not from the rule used to add variables to an equation, but rather from the improper alteration of one's original implicit priors.

The idea that is being introduced in this chapter is presimplification of models. Concept formation is interpreted as the decision to use a more complex model that was at least implicitly known all the time. Inference with presimplified models is discussed in Section 9.3. A presimplified model necessarily involves an uncertain misspecification error, which causes us to discount any evidence implied by it. Inference with models that are constructed after data are observed—data-instigated models—is discussed in Section 9.4. Quite simply, in using the simple model a researcher reveals certain things about his priors for the more complex models. We are merely suggesting that he stick with those judgments. The fifth section reports an example and the sixth some concluding remarks.

9.1 Concept Formation

The problem of concept formation may be illustrated in a simple example. A mythical kingdom is inhabited only by (green) parakeets, (green) crocodiles and (white) swans. A newly arrived visitor named Richard first meets two swans and two crocodiles. The latter, being in a nasty mood, proceed to bite Richard on the leg. That evidence suggests to Richard the slogan "green bite, white all right," a theory that seems to work well enough when he meets a third swan and a third crocodile. However, the seventh being he confronts is a friendly parakeet, who forces Richard to alter his slogan to "white all right, green usually bite." Being a good Bayesian with a uniform prior for p, the probability that a green being will bite, Richard assigns degree of belief $4/6$ to the proposition "The next green being I meet will bite me," and he furthermore anticipates that he will accumulate evidence about p, the proportion of green beings that bite. Richard's wife, who is little awed by the mathematical and logical bases of Richard's statement, proclaims "You fool! In truth, 'Four legs bite, two legs all right'."

Well, that is a theory that indeed "predicted" the data with certainty. The probability that three out of four green beings bite given p, the proportion of green beings that bite, is only $\binom{4}{3}p^3(1-p)=4p^3(1-p)$. Richard approximated his prior for p with a uniform distribution and computes the "Bayes factor" in favor of the "legs" hypothesis relative to the "color" hypothesis as

$$\frac{P\,(\text{data}\,|\,\text{legs theory})}{\int P\,(\text{data}\,|\,\text{color theory with proportion } p)f(p)\,dp}$$

$$=1\bigg/\left(4\int_0^1 p^3(1-p)\,dp\right)$$

$$=1\bigg/\left(1-\frac{4}{5}\right)=5.$$

He concludes that his wife's hypothesis is favored by the ratio five to one relative to his own. In fairness to his wife, Richard supposes that he had equal prior degrees of belief in each hypothesis, from which he calculates the probabilities of each hypothesis as

$$P\,(\text{legs hypothesis}|\text{data})=5/6$$

$$P\,(\text{color hypothesis}|\text{data})=1/6.$$

With these he can calculate his degree of belief $(4/6)(1/6)+1(5/6)=34/36$ in the proposition "The next green being with four legs that I meet will bite me."

At this point Richard, who is used to changing his degrees of belief only in response to data evidence, observes confusedly that his degree of belief in this proposition increased from $4/6$ to $34/36$ in response only to his wife's observation, "Four legs bite, two legs all right." "Can this be data evidence?" he asks himself. In retracing his steps, Richard discovers that the assumption that the legs hypothesis receives zero prior probability would mean that he would not change his degrees of belief. He tentatively suggests that what he has done is to alter the relative prior probability of the two hypotheses, and that more generally his current degrees of belief depend on that relative probability. Perplexed, he asserts, "If I alter the relative prior probability of the two hypotheses it is because my wife pointed out certain compelling regularities in the data evidence not because of any new experiments. My rules of inference are designed to prevent me from making inferential errors, in particular from double-counting the evidence. It is obvious double-counting to let the data first alter the prior odds ratio from zero to one as I change the prior odds in response to my wife's suggestion and secondly alter the odds ratio from one to five as I would if I applied Bayes' rule to the new prior odds. I shall stick to my original assessment." To this his wife replies, "You fool. Can you not see that your original odds ratio was a mistake? You surely never held degree of belief zero in the legs hypothesis. Better to admit your mistake now than to perpetuate your error." Richard sighs in response, "Yes, I suppose it was a mistake. But I don't see how I can without self-deceit assess any new prior odds ratio which is legitimately unpolluted by the data, and I don't see therefore how I can correct my old mistake without making a new one. Besides, you are so skilled at 'explaining' observations that, regardless of the data you would have come up with some plausible and compelling hypothesis. Why should I believe this one?"

This example aptly illustrates the dilemma of concept formation. In any real learning situation, data evidence strongly compels us to alter our prior, but if we do so, we risk double-counting the data and placing excessive faith in the current evidence. An appropriate model of inference would necessarily allow hypothesis discovery but would also discount the data evidence when it occurs. We propose a method that does just that. It rests on the observation that Richard's p, the probability that a green being will bite, is not conditional on all other features of the being. It is rather a marginal probability such as

$$\begin{aligned} p &= P(\text{bite}|\text{green}) \\ &= P(\text{bite}|\text{green and two legs})P(\text{two legs}|\text{green}) \\ &\quad + P(\text{bite}|\text{green and four legs})P(\text{four legs}|\text{green}). \end{aligned}$$

In particular, let us suppose that conditional on the two hypotheses, the probabilities of biting are as given in Table 9.1. Furthermore, let $1-f$ be the proportion of green beings with two legs, f the proportion with four legs. Assume also that Richard observes randomly selected beings.

Table 9.1
Probability of Bite

Hypothesis	H_0	H_1
Green, two legs	p_0	0
Green, four legs	p_0	1

Letting $\pi = P(H_0)$, $1-\pi = P(H_1)$, we can then calculate p as

$$p = P(\text{bite}|\text{green})$$
$$= \begin{cases} p_0 & \text{with probability } \pi \\ f & \text{with probability } 1-\pi \end{cases} \tag{9.1}$$

We suppose that Richard has a prior distribution for p_0 and f and also that he assigns a number to π, which, together, imply the mixed distribution (9.1) for p. When he first arrives in our mythical kingdom he makes the judgment that he will observe only color, partly because he is not so good at counting legs and partly because he does not have much faith in H_1 (π is close to one). His prior for p together with the likelihood function implied by three bites in four trials imply a posterior distribution for p. At this point, since his wife implicitly lowers the cost of counting the legs, Richard proceeds to observe the number of legs and therefore to condition on that data as well in applying Bayes' rule. He uses his prior for p_0 and the likelihood evidence of three bites in four trials to compute a posterior for p_0 conditional on H_0. He also uses the Bayes factor as above to compute the posterior probabilities of the two hypotheses.

What, then, is the problem of concept formation? We have just described a perfectly valid application of Bayes' rule with sequential construction of theories. Most of us would be suspicious of the new concept because it was "constructed" only when the first one failed to predict the data perfectly. But the decision to observe the number of legs when the color concept "fails" is what is known as a noninformative stopping rule that to a Bayesian has no implications for inference. (More is said about noninformative stopping rules shortly.) The problem of concept formation lies not in the stopping rule but rather in the failure to observe the constraints implied by the probability function (9.1). In particular, p and p_0 are not the same parameter unless $\pi = 1$, the trivial case in which the observation of the number of legs, is ignored.

Of course, in constructing his prior probability function for p, Richard did not consciously have in mind the alternative hypothesis H_1. It is thus not possible to treat the problem of concept formation strictly as described above with the distribution for p derived from distributions for f and p_0. Inference in the context of concept formation is, therefore, necessarily not a topic that can be handled by formal methods of statistical inference. Richard can, however, act *as if* he were deriving his distribution for p from the more basic distributions. Several arguments may be made that this is a desirable approach. The fact that Richard wants to change his mind when his wife makes her suggestion is evidence that Richard does not assign to H_0 a degree of belief equal to one. Unless π were one, a prior for p is necessarily a derivative distribution, and Richard could not apply Bayes' rule to make inferences about p unless he could implicitly derive the prior distribution for p from the more basic distributions. And finally, quite pragmatically, if he behaves as if he so derived his distribution for p, we can solve the inferential issues raised by the phenomenon of concept formation; otherwise they remain entirely beyond our reach.

9.2 Stopping Rules and Inference

At first blush the problem of concept formation appears to be associated with the fact that new hypotheses are data instigated. How could we possibly claim that the data favor H_1 relative to H_0 when the only reason H_1 is examined at all is that H_0 did not work? To be more specific, imagine a researcher who adds variables to his regression equation until a favorite coefficient is significantly positive. We would all chastise him for prejudicing his conclusions in such an obvious way, and we would want to discount his results because of his biased rule for observing the data.

For me, the greatest surprise of the Bayesian logic is that these instincts are simply wrong. This rule and, practically speaking, any rules for observing the data are noninformative stopping rules. They have no implications whatsoever for inference. This counterintuitive assertion needs considerable argument before it can be accepted. Thus we consider in this section the inferential problems implied by stopping rules.

As an example of the kind of problem raised by optional stopping, consider binomial sampling with a sample size equal to one if there is a success on the first trial and equal to two otherwise. The sampling distribution and a hypothetical estimator of p are given below.

Sample	Probability	Estimator
S	p	a
FS	$p(1-p)$	b
FF	$(1-p)^2$	c

If we wanted our estimator of p to be unbiased, we would have to choose a, b, and c to satisfy for all p

$$p \equiv ap + bp(1-p) + c(1-p)^2$$

or

$$0 \equiv c + (a+b-2c-1)p + (c-b)p^2.$$

This is satisfied in the interval $0 \leqslant p \leqslant 1$ if and only if all three coefficients in this polynomial are zero. The only solution is $c=b=0$, $a=1$. This is equivalent to throwing out the second observation.

The usual estimator—the number of successes r divided by the number of trials n—has $a=1$, $b=1/2$, and $c=0$. The expected value of this estimator

$$E\left(\frac{r}{n}\right) = p + \frac{p(1-p)}{2}$$

exceeds p. That is, the sampling scheme prejudices the sample in favor of high values of p, since there is a tendency to observe samples with too many successes.

But let us look at this from a Bayesian point of view. The posterior distribution of p is, of course, the product of the likelihood function times the prior. But the likelihood function (the second column in the table) is exactly the same for every sample as the likelihood function derived under a sampling rule with fixed sample size. Thus from a Bayesian point of view the meaning of the sample *FS* does not depend on the stopping rule, and the fact that you might have stopped on the first trial is quite irrelevant for the interpretation of this particular sample.

A more concrete example illustrates why the stopping rule should not matter. Suppose that boys are born with probability one-half, and that all families stop having children if their first is a boy, otherwise, they have two children. Family composition and probability would then be

Family	Probability
B	$p=1/2$
GB	$p(1-p)=1/4$
GG	$(1-p)(1-p)=1/4$

The average proportion of boys per family would be $1/2+1/4\cdot 1/2=5/8$, more than one-half. Apparently, the stopping rule has biased the population in favor of boys. But the proportion of boys in the whole population is still one-half. Apparently, the stopping rule has failed.

Concern over the stopping rule derives from the following proposition: the mean proportion of boys per family exceeds one-half; if you estimate p

by taking an average over all families of the family proportions, you will necessarily exceed one-half. The error being made, however, is not that families are reporting "biased" numbers; rather, it is that you have not allowed for family size. If you weight families by family size, you will obtain the right number. Exactly the same thing can be said about the estimator (r/n). The fact that (r/n) is a biased estimator is not because r/n is a "biased" summary of the data for any sample; rather, it is because the expected value operator does not weight by sample size.

It is easy to see that stopping rules dependent on the data only are noninformative. Let θ be the parameter of interest, let $X_1, X_2, \ldots$ be a sequence of observations, and let N be the sample size. Given the sample $X_1 = x_1, X_2 = x_2, \ldots X_n = x_n$ and $N = n$, we may write the likelihood function as

$$\begin{aligned} L(\theta; \mathbf{x}, n) &\propto P(X_1 = x_1, X_2 = x_2, \ldots, X_n = x_n, N = n|\theta) \\ &= P(N = n|X_1 = x_1, X_2 = x_2, \ldots, X_n = x_n, \theta) \\ &\quad P(X_1 = x_1, X_2 = x_2, \ldots, X_n = x_n|\theta). \end{aligned}$$

But the stopping rule is assumed to terminate sampling with probability one given the sample $x_1, x_2, \ldots, x_n$; thus $P(N = n|\mathbf{X} = \mathbf{x}, \theta) = 1$, independent of θ, and the likelihood function is proportional to $P(\mathbf{X} = \mathbf{x}|\theta)$ regardless of the stopping rule.[3]

[3]The apparent danger of a stopping rule is that a researcher can prejudice the sample in any way he sees fit. He may even be able to sample to a foregone false conclusion. Suppose that we sample from a normal population with mean zero and variance one. If we wish to prove that the mean is, in fact, not zero, we may want to continue sampling until the sample mean is "significantly different from zero." That is, let

$$m_n = \sum_{i=1}^{n} x_i / n,$$

the mean of the sample of size n, and adopt a stopping rule

$$\text{if } |m_n| n^{1/2} > 1.96, \text{ stop}$$
$$\text{otherwise,} \qquad \text{continue sampling.}$$

The statistic $m_n n^{1/2}$ is, of course, the normal statistic typically used to test the hypothesis $\mu = 0$ against the alternative $\mu \neq 0$. A value of $|m_n| n^{1/2} > 1.96$ is taken as evidence against the point-null hypothesis.

Surprisingly enough, this inequality will eventually be satisfied with probability "essentially one." It is apparently possible to sample to a foregone false conclusion. The paradox is completely resolved, however, by noting the discussion in Chapter 4.2, that from a Bayesian point of view a value of $m_n n^{1/2}$ equal to 1.96 may, in fact, be overwhelming evidence in favor of $\mu = 0$, if the sample size is large enough. That is the significance of a "statistically significant" result depends on sample size. If it takes a large sample to get the result, this should be taken as evidence in favor of the null hypothesis. For references and further discussion see Cornfield (1969).

To sum up, we have argued by analogy with the problem of estimating the proportion of boys in a population that the bias of the usual estimator when there is a stopping rule should cause uncomfortableness not with the estimator but with the concept of bias. Constructing an unbiased estimator is roughly equivalent to passing a law that families with one boy may tell the truth, whereas larger families must report to the census taker that they have no boys at all. Bias, being a property of a sampling distribution, is not of direct interest to a Bayesian. Concern over bias from a sampling theory point of view apparently derives from the following proposition. If $\hat{\theta}_i$ is a biased estimator of θ, if n independent values $\hat{\theta}_1,\ldots,\hat{\theta}_n$, are observed, and if a composite estimator $\theta_n^* = \Sigma\hat{\theta}_n/n$ is computed, then as n grows, θ_n^* will converge (in a probability sense) to a value different from θ. Thus accumulation of evidence will not lead to the truth. The counter-argument, as we have seen, is that the appropriate pooling of evidence does not lead to θ_n^*. Instead, we should maximize the composite likelihood function formed by multiplying the individual likelihoods together. Bias should, therefore, concern us only if for some peculiar reason we are compelled to pool information from different samples in this undesirable way.

Most practical stopping rules that lead to biased estimators are from the Bayesian point of view noninformative and therefore irrelevant to the inference problem. In particular, the class of rules dependent on the data alone is noninformative.

9.3 Inference with Presimplified Regression Models

The idea on which our solution to the problem of concept formation rests is that inferential models are highly simplified versions of the learner's inherent set of beliefs. In this section we discuss inference with regression models that are simplified versions of more complete models.

A regression function in a linear nonstochastic world may be written

$$\mathbf{Y} = \mathbf{X}\boldsymbol{\beta} + \mathbf{Z}\boldsymbol{\gamma} \tag{9.2}$$

where $\mathbf{X}$ and $\mathbf{Z}$ are observable matrices, $\mathbf{Y}$ an observable vector, and $\boldsymbol{\beta}$ and $\boldsymbol{\gamma}$ are unobservable parameter vectors. As in the example in the introduction, inferences about $\boldsymbol{\beta}$ may be made by observing $\mathbf{Y}$ and $\mathbf{X}$ alone,

$$P(\boldsymbol{\beta}|\mathbf{Y},\mathbf{X}) \propto \int_{\gamma}\int_{\mathbf{Z}} P(\mathbf{Y},\mathbf{X},\mathbf{Z},\boldsymbol{\beta},\boldsymbol{\gamma})\,d\mathbf{Z}\,d\boldsymbol{\gamma}.$$

The integral in $\mathbf{Z}$ is easily computed by writing the linear regression function (an assumption)

$$\mathbf{Z}|\mathbf{X} = \mathbf{X}\mathbf{R} + \mathbf{U}$$

where $\mathbf{U}$ is a matrix of random variables subjectively independent of $\mathbf{X}$ and where $\mathbf{R}$ is assumed known. The resulting working hypothesis is then

$$\mathbf{Y}|\mathbf{X}=\mathbf{X}\boldsymbol{\beta}+\mathbf{X}\mathbf{R}\boldsymbol{\gamma}+\mathbf{U}\boldsymbol{\gamma} \tag{9.3}$$

where we have written $\mathbf{Y}|\mathbf{X}$ to emphasize the point that other observables have been marginalized out.

The usual regression model

$$\mathbf{Y}=\mathbf{X}\boldsymbol{\beta}+\mathbf{u} \tag{9.4}$$

with $\mathbf{u}$ independent of $\mathbf{X}$ is derivable from the working hypothesis (9.3) only if $\mathbf{R}\boldsymbol{\gamma}=\mathbf{0}$. Thus the usual analysis involves a (well-known) specification assumption that left-out variables have either zero effect ($\boldsymbol{\gamma}=\mathbf{0}$) or are uncorrelated with included variables ($\mathbf{R}=\mathbf{0}$). Postdata model construction in response to peculiarities in the least-squares estimate of $\boldsymbol{\beta}$ constitutes a de facto rejection of this assumption.

The model

$$\mathbf{Y}=\mathbf{X}\boldsymbol{\beta}+\mathbf{X}\boldsymbol{\beta}^c+\mathbf{u} \tag{9.5}$$

with $\mathbf{u}\sim N(\mathbf{0},\sigma^2\mathbf{I})$ offers a closer approximation to the probability assignments implied by the working hypothesis (9.3). It admits the possibility of left-out variables ($\boldsymbol{\beta}^c$ plays the role of $\mathbf{R}\boldsymbol{\gamma}$) but does not require us actually to identify them. It also fits, with minor modification, into the traditional statistical theory.

The parameter vector $\boldsymbol{\beta}^c$ summarizes the bias in the information about $\boldsymbol{\beta}$ due to excluded variables. It is called either a *contamination vector* or an *experimental bias* vector. The usual regression model (9.4) is called a *false model*, since it unbelievably sets the contamination vector to zero and since it yields reasonable results only if that approximation is adequate. The amended model (9.5) is called the *working hypothesis*, indicating that degrees-of-belief are not allocated directly to it but rather are derived implicitly from a true model or "*world view*" such as (9.2). A working hypothesis includes a statement about the quality of the experiment (a prior on $\boldsymbol{\beta}^c$); a false model does not.

We may "identify" model (9.5) by specifying $\boldsymbol{\beta}^c$ and making inferences about the theoretical coefficient $\boldsymbol{\beta}$ or by specifying $\boldsymbol{\beta}$ and making inferences about the experimental bias $\boldsymbol{\beta}^c$. Informative priors offer a range of intermediate inferences. Analysis of this model from a Bayesian point of view is a straightforward generalization of Pratt, Raiffa, and Schlaifer's (1965) biased sampling. Let the prior be normal with mean and variance

$$E\begin{bmatrix}\boldsymbol{\beta}\\ \boldsymbol{\beta}^c\end{bmatrix}=\begin{bmatrix}\mathbf{b}^*\\ 0\end{bmatrix} \tag{9.6}$$

$$V\begin{bmatrix}\boldsymbol{\beta}\\ \boldsymbol{\beta}^c\end{bmatrix}=\sigma^2\begin{bmatrix}\mathbf{N}^* & 0\\ 0 & \mathbf{B}\end{bmatrix}^{-1}, \tag{9.7}$$

with $\sigma^2 = \operatorname{Var} u_t$. With $\mathbf{N}=\mathbf{X}'\mathbf{X}$, the posterior precision of the coefficient vector is

$$\sigma^{-2}\begin{bmatrix} \mathbf{N}^* & 0 \\ 0 & \mathbf{B} \end{bmatrix} + \sigma^{-2}\begin{bmatrix} \mathbf{N} & \mathbf{N} \\ \mathbf{N} & \mathbf{N} \end{bmatrix}$$

with posterior variance

$$V\left(\begin{bmatrix} \boldsymbol{\beta} \\ \boldsymbol{\beta}^c \end{bmatrix} | \mathbf{Y}\right) = \sigma^2 \begin{bmatrix} \mathbf{N}^* + \mathbf{N} & \mathbf{N} \\ \mathbf{N} & \mathbf{B}+\mathbf{N} \end{bmatrix}^{-1}$$

$$= \sigma^2 \begin{bmatrix} \mathbf{D}^{-1} & -\mathbf{D}^{-1}\mathbf{N}\,(\mathbf{B}+\mathbf{N}\,)^{-1} \\ -\mathbf{E}^{-1}\mathbf{N}\,(\mathbf{N}^*+\mathbf{N}\,)^{-1} & \mathbf{E}^{-1} \end{bmatrix} \qquad (9.8)$$

with

$$\mathbf{D} = \mathbf{N}^* + \mathbf{N} - \mathbf{N}(\mathbf{B}+\mathbf{N})^{-1}\mathbf{N}$$

and

$$\mathbf{E} = \mathbf{B} + \mathbf{N} - \mathbf{N}(\mathbf{N}^*+\mathbf{N})^{-1}\mathbf{N}.$$

Similarly, the posterior mean is (with $\mathbf{b}$ a solution to $\mathbf{X}'\mathbf{X}\mathbf{b}=\mathbf{X}'\mathbf{Y}$)

$$E\left(\begin{bmatrix} \boldsymbol{\beta} \\ \boldsymbol{\beta}^c \end{bmatrix} | \mathbf{Y}\right)$$

$$= \begin{bmatrix} \mathbf{D}^{-1} & -\mathbf{D}^{-1}\mathbf{N}\,(\mathbf{B}+\mathbf{N}\,)^{-1} \\ -\mathbf{E}^{-1}\mathbf{N}\,(\mathbf{N}^*+\mathbf{N}\,)^{-1} & \mathbf{E}^{-1} \end{bmatrix} \begin{bmatrix} \mathbf{N}^*\mathbf{b}^* + \mathbf{N}\mathbf{b} \\ \mathbf{N}\mathbf{b} \end{bmatrix}$$

$$= \begin{bmatrix} \mathbf{D}^{-1}\left(\mathbf{N}^*\mathbf{b}^* + \left[\mathbf{N} - \mathbf{N}(\mathbf{B}+\mathbf{N})^{-1}\mathbf{N}\right]\mathbf{b}\,\right) \\ \mathbf{E}^{-1}\mathbf{N}\,(\mathbf{N}^*+\mathbf{N}\,)^{-1}\mathbf{N}^*\,(\mathbf{b}-\mathbf{b}^*\,) \end{bmatrix}. \qquad (9.9)$$

Notice, first, that the posterior mean of $\boldsymbol{\beta}$ is, in the usual way, a weighted average of the prior mean $\mathbf{b}^*$ and the sample mean $\mathbf{b}$. Whereas ordinarily the sample mean receives weight $\mathbf{N}$, its weight is here reduced to $\mathbf{N}-\mathbf{N}(\mathbf{B}+\mathbf{N})^{-1}\mathbf{N}=\mathbf{N}(\mathbf{B}+\mathbf{N})^{-1}\mathbf{B}$. That is to say, we discount the evidence provided by contaminated (or potentially contaminated) experiments. The discount depends on $\mathbf{B}$, the prior precision of the experimental bias $\boldsymbol{\beta}^c$. As $\mathbf{B}$ grows the posterior parameters converge to their values in an uncontaminated experiment.

The posterior mean of $\boldsymbol{\beta}^c$ is a matrix-weighted average of zero and $(\mathbf{b}-\mathbf{b}^*)$. When $\mathbf{b}$ exceeds $\mathbf{b}^*$ we conclude in part that $\boldsymbol{\beta}$ exceeds $\mathbf{b}^*$ (in the matrix-weighted average sense) but we prejudice the posterior distribution toward $\mathbf{b}^*$ more than in the case of a true model. We adjust for this by moving the distribution of $\boldsymbol{\beta}^c$ from the origin; that is, part of the excess of $\mathbf{b}$ over $\mathbf{b}^*$ is attributed to experimental bias, part to large $\boldsymbol{\beta}$.

Given a diffuse prior for $\boldsymbol{\beta}$, with $\mathbf{N}^* = \mathbf{0}$, the posterior parameters are

$$E\left(\begin{bmatrix} \boldsymbol{\beta} \\ \boldsymbol{\beta}^c \end{bmatrix} \middle| \mathbf{Y}\right) = \begin{bmatrix} \mathbf{b} \\ \mathbf{0} \end{bmatrix}$$

$$V\left(\begin{bmatrix} \boldsymbol{\beta} \\ \boldsymbol{\beta}^c \end{bmatrix} \middle| \mathbf{Y}\right) = \sigma^2 \begin{bmatrix} \mathbf{N}^{-1} + \mathbf{B}^{-1} & -\mathbf{B}^{-1} \\ -\mathbf{B}^{-1} & \mathbf{B}^{-1} \end{bmatrix}$$

where we have used the fact that

$$\left[\mathbf{N} - \mathbf{N}(\mathbf{B}+\mathbf{N})^{-1}\mathbf{N}\right]^{-1} = \mathbf{N}^{-1} + \mathbf{B}^{-1}.$$

This result is of interest, since it suggests that the least squares estimator $\mathbf{b}$ is the best we can do when the direction of the bias is unknown $(E\boldsymbol{\beta}^c = \mathbf{0})$. The posterior variance matrix, however, differs importantly from the usual OLS result $\sigma^2\mathbf{N}^{-1}$. We must add to this matrix another matrix $\sigma^2\mathbf{B}^{-1}$, which is just the prior variance of $\boldsymbol{\beta}^c$. Most importantly, as sample size increases, this second term does not decay away and thus becomes a lower bound to the variance of $\boldsymbol{\beta}$. In words, if you arrive at a sampling experiment with no knowledge of $\boldsymbol{\beta}$, you can never know more about $\boldsymbol{\beta}$ than you claim to know about $\boldsymbol{\beta}^c$. Of course, you cannot measure more accurately than your measuring instrument is capable of. More than that, the capability of the measuring instrument is not disclosed in the process of measuring (since $\mathbf{B}$ is fixed before measurement commences).

Once the sampling uncertainty $\sigma^2\mathbf{N}^{-1}$ becomes small relative to the misspecification uncertainty $\sigma^2\mathbf{B}^{-1}$, continued sampling of this process is, essentially, a waste of time. Additional information may be gathered only by improved experimentation, that is, by smaller σ^2 or larger $\mathbf{B}$. Larger $\mathbf{B}$ is a pure prior concept, whereas reduction in σ^2 is evidenced through smaller error sum of squares; thus the latter, when it is not offset by smaller $\mathbf{B}$, seems to be the only unambiguous method of improving our knowledge of the process parameters.

In the nonexperimental sciences, the possibility of improving an "experiment" is, by definition, excluded. Researchers implicitly do what they regard to be the next best thing: they treat the R^2 as an indicator of the quality of experimental control and discount results when R^2s are small. The extent to which this discounting is appropriate depends on how it is done. Since R^2 does map into an estimate of σ^2, R^2 may give an indication of the absolute misspecification uncertainty $\sigma^2\mathbf{B}$. However, the percentage understatement of the uncertainty is a function of sample size $(\mathbf{N}^{-1})$ and not of σ^2. In this sense, the R^2 is not an indicator of experimental control. Independent of R^2, the OLS variance $\sigma^2\mathbf{N}^{-1}$ accurately summarizes the uncertainty for small samples but understates the uncertainty for large samples. This is simply because it ignores the misspecification uncertainty

$\sigma^2\mathbf{B}^{-1}$, which is negligible compared to the sampling uncertainty $\sigma^2\mathbf{N}^{-1}$ in small samples, but not in large.

9.4 Inference with Data-instigated Models

In this section we discuss the inferences that are legitimate when new variables are added to regression equations. As we have suggested previously, it is not the stopping rule that should cause suspicion. Rather, the error that is potentially made is that in adding a new variable to the equation the researcher implicitly changes his priors about various parameters. He left the variable out in the first place because he thought it did not belong, and to be consistent he must have a prior on the new coefficient that concentrates the probability in the neighborhood of zero. This automatically "discounts" the evidence implied by the new regression model in the sense that the posterior distribution of the new regression coefficient is pushed toward the origin.

Consider the following hypothesis:

$$\mathbf{Y}|\mathbf{X},\mathbf{Z}=\mathbf{X}\boldsymbol{\beta}+\mathbf{X}\boldsymbol{\beta}^c+\mathbf{Z}\boldsymbol{\gamma}+\mathbf{Z}\boldsymbol{\gamma}^c+\boldsymbol{\varepsilon} \tag{9.10}$$

with $\boldsymbol{\varepsilon}\sim N(\mathbf{0},\sigma_\varepsilon^2\mathbf{I})$ and where $\mathbf{X}$ and $\mathbf{Z}$ may be matrices. A simpler working hypothesis may be obtained by marginalizing out $\mathbf{Z}$. Assume a linear regression function

$$\mathbf{Z}=\mathbf{XR}+\mathbf{U} \tag{9.11}$$

with $\mathbf{U}$ having a matrix-normal distribution such that if $\mathbf{u}_i$ is a row of $\mathbf{U}$, $\mathbf{u}_i\sim N(0,\boldsymbol{\Sigma}_{\mathbf{UU}})$, $\mathbf{u}_i$ independent of $\mathbf{u}_j$ and $\boldsymbol{\varepsilon}$. The model conditioned on $\mathbf{X}$ would then be

$$\begin{aligned}\mathbf{Y}|\mathbf{X}&=\mathbf{X}\boldsymbol{\beta}+\mathbf{X}\boldsymbol{\beta}^c+\mathbf{XR}\boldsymbol{\gamma}+\mathbf{XR}\boldsymbol{\gamma}^c+\mathbf{U}\boldsymbol{\gamma}+\mathbf{U}\boldsymbol{\gamma}^c+\boldsymbol{\varepsilon}\\&=\mathbf{X}\boldsymbol{\beta}+\mathbf{X}\boldsymbol{\Gamma}+\mathbf{e}\end{aligned} \tag{9.12}$$

which is in the form of the usual contaminated model but with the constraints

$$\boldsymbol{\Gamma}=\boldsymbol{\beta}^c+\mathbf{R}\boldsymbol{\gamma}+\mathbf{R}\boldsymbol{\gamma}^c \tag{9.13}$$

$$\mathbf{e}=\mathbf{U}\boldsymbol{\gamma}+\mathbf{U}\boldsymbol{\gamma}^c+\boldsymbol{\varepsilon}, \tag{9.14}$$

with parameters

$$\sigma_e^2=(\boldsymbol{\gamma}+\boldsymbol{\gamma}^c)'\boldsymbol{\Sigma}_{\mathbf{UU}}(\boldsymbol{\gamma}+\boldsymbol{\gamma}^c)+\sigma_\varepsilon^2 \tag{9.15}$$

where $\boldsymbol{\Sigma}_{\mathbf{UU}}$ is a contemporaneous covariance matrix of the (matrix) random variable $\mathbf{U}$.

Inferences about the parameters may be made as implied by either (9.10) or (9.12), depending on whether $\mathbf{Z}$ is observed. This is true even when the analysis proceeds sequentially with **Z** observed depending on the least-squares outcome $(\mathbf{X}'\mathbf{X})^{-1}\mathbf{X}'\mathbf{Y}$. [There is a tendency to think that the data evidence is contaminated by the stopping rule. Suppose, for example, we decide to observe **Z** if $(\mathbf{X}'\mathbf{X})^{-1}\mathbf{X}'\mathbf{Y}$ has any negative elements. Since this rule depends only on the data and not at all on the parameters, it is noninformative. Classically, of course, the resulting estimator is biased, and the stopping rule would be regarded as a source of contamination.]

If we begin our analysis with the simple model (9.12) and therefore a probability assignment to $(\boldsymbol{\beta},\boldsymbol{\Gamma},\sigma_{\mathbf{e}}^2)$, the fact that the more complex working hypothesis (9.10) is lurking in the background is quite irrelevant. Within the confines of noninformative stopping rules, we may decide to observe **Z**, and to expand to the fuller model at any time. We are *not* free, however, to assign any distributions to $(\boldsymbol{\beta},\boldsymbol{\beta}^c,\gamma,\gamma^c,\sigma_\varepsilon^2)$, since we already have an assignment on $(\boldsymbol{\beta},\boldsymbol{\Gamma},\sigma_e^2)$, functions (9.13) and (9.15) defined on the expanded parameter space. As long as we satisfy these constraints, we *are* free to alter the model as we choose. Thus postdata model construction becomes fully legitimate.

Although these constraints are conceptually straightforward, they are not easy to implement. When **R** is known, however, the implication of constraint (9.13) is straightforward. Under a normality assumption, only the first two moments are of interest.

$$E(\boldsymbol{\Gamma})=E(\boldsymbol{\beta}^c)+\mathbf{R}E(\gamma+\gamma^c) \tag{9.16}$$

$$V(\boldsymbol{\Gamma})=V(\boldsymbol{\beta})+\mathbf{R}V(\gamma+\gamma^c)\mathbf{R}' \tag{9.17}$$

where we have assumed the independence of $\boldsymbol{\beta}^c$ and (γ,γ^c). We would typically set $E(\boldsymbol{\Gamma})$, $E(\boldsymbol{\beta}^c)$ and $E(\gamma^c)$ to zero to indicate expected unbiased experiments, and (9.16) implies that γ must have prior mean zero. The extent to which we allow γ to wander from zero in response to the data evidence is determined by the variance $V(\gamma)$, which is constrained by (9.17). The larger $V(\boldsymbol{\Gamma})$ is, the more we discount the evidence derived from the regression of **Y** on **X** (see Section 9.3). But large $V(\boldsymbol{\Gamma})$ also allows us to assign large $V(\gamma)$, and this allows γ to wander from zero when **Y** is regressed on both **X** and **Z**. Loosely speaking, if we are willing to discount the evidence collected when we regress **Y** on **X**, we may believe the evidence collected when we regress **Y** on **X** and **Z**. Note, by the way, that the constraints become inoperative for orthogonal data, $\mathbf{R}=\mathbf{0}$, and we are thus completely free to add in orthogonal variables.

See Leamer (1974) for further discussion of the implications of these constraints.

9.5 An Example: Bode's Law

An interesting example of a data-instigated model is the numerological relationship discovered by Titius describing the mean distance of a planet from the sun as a simple function of the planet order, specifically, by the simple geometric progression

$$d_n = 4 + 3(2^n) \tag{9.18}$$

where d_n is the distance from the sun to the nth planet from the sun. For the first eight planets this implies the mean distances of 4, 7, 10, 16, 28, 52, 100, 196 (using $n = -\infty, 0, 1 \ldots$). The seven planets known in 1800 had mean distances of 3.9, 7.2, (10), 15.2, 52, 95, 192, with the earth's distance arbitrarily set to 10. These numbers fit the numerical sequence remarkably well, with the exception of a missing planet 28 units from the sun. The very real decision problem of astronomers at that time was whether this evidence was compelling enough to warrant a search for the missing planet in the region suggested by the relationship.

Surprisingly enough, Bode and five other German astronomers, searching the heavens at roughly 28 units from the sun, on January 1, 1801, did indeed find the small planet, Ceres, and since then dozens of other small planets have been found and are hypothesized to be the fragments of a single larger planet.[4] The "law" was given Bode's name perhaps because his discovery of the missing planet makes the law ever so much more believable than it would have been otherwise. The law was instigated by the observations available up to 1800, and we tend, properly I think, to discount the evidence implied by those observations. The single observation that was not known at the time and that could not have instigated the model is taken as essentially the only data point relevant for testing. It intuitively lends considerable believability to the law, and application of Bayes rule leads to the same conclusion (using as an alternative hypothesis almost any other plausible hypotheses about the dispersion of the planets around the sun).[5]

But it is not enough to observe that Bode's discovery considerably adds to the believability of the law. Beliefs prior to the observation of Ceres also partly determined the posterior belief. In order to determine if Bode's Law is believable today we must determine what degree of believability it had prior to 1801. We must assess the uncertainty, allowing for the fact that the law was instigated by the first seven observations.

It is interesting to observe that the statisticians Good (1969) and Efron (1971) seem to be concerned primarily with the construction of interesting alternative hypotheses, with little argument over the appropriateness of statistical theory in general. Blyth (1971, p. 566) comments on this, "The Efron and Good tests seem to me invalid because they are based on the same data that suggested both hypotheses." He takes a pessimistic posi-

[4]This is Polanyi's (1964) version of the facts. Good (1972) attributes the discovery to Piazzi and describes Bode as merely a publicist.

[5]Almost any other hypothesis places low probability on finding a planet in this region. See Good (1969) or Efron (1971).

tion, "And it would appear that any real test of this would have to be based on future observations."

We claim to have a way of characterizing the uncertainty about Bode's law that allows for the fact that it was data instigated. We may include all the observations available today, Bode's eight planets plus Neptune and Pluto. As it turns out, neither of these last two planets obeys Bode's law very well.

To perform the analyses described in the previous section, we must phrase the postdata model construction aspect of Bode's law as the addition of variables to a linear model. There are, apparently, three "discoveries" from the data set that are candidates for the postdata label:

1. That distance depends on the order
2. That distance depends nonlinearly on order
3. That there are three "outliers"

It seems to me that any model of planetary distance would include order as an explanatory variable. The significant postdata discovery is that distance depends nonlinearly on order. The discarding of the three outliers represents a second step we do not explore here.

Let us phrase the model in terms of our linear regression parameters of the previous section as

$y_n =$ distance from planet $n-1$ to planet n (in units of Sun-Earth distance)

$x_0 =$ constant

$x_1 = n$, the planet order

$z = n^2$

The first and second phase regressions are

Phase I: $y_n = \beta_0 + \beta_1 n + \Gamma_0 + \Gamma_1 n + e_n$

Phase II: $y_n = \beta_0 + \beta_1 n + \beta_0^c + \beta_1^c n + \gamma n^2 + \gamma^c n^2 + \varepsilon_n$

with the regression of Z on X being

$$n^2 = r_0 + r_1 n + u, \qquad n = 1, \ldots, 10. \tag{9.19}$$

Note, of course, that Bode's law is distorted to fit it into our framework. I do not think that the distortion has important substantive implications, however. I also substitute sample estimates for σ_e^2 and σ_ε^2.

In this case, it is possible to calculate r_0 and r_1 with certainty, and the linear system (9.13) becomes

$$\begin{bmatrix} \Gamma_0 \\ \Gamma_1 \end{bmatrix} = \begin{bmatrix} \beta_0^c \\ \beta_1^c \end{bmatrix} + \begin{bmatrix} r_0(\gamma + \gamma^c) \\ r_1(\gamma + \gamma^c) \end{bmatrix} \tag{9.20}$$

where $r_0 = -22$, $r_1 = 11$.

I choose to ignore the constraint on the variances, Equation (9.15), on the basis that the assumed vagueness of the prior distribution of σ_e^2 effectively eliminates the constraint. If we take all prior means to be zeroes, the constraints (9.20) under a normality and an independence assumption are satisfied when

$$V(\Gamma_0)=V(\beta_0^c)+r_0^2(V(\gamma)+V(\gamma^c))$$
$$V(\Gamma_1)=V(\beta_1^c)+r_1^2(V(\gamma)+V(\gamma^c)). \tag{9.21}$$

We require that the researcher who employs the phase I regression must at that time select $V\Gamma_0$ and $V\Gamma_1$, where Γ_0 and Γ_1 are the first-phase experimental-bias coefficients. Relatively small values of these variances imply relatively small discounting of the first phase result but also imply through (9.21) relatively tight distributions of γ and γ^c, or equivalently, relatively large "discounting" of the second-phase regression.

To begin, let us take a look at the unadorned regressions (with standard errors in parentheses)

$$\begin{array}{llll} y_n = -32.01 & +13.0n & & \\ \quad\;\;(13.5) & \;\;(2.18) & & \\ R^2 = .82 & \bar{R}^2 = .80 & \text{d.f.}=8 & \text{D.W.}=.91 \end{array} \tag{9.22}$$

$$\begin{array}{llll} y_n = 3.25 & -4.63n + 1.6n^2 & & \\ \quad\;\;(18.7) & \;\;(7.8) \quad\;\; (.7) & & \\ R^2 = .90 & \bar{R}^2 = .87 & \text{d.f.}=7 & \text{D.W.}=1.6 \end{array} \tag{9.23}$$

where D.W. indicates the Durbin-Watson statistic.

We assume that the researcher first runs (9.22) and then notes peculiarities in the residual pattern, indicated especially by the Durbin-Watson statistic of .91. To rid his model of those peculiarities, he adds the n^2 term and refits. It is pretty clear from (9.23) that he obtains a substantially improved fit. The variable n^2 effectively wipes out any apparent influence of the variable n. Is this, however, real or manufactured evidence?

I claim that before the regression equation (9.22) is estimated, one must decide how much he will believe the result. He does this by selecting $V(\beta_0)$, $V(\beta_1)$, $V(\Gamma_0)$, $V(\Gamma_1)$. Consider the following three cases:

	$V(\beta_0)$	$V(\beta_1)$	$V(\Gamma_0)$	$V(\Gamma_1)$
Case 1	10^4	10^4	10^4	10^4
Case 2	10^4	10^4	10^2	10^2
Case 3	10^4	10^4	10	10

In all cases the researcher is very uncertain about the theoretical coefficients β_i. As we proceed from case 1 to case 3, he is increasingly confident about the quality of the experiment.

The phase I posterior means and standard errors implied by these priors may be found in Table 9.2. One is expected at this stage to choose one of these cases. If you wish to believe the sample result you must select Case 3. At the other extreme, you may select Case 1 and discount the sample very significantly.

Having committed oneself during phase I to one of the three cases, one has a restricted menu of things he can believe following phase II. These are given for the three cases in Table 9.3. The constraints (9.20) imply

$$V(\gamma)+V(\gamma^*)\leqslant m=\min\left(\frac{V\Gamma_0}{r_0^2},\frac{V\Gamma_1}{r_1^2}\right).$$

The small letters in Table 9.3 indicate

$$\begin{aligned}&(\mathrm{a})\,V(\gamma)+V(\gamma^*)=.01m\\&(\mathrm{b})\,V(\gamma)+V(\gamma^*)=.5m\\&(\mathrm{c})\,V(\gamma)+V(\gamma^*)=.99m.\end{aligned}$$

That is, for distribution (c), the coefficient $(\gamma+\gamma^*)$ has the largest variance and therefore the greatest freedom to vary from zero. In all cases we have set $V(\gamma)=99V(\gamma^*)$, that is, we are allocating almost all the evidence to the theoretical coefficient γ. (More on this point shortly.)

Notice in Table 9.3 that the phase II posterior distributions for case 1 are effectively the phase II sample regression function, whereas the distributions for case 3 are effectively the phase I sample regression function. In words, if you were willing to completely discount the evidence generated in phase I (case 1) you may now believe in the nonlinearity of the function. If, on the other hand, you thought you were getting evidence about the linear term in the first phase, no significant evidence about the nonlinearity of the function was generated during the second phase. The reason for

Table 9.2

Posterior Means and Standard Errors
(Standard errors in parentheses)

	β_0	β_1	Γ_0	Γ_1
Sample	−32.01	13.0		
	(13.5)	(2.18)		
Case 1	−15.8	6.4	−15.8	6.4
	(71)	(71)	(71)	(71)
Case 2	−31.1	12.8	−.31	.13
	(16.6)	(10.2)	(10.0)	(10.0)
Case 3	−31.4	12.9	−.03	.013
	(13.7)	(3.8)	(3.16)	(3.16)

Table 9.3
Phase II Posterior Means and Standard Errors

		β_0	β_1	γ	β_0^c	β_1^c	γ^c
Sample		3.25	−4.63	1.6			
		(18.7)	(7.8)	(.7)			
Case 1	*a*	−1.7	−.67	1.3	−1.6	−.64	.013
		(69)	(71)	(.63)	(69)	(71)	(.14)
	b	1.1	−2.1	1.5	.54	−1.79	.015
		(58)	(69)	(.74)	(58)	(69)	(.32)
	c	2.3	−2.4	1.55	0	−1.77	.016
		(18.3)	(66.8)	(.81)	(.089)	(66)	(.045)
Case 2	*a*	−29	12.1	.067	−.27	.12	0
		(15)	(10)	(.14)	(9.4)	(9.8)	.014
	b	−25	9.7	.29	−.13	.085	.003
		(14)	(9.9)	(.29)	(7.1)	(9.3)	(.003)
	c	−21	7.5	.48	0	.06	.005
		(13)	(9.7)	(.38)	(.006)	(8.6)	(.05)
Case 3	*a*	−31	12.9	.007	−.03	.01	0
		(11)	(3.6)	(.045)	(2.99)	(3.12)	(.005)
	b	−31	12.5	.03	−.015	.011	0
		(11.2)	(3.6)	(.10)	(2.2)	(2.96)	(.01)
	c	−30	12.2	.07	0	.009	0
		(11)	(3.6)	(.14)	(.0009)	(2.7)	(.014)

this seems clear. Since n and n^2 are highly correlated, the only way a regression of y on n alone could give us evidence about how n affects y is if n^2 simply does not belong in the equation. The sample evidence that n^2 does belong simply is not enough to overcome that prejudice.

Let us now suppose that you did discount the evidence in phase I, that is, that you selected case 1. As I have indicated, the phase II posteriors assign almost all of the evidence to γ rather than γ^*. Thus although it is possible to believe in the nonlinearity of the function you may instead decide to discount this evidence by reallocating the prior variance from $V(\gamma)$ to $V(\gamma^*)$. The advantage of doing this is clear: it greatly increases your flexibility in phase III. In the absence of a "deep" model that encourages me to commit myself to this peculiar nonlinearity, it seems wise to maintain as much flexibility as possible.

9.6 Conclusion

It is possible to construct a formal decision-theoretic solution to the problem of choice of variables that allows for reconsideration of that choice *after* data have been observed. Such a solution requires us first to identify all the potential variables. We then must provide subjective probability distributions for both the parameters that govern the generation of these variables and also for the parameters that link these variables to the dependent variable under study. For many problems, unfortunately, the identification and assessment problems jointly constitute the most significant costs of dealing with other variables. The observation costs are trivial in comparison, and once we bear the former costs, we almost certainly want to observe and process the complete data set.

I am proposing, therefore, that we behave only *as if* we were formally solving this decision problem. We identify through economic theory and/or introspection certain variables that are potentially important. This is, essentially, the first phase in the formal decision problem. The left-out variables are not, however, formally identified. Instead, we summarize their influence in a contamination parameter $\boldsymbol{\beta}^c$, the prior on which essentially determines the extent to which we are committed to "believe" the regression result.

Just as if we were solving the formal decision problem, we may decide to observe other variables because of either low R^2, peculiar residuals, or peculiar coefficient estimates. At the very least, the probability distribution over the new parameters must imply the original on $\boldsymbol{\beta}^c$.

This constraint prejudices the coefficients on the new variables to zero; that is, by leaving the variable out of the equation to begin with, thereby expressing interest in a "false" model, we have revealed that we think the variable will not significantly distort inferences on the other parameters. This is the case if the regression coefficients are negligible or if the added variables are orthogonal to the original set (or a combination).

This analysis can obviously be carried on to additional stages. At each stage the constraints on the new variables become more severe. Incidentally, the order in which variables are added to the equation influences the interpretation of the evidence. For example, two researchers may end up with the same set of explanatory variables. If these variables have been added to the equations in different orders, then the researchers have revealed different priors and must also make different interpretations of the data evidence.

CHAPTER **10**

SYSTEMATIC JUDGMENTAL ERRORS

A theme of this book is that judgment is the critical input into the analysis of nonexperimental data. Systematic errors in the formation of judgment may lead to significant systematic errors in the interpretation of evidence. The elimination of systematic judgmental errors is thus highly desirable. As a first step in that direction, we may identify in this chapter what seem to be the more consequential systematic errors.

10.1 "Explaining Your Results" as Access-Biased Memory

QUESTION. What do the following quotations have in common?

> The stock market reacted today to the favorable news released by the Commerce Department that our fourth-quarter trade surplus established a new record.
>
> Casey Stengel demonstrated again his lack of managerial talent by replacing pitcher Whitey Ford by a wild Ryne Duren, who proceeded to walk in the winning run.
>
> The negative estimated effect of the price of butter on the consumption of wheat is fully consistent with the fact that bread and butter are jointly consumed.

Answer. (a) All three statements are "explanations" of certain events. In the terminology of probability, where A and B are events, an explanation of an event

A is another event B such that $P(B)$ is large and $P(A|B)$ is significantly greater than $P(A)$.[1] (b) All four explanations were offered after the events occurred. More formally, the statement "$P(A|B)$ is close to one" is made only after both A and B are known to occur.

Suspicion is cast upon *ex post* explanations by the popular English phrases "20-20 hindsight" and "Monday-morning quarterbacking." These phrases allude to the tendency to think in retrospect that events were perfectly predictable, whereas, in fact, the events could not have been foreseen. When probabilities are computed incorrectly, inferential errors are certain to occur. In commenting on the "silly certainty of hindsight," Fischhoff (1975) observes that "if we believe, because of creeping determinism, that the past holds few surprises for us, then we fail to realize that we have anything to learn from it.... A surprise-free past is prologue to a surprise-full future."

It may be useful to construct explicit statistical models to illustrate the inferential errors. Actually, one has already been constructed. The data-instigated models discussed in the previous chapter tend to over-explain the data, and the inferences implied by such models should be discounted. In contraposition to the Fischhoff quotation, the potential error is excessive learning from the given experiment, not insufficient learning.

Another model of hindsight—"access-biased" memory—is now discussed. According to this model a great wealth of prior information is available to aid in the interpretation of current observations. Unfortunately, prior information is not accessed costlessly. As a result, the information that is remembered may bear only a fuzzy relationship to the actual past. The recalled information may even be a version of the past that is distorted to suit the present purposes. If the present purposes are explaining some event A, it is possible that events B, favorable to the outcome A, will be remembered; whereas events C, which are also stored in memory but are unfavorable to A, will be forgotten. It is then asserted that A was inevitable since B is true, and the fact that C is also true is ignored. This we call access-biased memory.

A formal model of access-biased memory consists of the following elements:[2]

a. An uncertain parameter p, $0 \leqslant p \leqslant 1$.
b. A beta prior distribution with parameters r and n.

[1]Good (1975) proposes as a measure of the degree of explanation the number $\log P(A|B) - \log P(A) + \gamma \log P(B)$ where $0 \leqslant \gamma \leqslant 1$. The first two terms measure the increase in probability, and the last measures the plausibility of B.

[2]This material is from Leamer (1975).

c. T independent random variables $X_1,\ldots,X_T$ each distributed binomially with parameters N and p.
d. An observation consisting of the random variable S_k where S_k is the number of the random variables X_i that assume the value k. $(S_0+S_1+\cdots+S_N=T)$.

The parameters of the prior, r and n, are considered to be the outcome of a current binomial experiment. The random variables $X_1,\ldots,X_T$ are considered to be the outcomes of T previous independent sampling experiments. These are assumed to be stored in memory that consists of $N+2$ counters: one that counts the total number of experiments and $N+1$ that count the number of experiments involving exactly k successes, $k=0,1,\ldots,N$. Memory is accessed by observing the total number of experiments T and also *one* of the other counters, S_k.

This is intended primarily to model the phenomenon that may be called "explaining your results." It is common practice first to estimate parameters from sample information and then to give reasons why these estimates are correct. The sample information is thereby supported by nonsample information in a way qualitatively in agreement with formal Bayesian analysis, and the "posterior distribution" that informally pools the quantitative sample information with the qualitative nonsample information is implicitly suggested to be more concentrated than the likelihood function alone.

Examples of this sort of thing are abundant; one suffices to illustrate the phenomenon. In a study of the effects of ability on earnings and schooling, Griliches (1974) obtains the "wrong" sign for a coefficient and reports, "This is an unexpected and strange result, which leads us to reexamine our model. Before we do that however, it is worth noting that the results may not be all that foolish (The human facility for rationalization is boundless)..." He then proceeds to explain why this wrong sign might, in fact, be correct.

The essential feature of this example is that Griliches' access to qualitative nonsample information is *selective*. We may assume, as does Griliches, that any result can be explained. But reasons why a coefficient can assume a "wrong" sign are constructed (remembered) only given the data signal of an estimated coefficient with the wrong sign. The signal of the wrong sign triggers not a general reaccess of memory but a selective one, aimed at remembering previous (qualitative) "experiments" that favor the "wrong sign."

The analogous situation in our formal model is that instead of remembering a random selection of the previous binomial experiments, only those experiments that led to exactly k successes are remembered. Each of these experiments necessarily favors the value $p=k/N$, just as reasons why a sign may be "wrong" necessarily support the hypothesis of wrong signs. But as we shall see, this is information that should be discounted, depending especially on T, the total number of previous experiments. It may turn out that the remembered previous experiments that apparently support the value $p = k/N$, in fact should be taken as evidence against this value.

Attention should also be drawn to the fact that Griliches' access to at least some of the qualitative nonsample information occurs *after* a current data set is observed. If this access to memory were random, we may justifiably ask why it did not occur *before* the data were observed. The only apparent answer is no answer at all—an accident of no importance occurred. The hypothesis of selective access to memory, on the other hand, has as a consequence the fact that it is optimal to search memory after the data are observed. We may, therefore, conclude that when we observe memory access after data analysis we have evidence in favor of the hypothesis of selective access relative to the hypothesis of random access.

The selective access model can itself be explained in terms of the library or computer storage models of memory used by psychologists (Howe, 1970). Experiments rather than sufficient statistics are stored in memory, because the computation of a sufficient statistic requires retrieval, computation, and storage operations each time a new experiment is observed. Storage is conceptually limited only by retrieval costs, and the decision to store experiments rather than sufficient statistics can be interpreted as the economic decision to save computation and some retrieval costs at the expense of greater retrieval costs when the information is actually to be used. In other words, there is a configuration of retrieval costs, computation costs, and information use patterns that discourages the computation and storage of sufficient statistics.

The other feature of our model—selective retrieval—has been the implicit study of numerous psychologists under the headings of secondary organization and associationism (Howe, 1970, p. 60). Events are clustered or categorized for later retrieval on the basis of some contentful categorization. For example, events contiguous in time may be accessed in blocks by retrieval questions such as "what happened after that?" It is here assumed that the relevant previous experiments were not conducted seriatim and that categorization by time would not aid retrieval. Instead, experiments are categorized by their meaning or implications for inference: those that favor one hypothesis are stored in one file, those that favor another are stored in another file.

To summarize, we are suggesting the two-part hypothesis: (1) events, not sufficient statistics or their qualitative equivalent, are stored in memory; (2) events are categorized in memory for later retrieval depending on their implications for inference. The intent of this section is not to test this hypothesis but only to explore its implications. We have, however, already offered some qualitative evidence—the Griliches quotation. We may add to this the author's (and perhaps the readers') casual reading of numerous papers in econometrics and casual observation of economics seminars too numerous and too dull to recount in detail here.

The principal implication of this model is that the information attained from memory ought to be discounted. When it is not, and when a researcher retrieves from memory only those experiments similar in content to his current experiment, he greatly understates his uncertainty and places excessive faith in the validity of the current evidence.

This leads to the second question addressed in this section: given this form of memory, which of the categories is it optimal to access? Which is better: to retrieve experiments that tend to support the current experiment or ones that tend to cast doubt on it? In terms of the formal model, given some loss function for estimating p, which value of k is optimal? For example, a value of $k=N(r/n)$ necessarily accesses experiments that favor the same values of p as the current experiment, that is, $p=r/n$. It turns out that with squared error loss it is better in the sense of minimizing Bayes risk to access experiments that slightly contradict the current experiment.

Parenthetically, it may be observed that this model applies not only to personal memory but also to social information processes. The complaint that journals publish only extreme results is quite common. Newspapers publish "bad" news. My friends transmit to me only their most titillating stories. Casual observation thus suggests that information is categorized by its implications for inference and that social information transmission tends to emphasize the extremes. It is easy to construct a model of information transfer that makes this desirable, if it is understood by the participants that the information is selectively transmitted. Failure to understand the selective nature of the transmission results in erroneous inferences—the conclusion that the newspaper's man-bites-dog story accurately portrays the average relationship between men and dogs.

The structure of the formal model is the following:

1. An infinite population having a proportion, p, of its elements that possess a given attribute.
2. An "experience-free" prior for p in the beta family, which for convenience we take to be the diffuse prior
$$f_0(p) \propto p^{-1}(1-p)^{-1}. \tag{10.1}$$
3. A binomial sample consisting of r successes in n independent trials with likelihood function
$$f(r,n|p) \propto p^{r}(1-p)^{n-r}. \tag{10.2}$$

4. A set of T experiences stored in memory, each of which consists of a binomial sample of size N with r_t successes, $t=1,\ldots,T$.
5. A rule for accessing memory, which we take to be the following: select an integer k and memory reports the number of experiences S_k for which $r_t=k$.

It is assumed for simplicity that the cost function for accessing memory allows for one choice of k essentially for free but disallows any further interrogation of memory.[3]

In principle, memory could be interrogated before the sample values (r,n) are observed. The strategy of letting k be a function of r and n includes as a special case k independent of r and n. It cannot, therefore, increase expected loss, and does, in fact, decrease it, except in unusual circumstances. We therefore search memory after observing r and n. Combining the experience-free prior with the likelihood function to form a prememory distribution, we obtain

$$f(p|r,n) \propto p^{r-1}(1-p)^{n-r-1} \propto f_\beta(p|r,n) \tag{10.3}$$

where f_β indicates a beta distribution

$$f_\beta(p|r,n) = B^{-1}(r,n-r)p^{r-1}(1-p)^{n-r-1}, \qquad 0 \leqslant p \leqslant 1, 0<r<n$$

where

$$B(r,n-r) = \frac{(r-1)!(n-r-1)!}{(n-1)!}.$$

A postmemory distribution is formed by multiplying the prememory distribution (10.3) times the likelihood function of p depending on s, the memory output. This function can be derived by some straightforward probability manipulations. The number of successes in each of the experiences is assumed to be binomially distributed with parameter p

$$\pi(k,p) = P(r_t = k|p) = \binom{N}{k} p^k (1-p)^{N-k}.$$

Conditional on p each experience is independent and contains k successes with probability $\pi(k,p)$. The number of experiences with k successes is, therefore, binomially distributed with parameter π

$$P(S_k = s|p,k) = \binom{T}{s} \pi^s (1-\pi)^{T-s} \tag{10.4}$$

$$\propto \left[p^{ks}(1-p)^{(N-k)s}\right]\left[1-\binom{N}{k}p^k(1-p)^{N-k}\right]^{T-s}$$

[3] An alternative reasonable assumption is that extreme events are the most memorable, i.e., accessed at least cost.

which, given s, is the likelihood function of p. Note that the first term in the brackets in this expression is the likelihood function that we would usually use to characterize the information in our s-remembered experiences, each of which consists of a sample of size N with k successes. The second term is a discount factor that should apply to this information because of the way it was accessed.

If we let the discount factor be

$$d = d(N,k,T,s,p) = \left[1 - \binom{N}{k} p^k (1-p)^{N-k}\right]^{T-s},$$

the postmemory distribution of p formed by multiplying the prememory distribution (10.3) times the memory likelihood function (10.4) is

$$f(p|r,n,k,T,s,N) \propto f_\beta(p|r+sk, n+Ns)\cdot d \qquad (10.5)$$

which is the usual beta distribution times the discount factor d. The discount factor can be written as

$$d = \left[1 - \binom{N}{k} B(k+1, N-k+1) f_\beta(p|k+1, N+2)\right]^{T-s}$$

$$= \left[1 - (N+1) f_\beta(p|k+1, N+2)\right]^{T-s}$$

which assumes a minimum value at $p = k/N$, thereby discounting the memory evidence that would otherwise necessarily favor the value of $p = k/N$.

To illustrate the effect of the discount factor, let us consider the case when $k=0$, that is, when memory is accessed by the question "are there any previous experiences that involved no successes?" The factor then becomes

$$d = \left[1 - (1-p)^N\right]^{T-s}$$

which assumes a minimum of zero at $p=0$ and a maximum of one at $p=1$. This has the effect of pushing the posterior distribution away from $p=0$, depending positively on the number of forgotten experiences, $T-s$. Of course, any experiences that are remembered necessarily favor $p=0$. The net effect on the distribution thus depends on both the remembered experiences and the forgotten experiences. There is an informal lesson to be drawn from this. Old men with many experiences must tell more stories in support of their theories if they expect to generate the same amount of believability as a young man. Or to put it differently, the wisdom of age is greatly exaggerated if memory failures are ignored.

We now turn to the problem of optimal memory interrogation. Consider a researcher who has current experimental support for some proposition x against an alternative y. If he searches memory for further evidence in favor of x, he expects to find it but "hopes" he doesn't, since the absence of previous similar results will informatively cast doubt on the proposition while the presence of previous similar results is unsurprising and relatively uninformative. Similarly, if he searches memory for evidence in favor of the alternative y, he expects not to find it but "hopes" that he does. In both cases he anticipates obtaining relatively uninformative information. This symmetry makes ambiguous the decision whether to search memory for experiments in favor of x or in favor of y, an ambiguity that can be resolved only in the context of specific problems.

For our problem, optimal choice of k, the memory-accessed value, is not obvious but depends on a conceptually straightforward preposterior analysis. Let us take the variance of p as the measure of uncertainty and seek to find the value of k that minimizes the expected posterior variance.

The expected posterior moments can be written as

$$E(Ep|s) = Ep = \frac{r}{n}$$

$$E(V(p|s)) = E(E(p^2|s)) - E\{[E(p|s)]^2\}$$

$$= Ep^2 - E\{[E(p|s)]^2\}$$

$$= \frac{r(r+1)}{n(n+1)} - E\{[E(p|s)]^2\}.$$

That is, the expected mean is just the prior mean (r/n), and the expected variance depends on k only through the factor $E\{[E(p|s)]^2\}$. These can be computed by appropriately defining the joint distributions of p and s, as in Leamer (1975).

The expected posterior variance as a function of k is illustrated for one case in Figure 10.1. It is assumed that the current sample size n is equal to 10, that this sample involves five successes, that there are $T=5$ previous experiments stored in memory, and that each of these five previous experiments involve $N=10$ trials. Note that although $N=10$, the least desirable ($EV(ps)$ maximum) memory accessed value is $k=5$. Past experiences that involve k successes favor the value $p=k/N$. Thus the least desirable way to interrogate memory is to ask if there are any previous experiences that favor the value $p=k/N=.5$, the very value that is most favored by the current sample, $r/n=.5$. It is almost as undesirable to search memory for extreme experiences, $k=0$ or $k=10$. The best thing to do is to ask for experiences that slightly contradict the current sample in the sense of favoring the values $p=.8$ or $p=.2$.

This resolves the ambiguity referred to previously. Since the value r/n is currently the most favored value of p, the researcher regards it to be highly

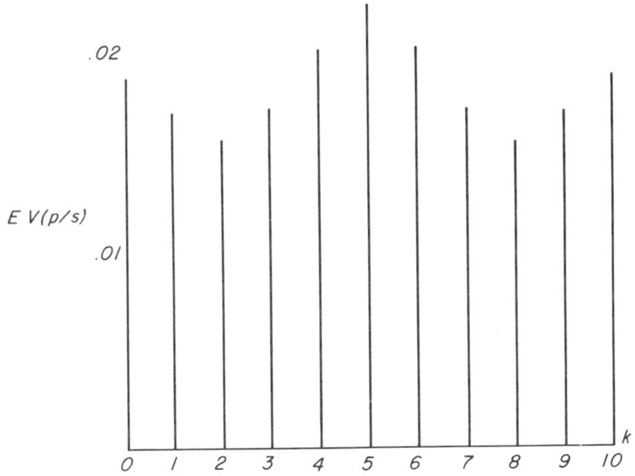

Fig. 10.1 Expected posterior variance; $n=10, r=5, N=10, T=5$.

likely that there will be previous experiments that favor this same value. Counterbalancing this influence is the fact that these experiments when remembered are not particularly informative. Slightly contradictory experiments, however, are much more informative and also are reasonably likely to be in memory. Greatly contradictory experiments, although especially informative, are highly unlikely to be in memory and therefore should not be searched out.

With the *caveat* that our inferences do not necessarily generalize to other situations, especially those free of memory failures, we conclude with the following homilies:

a. A young man with one good story may merit your attention more than an old man with several.
b. Spend your time thinking of reasons why you are slightly wrong. Waste not your time thinking why you are greatly wrong—you are likely to go home with an empty basket. Least of all think of reasons why you are right—though you are likely to fill your basket, the fruits you bring home will contain little knowledge.

10.2 Biases in Personal Probabilities

We have discussed in the previous section an error that is potentially made when prior probabilities are formed after the data are observed. This is the error of failing appropriately to discount prior information that otherwise necessarily is supportive of the current sample. There are other interesting "errors" that are often made and that have important implications for our study of inference with nonexperimental data. Many such errors are discussed by Tversky and Kahneman (1974) and by Slovic (1972), and Hogarth (1975). In this section we briefly discuss some of the more interesting of these phenomena.

Memory Failures

According to the model of the previous section, events are memorable because they have implications for inference, or to put it differently, storage is arranged in anticipation of a well-defined inferential problem. As an illustration of a similar memory error without an anticipated inferential problem, glance over the following list of names without reading further the textual materials:

Gerald Ford
Betty Jackson
Maria Muldaur
Ralph Nader
Ronald Reagan
Barbara Wilson
William Shakespeare
Katherine Burgoyne
Al Jolson
Deborah Hirsch

Now cover up the list of names and answer the question, "Are there more men or more women on this list?"

Tversky and Kahneman (1974) report that in such experiments there is a tendency to overstate the proportion of men, when the listed men are relatively more famous (as above), and to overstate the proportion of women, when the listed women are relatively more famous. The important point of the example is that the more memorable events are most likely used to form conscious opinions. The resulting opinions may significantly distort the past when the ease of memory is related in some way to the evidential content of the event. Somewhat the same point has been made in our discussion of access-biased memory: the current instance of an event makes more memorable similar past instances.

OPTIMISM/PESSIMISM

Another error in forming opinions is the confusion of a utility function with prior information. Optimists are people who think desirable events will likely happen; pessimists think they won't. Although it does not involve a logical contradiction to condition opinions on utilities, most thoughtful observers would argue against such a practice. The existence of optimists and pessimists has a firm foundation in folklore and has been more scientifically established by Slovic (1966).

The confusion between prior information and utility functions leads, in the language of this book, to a confusion between interpretive searches, which require prior information, and simplification searches, which require utility functions. As an example, consider the (constant elasticity of substitution) function

$$y(x_1,x_2)=\alpha\left[x_1^{-\beta}+(\gamma x_2)^{-\beta}\right]^{-\delta/\beta}. \tag{10.6}$$

In the limit, as β goes to zero, this function goes to

$$y^*(x_1,x_2)=\alpha x_1^{\gamma_1}x_2^{\gamma_2}, \tag{10.7}$$

which, conveniently, is linear in the logarithms. A common practice in economics is to test the restriction $\beta=0$, hoping that the restriction is accepted. We may ask if this is an interpretive search or a simplification search. My own opinion is that the restriction derives originally from a utility function—it is ever so much neater to work with the second function than the first. With the passage of time, however, what was first a simplification search has now become—inappropriately, I think—an interpretive search. The many studies that failed to reject the restriction $\beta=0$ generate a feeling among economists that the restriction is an hypothesis that is favored by prior information. Although these studies do, indeed, contain information about the parameter β, the fact that many of them fail to reject $\beta=0$ is only remotely connected to the accumulated probability in a region around $\beta=0$. It is important, as we have argued, to distinguish the output of a simplification search from the interpretation of the data.

EGOCENTRISM

In formulating opinions, about other human beings especially, you are necessarily forced to use yourself and your experiences as a norm. In so doing there is a clear tendency to regard yourself to be overly representative, that is, to fail to realize appropriately the diversity of humanity. An experiment I have done in my own classes is to have students estimate the weight of a relatively heavy person and a relatively light person. Not only do light people consistently underestimate the weight of the heavy subject, moreover, they fail to allow for their increased uncertainty in their choice of confidence intervals. Not only do light people have no concept of what it means to be heavy, but also they fail to realize that they have no such knowledge.

The importance of this phenomenon in social research could be overstated as follows: judgment is critically important in the analysis of nonexperimental data, and the diversity of judgments of academics is limited by the sameness of their lives. Thus the interpretation of evidence by academics is a class phenomenon: relative unanimity within the class and potentially sharp disagreements with other classes.

LAW OF SMALL NUMBERS

Tversky and Kahneman (1971a) in a study of psychologists observe that even people with some formal training in statistical inference have a deep belief in the "law of small numbers," according to which a sample is necessarily representative of the population as a whole. As an example of the law, the probability of a tail after five heads in a row is quite a bit larger than one-half because a tail is "due" or, if you like, because the sequence of six coin flips is strongly thought to be representative of the universe of coin flips consisting of 50% heads and 50% tails. A more important example is the unwarranted belief in the estimates generated by a relatively small sample, as demonstrated, first, by the researcher's willingness to go to great lengths to "explain" the sample result and, second, by the researcher's surprise at the extent to which estimates can change as evidence accumulates.

OVERCONFIDENCE

Heretofore we have been concerned largely with the location of prior distributions. A few things may also be said about the dispersions of these distributions. One phenomenon is overconfidence. In a study of students at the Harvard Business School, Alpert and Raiffa (1969) found that 426 out of 1000 98%-confidence intervals failed to capture the true value of the item being estimated. You might have expected approximately 20 misses, and the fact that as many as 426 intervals were "wrong" has to be considered strong evidence of overconfidence. However, the extent to which this can be considered to be a general phenomenon is subject to considerable doubt. In particular, it may be very sensitive to the method of eliciting the interval (Hogarth, 1975, provides many references).

CONSERVATISM

The counterpart of overconfidence with respect to prior information is undervaluation of current information (Edwards, 1968). To give a simple example, a coin is flipped to decide from which of two urns to select a ball. The first urn contains 80% red balls and the second contains 20%. If a red ball is selected, the conditional probability that it came from the first urn is "objectively" .8. Subjects consistently underestimate this probability, that is, they fail to revise their prior (.5) as much as would be suggested by Bayes' rule. There is a considerable literature in psychology, referenced by Hogarth (1975), that explores the circumstances in which this phenomenon occurs.

TRIVIALITY

Another error that is made is the confusion of the mass of information with the value of the information. A researcher sometimes attempts to convince his reader of the validity of his study by inundating the reader with great masses of information. Although it is obvious that the "informational weight" of a study is not equal to its physical weight, there is a tendency to unduly correlate the two.

For example, Oskamp (1965) provided psychological clinicians with written descriptions of the personalities of some subjects and then asked the clinicians to answer 25 questions about the subjects. He found that although the ability of the clinicians to answer the questions correctly was little affected by increases in the length of the written description, the confidence ascribed to the answers rose dramatically.

METHODOLOGICAL PATTERNING

Without comment, I would like to describe (with some literary license) Skinner's (1948) experiment of randomly induced behavior. Hungry birds fed at *random* intervals are observed to adopt the peculiar behavior of odd head movements, hopping from side to side, and the like. The apparent explanation is that on receipt of the seed, the bird hypothesizes that the seed is a reward for the most recently antecedent trick. If the bird happened to have twitched his head just before the seed arrived, the bird naturally tries twitching his head again. The increased frequency of head twitching makes more likely the event that a twitch will precede a seed, and eventually the bird is twitching frequently with seeds always following twitching. The belief in this relationship is so strong that even after the seed stops arriving altogether the bird may twitch as many as 10,000 times. Slovic (1972) sees parallels between pigeons and stockbrokers. I certainly would not want to draw parallels between pigeons and social scientists.

10.3 Social Learning Processes

The opinions and decisions of an individual are only partly dependent on his own observations. The advice and comments of other individuals is another highly important source of information and decision rules. It is embarrassing, on almost the last page of a book dealing with learning, to make the observation that the personal learning heretofore discussed constitutes only a small part of the learning process. The space allocated to social-learning processes reflects my knowledge, not my assessment of their importance.

IMPROPER POOLING OF INFORMATION ACROSS INDIVIDUALS

Two or more individuals who have opinions based on the same information should not change their opinions when confronted by the (necessarily coincident?) opinions of others. In practice, people are unduly affected by the opinions of others, to the point where unanimity is sometimes confused with certainty.

An experiment, first performed by Asch (1952), demonstrates the remarkable tendency of conformity of opinion in groups. Seven individuals are asked to identify which of three lines is the same length as some standard line. The first six individuals are, in fact, collaborators of the experimenter, and they deliberately and uniformly select (aloud) a line that is not the right one. The subject, who is unaware of the collusion, is then forced to choose between either giving what he feels is the right answer or conforming with the group. Faced with this decision, in 33.2% of the cases the subjects conformed to the group and gave the incorrect response. This contrasts with subjects not exposed to group pressure, who answered incorrectly only 7.4% of the time.

It may be argued, of course, that this is not improper pooling of information. Subjects may legitimately be influenced by the information revealed by others. What the experiment fails to distinguish is conformity due to information transfer from conformity merely to please the other members of the group. If a correct answer were sufficiently well rewarded, would the subject continue to conform? Yes, if he gets information from the group. No, if he is merely trying to please his colleagues.

ADVOCACY ABILITY

Under the heading of triviality we observed the confusion of the physical mass of information with its real content. Other features of the reporting style are also likely to greatly influence our willingness to believe results and arguments. Especially when the data evidence is relatively weak, it is possible to end up with a uniform professional opinion about some empirical issue, merely because one scholar has unusually fine advocacy abilities. We may even think that our judgment in his favor is heavily influenced by observed phenomena.

In a study designed to score advocates to eliminate ability bias, Warner (1975) had two teams of advocates each write pro and con briefs about a proposed expressway in Toronto. One pro and one con case was mailed to 1360 randomly selected electors. Warner found that the proportion of electors who favored the expressway after having received the briefs varied from .53 to .85, depending on which pro and which con case they received. The empirical facts are the same in all cases. The variability of the proportion in favor of the expressway is partly attributable to sampling error but mostly to advocacy ability.

CONSENSUS PRECEDES CERTAINTY

Divergent opinions when confronted with observations tend to converge to each other as well as to the "truth." Dickey and Fischer (1975) observe that for a broad class of sampling contexts, the posterior mean converges to the true value at the rate of $T^{-1/2}$, but the dispersion of posterior means across individuals goes to zero at the rate T^{-1} (T is the sample size).

OTHER

A consensus model is given by DeGroot (1974). See also Pruitt (1971) and Stone (1961).

APPENDIX 1

PROPERTIES OF MATRICES

Properties of matrices are reported in this appendix. Familiarity with the elementary algebra of matrices is assumed, and properties are stated but not proved. For proofs of most of the propositions, consult Graybill (1969).

Definitions

A *matrix* is a two-dimensional array of numbers denoted by capital boldface letters such as **A**. If the array has only one column, it is called a *vector* and is indicated (usually) by lowercase boldface letters such as **a**. The *ij*th element of **A** is indicated by A_{ij}, and the symbol $\{A_{ij}\}$ stands for a matrix with elements A_{ij}. An $m \times n$ matrix has m rows and n columns.

(D1) If **A** and **B** are $m \times n$ matrices, the *sum* of **A** and **B** is defined by

$$\mathbf{A}+\mathbf{B}=\{A_{ij}+B_{ij}\}.$$

(D2) If **A** is $m \times n$ and **B** is $n \times k$, then the *product* of **A** and **B** is defined by

$$\mathbf{AB}=\left\{\sum_{p=1}^{n} A_{ip}B_{pj}\right\}$$

(D3) *Multiplication* of a matrix **A** by a *scalar* α is defined by

$$\alpha\mathbf{A}=\{\alpha A_{ij}\}.$$

(D4) The *transpose* of a matrix $\{A_{ij}\}$ is the matrix

$$\mathbf{A}'=\{A_{ji}\}$$

(D5) A matrix is *symmetric* if it is equal to its transpose $\mathbf{A}=\mathbf{A}'$.

(D6) A *diagonal matrix* is a square matrix with zeroes off the main diagonal, $A_{ij}=0$, for $i\neq j$. It is indicated by $\mathbf{A}=\text{diag}\{d_1,d_2,\ldots,d_n\}$ where $A_{ii}=d_i$.

(D7) An *identity matrix* of order p is the $p\times p$ diagonal matrix

$$\mathbf{I}_p=\text{diag}\{1,1,\ldots,1\}.$$

If the order is obvious, the notation $\mathbf{I}$ suffices.

(D8) The notation $\mathbf{1}_n$ indicates an $n\times 1$ *vector of ones*. Where convenient, the subscript n is suppressed.

(D9) The *trace* of a square $n\times n$ matrix $\mathbf{A}$ is the sum of its diagonal elements

$$\text{tr}\mathbf{A}=\sum_{i=1}^{n} A_{ii}$$

(D10) Matrices $\mathbf{A}$ and $\mathbf{B}$ are said to *commute* if $\mathbf{AB}=\mathbf{BA}$.

(D11) A real symmetric matrix $\mathbf{A}$ is said to be *positive definite* if for any vector $\mathbf{x}\neq\mathbf{0}$, $\mathbf{x}'\mathbf{A}\mathbf{x}>0$, and is said to be *positive semi-definite* if $\mathbf{x}'\mathbf{A}\mathbf{x}=0$ for at least one $\mathbf{x}\neq\mathbf{0}$ and $\mathbf{x}'\mathbf{A}\mathbf{x}>0$ otherwise. (*Negative definite* and *negative semi-definite* are defined analogously.)

(D12) A square matrix $\mathbf{A}$ is said to be *orthogonal* if $\mathbf{AA}'=\mathbf{I}$. (As a consequence, $\mathbf{A}'=\mathbf{A}^{-1}$ and $\mathbf{A}'\mathbf{A}=\mathbf{I}$.)

(D13) A square symmetric matrix $\mathbf{A}$ is said to be *idempotent* if $\mathbf{AA}=\mathbf{A}$. (Sometimes symmetry is not included in the definition.)

(D14) The determinant of an $n\times n$ matrix $\mathbf{A}$ is

$$|\mathbf{A}|=\sum(\pm)A_{1i_1}A_{2i_2}\cdots A_{ni_n}$$

where $(i_1,i_2,\ldots,i_n)$ is a permutation of the first n integers, the summation extends over all $n!$ permutations, and the sign is + if the permutation is even or − if the permutation is odd.

(D15) The *inverse* of a square matrix $\mathbf{A}$ is a matrix $\mathbf{A}^{-1}$ such that $\mathbf{AA}^{-1}=\mathbf{A}^{-1}\mathbf{A}=\mathbf{I}$.

(D16) The *rank* of a matrix is the number of linearly independent columns (also the number of linearly independent rows).

(D17) An $n \times n$ matrix $\mathbf{A}$ is said to be *nonsingular* if and only if the rank of $\mathbf{A}$ is n.

(D18) If $\mathbf{A}$ is $(m \times n)$ and $\mathbf{B}$ is $(p \times q)$ the *Kronecker product* of $\mathbf{A}$ and $\mathbf{B}$ is the $(mp \times nq)$ matrix

$$\mathbf{A} \otimes \mathbf{B} = \begin{bmatrix} A_{11}\mathbf{B} & A_{12}\mathbf{B} & \cdots & A_{1n}\mathbf{B} \\ A_{21}\mathbf{B} & A_{22} & B & A_{2n}\mathbf{B} \\ \vdots & & \vdots & \\ A_{m1}\mathbf{B} & A_{m2}\mathbf{B} & \cdots & A_{mn}\mathbf{B} \end{bmatrix}$$

Properties

The following properties are easily verifiable, assuming that the operations are well defined.

(T1) $$\operatorname{tr}\mathbf{AB} = \operatorname{tr}\mathbf{BA}$$

(T2) $$\operatorname{tr}(\mathbf{A}+\mathbf{B}) = \operatorname{tr}\mathbf{A} + \operatorname{tr}\mathbf{B}$$

(T3) $$|\mathbf{AB}| = |\mathbf{A}||\mathbf{B}|$$

(T4) $$|\alpha\mathbf{A}| = \alpha^n|\mathbf{A}| \qquad (\mathbf{A} \text{ is } (n \times n))$$

(T5) $$(\mathbf{A}')^{-1} = (\mathbf{A}^{-1})'$$

(T6) $$(\mathbf{AB})' = \mathbf{B}'\mathbf{A}'$$

(T7) $$(\mathbf{AB})^{-1} = \mathbf{B}^{-1}\mathbf{A}^{-1}$$

(T8) $$(\mathbf{A}+\mathbf{BCB}')^{-1} = \mathbf{A}^{-1} - \mathbf{A}^{-1}\mathbf{B}(\mathbf{B}'\mathbf{A}^{-1}\mathbf{B}+\mathbf{C}^{-1})^{-1}\mathbf{B}'\mathbf{A}^{-1}$$

(T9) $$(\mathbf{A}+\mathbf{B})^{-1} = \mathbf{A}^{-1}(\mathbf{A}^{-1}+\mathbf{B}^{-1})^{-1}\mathbf{B}^{-1}$$

Let the quadratic form $\mathbf{x}'\mathbf{A}\mathbf{x}$ be indicated by $Q(\mathbf{x}, \mathbf{A})$; then assuming $\mathbf{H}$ and $\mathbf{H}^*$ invertible

(T10) $$Q((\boldsymbol{\beta}-\mathbf{b}), \mathbf{H}) + Q((\boldsymbol{\beta}-\mathbf{b}^*), \mathbf{H}^*)$$
$$= Q(\boldsymbol{\beta}-\mathbf{b}^{**}, \mathbf{H}+\mathbf{H}^*) + Q\left(\mathbf{b}-\mathbf{b}^*, \mathbf{H}^*(\mathbf{H}+\mathbf{H}^*)^{-1}\mathbf{H}\right)$$

where $$\mathbf{b}^{**}=(\mathbf{H}+\mathbf{H}^{*})^{-1}(\mathbf{H}\mathbf{b}+\mathbf{H}^{*}\mathbf{b}^{*}).$$

Theorems concerning the rank of a matrix are

(T11) $$\text{rank}(\mathbf{AB})\leqslant\min[\text{rank}(\mathbf{A}),\ \text{rank}(\mathbf{B})]$$

(T12) $$\text{rank}(\mathbf{A})=\text{rank}(\mathbf{A}')=\text{rank}(\mathbf{AA}').$$

The existence of an inverse is implied by the following theorem.

(T13) If $\mathbf{A}$ is an $n\times n$ matrix, its inverse exists if and only if $|\mathbf{A}|\neq 0$.

A square matrix may be partitioned as

$$\mathbf{A}=\begin{bmatrix}\mathbf{E} & \mathbf{F}\\ \mathbf{G} & \mathbf{H}\end{bmatrix}$$

where $\mathbf{E}$ and $\mathbf{H}$ are themselves square. Then

(T14) $$\begin{bmatrix}\mathbf{E} & \mathbf{F}\\ \mathbf{G} & \mathbf{H}\end{bmatrix}^{-1}=\begin{bmatrix}\mathbf{E}^{-1}(\mathbf{I}+\mathbf{F}\mathbf{C}^{-1}\mathbf{G}\mathbf{E}^{-1}) & -\mathbf{E}^{-1}\mathbf{F}\mathbf{C}^{-1}\\ -\mathbf{C}^{-1}\mathbf{G}\mathbf{E}^{-1} & \mathbf{C}^{-1}\end{bmatrix}$$

where $\mathbf{C}=\mathbf{H}-\mathbf{G}\mathbf{E}^{-1}\mathbf{F}$.

(T15) $$\left|\begin{bmatrix}\mathbf{E} & \mathbf{F}\\ \mathbf{G} & \mathbf{H}\end{bmatrix}\right|=|\mathbf{E}||\mathbf{H}-\mathbf{G}\mathbf{E}^{-1}\mathbf{F}|=|\mathbf{H}||\mathbf{E}-\mathbf{F}\mathbf{H}^{-1}\mathbf{G}|.$$

An implication of (T15) is $|\mathbf{E}+\mathbf{xx}'|=|\mathbf{E}|(1+\mathbf{x}'\mathbf{E}^{-1}\mathbf{x})$ for $\mathbf{E}$ a matrix and $\mathbf{x}$ a vector.

The following theorems apply to positive definite (p.d.) and positive semi-definite (p.s.d.) matrices

(T16) If $\mathbf{A}$ is $n\times n$ p.d., $\mathbf{P}$ is $n\times m$ with rank $m\leqslant n$, then $\mathbf{P}'\mathbf{AP}$ is p.d.

(T17) If $\mathbf{A}$ is p.d., then $\mathbf{A}^{-1}$ is p.d.

(T18) If $\mathbf{A}$ is p.d., then $|\mathbf{A}|>0$.

(T19) If $\mathbf{A}$ is p.d., and $\mathbf{B}$ p.s.d., then $\mathbf{A}+\mathbf{B}$ is p.d.

The following results apply to idempotent matrices.

(T20) If $\mathbf{A}$ is idempotent, the values of λ satisfying $|\mathbf{A}-\lambda\mathbf{I}|=0$ are either one or zero.

(T21) If $\mathbf{A}$ is idempotent, $\text{tr}(\mathbf{A})=\text{rank}(\mathbf{A})$.

On the assumption that the operations are well defined, the Kronecker product has the following properties:

(T22) $$\mathbf{A}\otimes(\mathbf{B}+\mathbf{C})=\mathbf{A}\otimes\mathbf{B}+\mathbf{A}\otimes\mathbf{C}$$

(T23) $$(\mathbf{A}\otimes\mathbf{B})\otimes\mathbf{C}=\mathbf{A}\otimes(\mathbf{B}\otimes\mathbf{C})$$

(T24) $$(\mathbf{A}\otimes\mathbf{B})'=\mathbf{A}'\otimes\mathbf{B}'$$

(T25) $$(\mathbf{A}\otimes\mathbf{B})(\mathbf{C}\otimes\mathbf{D})=\mathbf{AC}\otimes\mathbf{BD}$$

(T26) $$(\mathbf{A}\otimes\mathbf{B})^{-1}=\mathbf{A}^{-1}\otimes\mathbf{B}^{-1}$$

(T27) $$|\mathbf{A}\otimes\mathbf{B}|=|\mathbf{A}|^m|\mathbf{B}|^n, \qquad (\mathbf{A} \text{ is } (n\times n), \mathbf{B} \text{ is } (m\times m))$$

(T28) $$\mathrm{tr}|\mathbf{A}\otimes\mathbf{B}|=\mathrm{tr}(\mathbf{A})\cdot\mathrm{tr}(\mathbf{B})$$

Matrix Differentiation

Let $y=f(\mathbf{x})$ be a scalar function of the vector $\mathbf{x}$, and let the vector of derivatives of y with respect to $\mathbf{x}$ be denoted by

$$\frac{\partial y}{\partial \mathbf{x}}=\left\{\frac{\partial y}{\partial x_i}\right\}.$$

Let $\mathbf{y}=f(\mathbf{x})$ be a vector function of the vector $\mathbf{x}$, and let the *matrix* of derivatives of $\mathbf{y}$ with respect to $\mathbf{x}$ be

$$\frac{\partial \mathbf{y}}{\partial \mathbf{x}}=\left\{\frac{\partial y_i}{\partial x_j}\right\}.$$

Let y be a scalar function of the matrix $\mathbf{A}$, and let the *matrix* of derivatives of y with respect to $\mathbf{A}$ be

$$\frac{\partial y}{\partial \mathbf{A}}=\left\{\frac{\partial y}{\partial A_{ij}}\right\}.$$

Let $\mathbf{A}$ be a matrix function of the scalar t, and let the matrix of derivatives of $\mathbf{A}$ with respect to t be

$$\frac{\partial \mathbf{A}}{\partial t}=\left\{\frac{\partial A_{ij}}{\partial t}\right\}.$$

The following formulas taken from Dwyer (1967), are straightforwardly verified where $\mathbf{A}$ and $\mathbf{B}$ are matrices, $\mathbf{x}$ is a vector, and t is a scalar.

(T29) $$\frac{\partial \mathbf{Ax}}{\partial \mathbf{x}} = \mathbf{A}$$

(T30) $$\frac{\partial \mathbf{x}' A \mathbf{x}}{\partial \mathbf{x}} = 2\mathbf{Ax} \qquad (\text{with } \mathbf{A} = \mathbf{A}')$$

(T31) $$\frac{\partial \operatorname{tr} \mathbf{A}}{\partial \mathbf{A}} = \mathbf{I}$$

(T32) $$\frac{\partial |\mathbf{A}|^r}{\partial \mathbf{A}} = r|\mathbf{A}|^r (\mathbf{A}^{-1})'$$

(T33) $$\frac{\partial \, 1n|\mathbf{A}|}{\partial \mathbf{A}} = (\mathbf{A}^{-1})'$$

(T34) $$\frac{\partial \mathbf{x}' \mathbf{Ax}}{\partial \mathbf{A}} = \mathbf{xx}' \qquad (\mathbf{A} \text{ symmetric})$$

(T35) $$\frac{\partial \mathbf{xA}^{-1}\mathbf{x}}{\partial \mathbf{A}} = -\mathbf{A}^{-1}\mathbf{xx}'\mathbf{A}^{-1} \qquad (\mathbf{A} \text{ symmetric})$$

(T36) $$\frac{\partial \operatorname{tr} \mathbf{AB}}{\partial \mathbf{A}} = \mathbf{B}'$$

(T37) $$\frac{\partial \mathbf{AB}}{\partial t} = \mathbf{A}\frac{\partial \mathbf{B}}{\partial t} + \frac{\partial \mathbf{A}}{\partial t}\mathbf{B}$$

(T38) $$\frac{\partial \mathbf{A}^{-1}}{\partial t} = -\mathbf{A}^{-1}\frac{\partial \mathbf{A}}{\partial t}\mathbf{A}^{-1}$$

Gradients, Normals, and Tangent Hyperplanes

Let $f(\mathbf{x})$ be a scalar valued function of the vector $\mathbf{x}$. The differential of f is

$$df = \sum_i \frac{\partial f}{\partial x_i} dx_i$$

$$= \left(\frac{\partial f}{\partial \mathbf{x}}\right)' d\mathbf{x}.$$

Setting this differential to zero and solving for $d\mathbf{x}$ determines a direction in which there is no differential change in the function f. Geometrically, the equation $0 = (\mathbf{x} - \mathbf{x}_0)'(\partial f / \partial \mathbf{x}|_{\mathbf{x}=\mathbf{x}_0})$ thus defines a hyperplane tangent to the surface $f(\mathbf{x}) = f(\mathbf{x}_0)$ at $\mathbf{x}_0$. The direction $\partial f / \partial \mathbf{x}$ is orthogonal to this hyperplane and is called the *gradient vector* of the function or the (inward or outward) normal of the surface. The quadratic form $f(\mathbf{x}) = \mathbf{x}'\mathbf{Ax}$ has the tangent hyperplane at $\mathbf{x}_0$ given by $\mathbf{x}_0'\mathbf{A}\mathbf{x}_0=\mathbf{x}'\mathbf{A}\mathbf{x}_0$, and $\mathbf{A}\mathbf{x}$ is the normal of the surface.

Eigenvectors and Ellipsoids

The first result of this section is illustrated in Figure A1: If $\mathbf{A}$ is a real symmetric positive definite matrix, the equation $\mathbf{x}'\mathbf{A}\mathbf{x}=r^2$ defines an ellipsoid. The directions of tangency between a unit sphere and an ellipsoid, $\mathbf{c}_1$ and $\mathbf{c}_2$ in the figure, are the eigenvectors of the matrix $\mathbf{A}$, and are also the major axes of the ellipsoid. The corresponding eigenvalues are the values of r^2 needed to produce a tangency between the unit sphere and an ellipsoid $\mathbf{x}'\mathbf{A}\mathbf{x}=r^2$. The relative eigenvalues are just the relative lengths of the major axes of the ellipse, the longer axis having the smaller eigenvalue.

If $\mathbf{A}$ is a fixed square $(k\times k)$ matrix, $\mathbf{x}$ is a $(k\times 1)$ vector of variables, and r is a fixed scalar, then $\mathbf{x}'\mathbf{A}\mathbf{x}=r^2$ is the equation of a *quadratic surface*. For example, if $\mathbf{A}=\mathbf{I}_k$, then $\mathbf{x}'\mathbf{A}\mathbf{x}=\mathbf{x}'\mathbf{x}=r^2$ is the equation of a sphere located at the origin with radius r. A *family of concentric quadratic surfaces* includes all surfaces $\mathbf{x}'\mathbf{A}\mathbf{x}=r^2$ for a given $\mathbf{A}$ and for any $r\geqslant 0$.

The direction of tangencies between a family of concentric quadratic surfaces and the unit sphere $\mathbf{x}'\mathbf{x}=1$ can be found by maximizing or

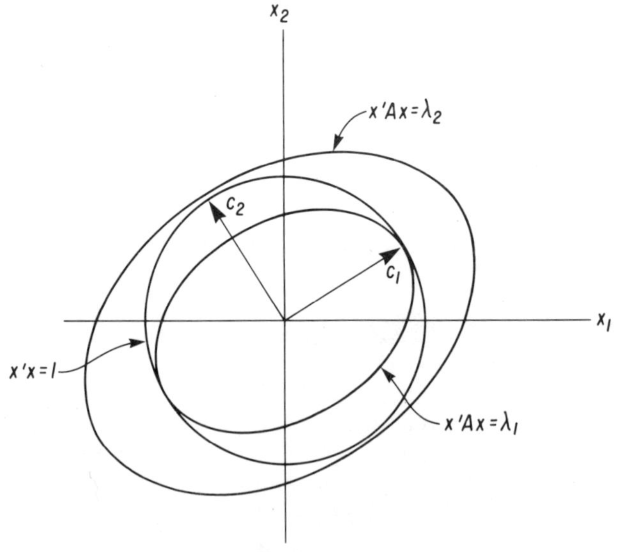

Fig. A.1 Eigenvectors and eigenvalues.

minimizing the quadratic form $\mathbf{x}'\mathbf{A}\mathbf{x}$ subject to the constraint $\mathbf{x}'\mathbf{x}=1$. This is a simple Lagrangian problem:

$$\mathbf{0}=\frac{\partial\left(\mathbf{x}'\mathbf{A}\mathbf{x}-\lambda[\mathbf{x}'\mathbf{x}-1]\right)}{\partial\mathbf{x}}=2\mathbf{A}\mathbf{x}-2\lambda\mathbf{x},$$

where λ is the Lagrange multiplier. A tangency direction is thus a direction $\mathbf{x}$ satisfying $\mathbf{A}\mathbf{x}=\lambda\mathbf{x}$, or

$$(\mathbf{A}-\lambda\mathbf{I}_k)\mathbf{x}=\mathbf{0}_k.$$

If the matrix $(\mathbf{A}-\lambda\mathbf{I}_k)$ were invertible, the only solution to this set of equations would be $\mathbf{x}=(\mathbf{A}-\lambda\mathbf{I}_k)^{-1}\mathbf{0}_k=\mathbf{0}_k$. Thus a condition for a nontrivial solution is that this matrix is not invertible and hence that the determinant is zero:

$$|\mathbf{A}-\lambda\mathbf{I}_k|=\sum_{i=0}^{k}p_i\lambda^i=0.$$

This polynomial in λ of degree k is known as the *characteristic equation* of the matrix $\mathbf{A}$, and the roots of the polynomial are known as the *characteristic* (or eigen, or latent) *values* of the matrix $\mathbf{A}$. If λ_i is a root of the polynomial $|\mathbf{A}-\lambda\mathbf{I}_k|=0$, the tangency direction $\mathbf{c}_i$ satisfying $(\mathbf{A}-\lambda_i\mathbf{I}_k)\mathbf{c}_i=\mathbf{0}_k$ and normalized such that $\mathbf{c}_i'\mathbf{c}_i=1$, is known as a *characteristic* (or eigen, or latent) *vector* of the matrix $\mathbf{A}$.

Example. Let

$$\mathbf{A}=\begin{bmatrix}4 & 2\\ 2 & 4\end{bmatrix}.$$

The characteristic polynomial is $|\mathbf{A}-\lambda\mathbf{I}_2|=(4-\lambda)^2-4$, with roots $\lambda_1=2$, $\lambda_2=6$. The characteristic vector corresponding to $\lambda_1=2$ is a solution to

$$\begin{bmatrix}4-2 & 2\\ 2 & 4-2\end{bmatrix}\begin{bmatrix}c_{11}\\ c_{12}\end{bmatrix}=\begin{bmatrix}0\\ 0\end{bmatrix}.$$

Thus $c_1'=(1,-1)/\sqrt{2}$, and similarly $c_2'=(1,1)/\sqrt{2}$.

If a matrix $\mathbf{B}$ is not symmetric, there is a symmetric matrix $\mathbf{A}$ such that $\mathbf{x}'\mathbf{B}\mathbf{x}=\mathbf{x}'\mathbf{A}\mathbf{x}$ for all $\mathbf{x}$. Merely set $A_{ij}=A_{ji}=(B_{ij}+B_{ji})/2$. Thus for the analysis of quadratic surfaces we need consider only symmetric matrices.

THEOREM 30. *Given a symmetric matrix* $\mathbf{A}$, *the characteristic vectors corresponding to unequal characteristic values are orthogonal.*

Proof: $0=\mathbf{c}_i'\mathbf{A}\mathbf{c}_j-\mathbf{c}_i'\mathbf{A}\mathbf{c}_j=\mathbf{c}_i'\mathbf{A}\mathbf{c}_j-\mathbf{c}_j'\mathbf{A}'\mathbf{c}_i=\mathbf{c}_i'\mathbf{A}\mathbf{c}_j-\mathbf{c}_j'\mathbf{A}\mathbf{c}_i=\mathbf{c}_i'\mathbf{c}_j\lambda_j-\mathbf{c}_j\mathbf{c}_i=(\lambda_j-\lambda_i)\mathbf{c}_i'\mathbf{c}_j$. Thus $\lambda_j\neq\lambda_i$ implies $\mathbf{c}_i'\mathbf{c}_j=0$.

The problem associated with multiple roots occurs, for example, when $\mathbf{A}=\mathbf{I}$. *Any* direction $\mathbf{x}$ is a direction of tangency between the unit sphere $\mathbf{x}'\mathbf{x}=1$ and the family of concentric spheres $\mathbf{x}'\mathbf{x}=r^2$. Thus any vector $\mathbf{x}$ is a characteristic vector. [The characteristic polynomial in this case is $(\lambda-1)^k=0$, which has k roots all equal to one.] Since any vector is a characteristic vector, it is possible to select a set of k orthogonal characteristic vectors. In general, the existence of multiple roots implies some freedom in choosing characteristic vectors, but it is always possible to choose them to be orthogonal. If there are no multiple roots, then the set of characteristic vectors is unique.

THEOREM 31. *The characteristic values of a real symmetric matrix are real.*

Proof: Suppose that the characteristic polynomial has the complex root $\lambda=a+b(-1)^{1/2}$ with characteristic vector $\mathbf{c}=\mathbf{x}+\mathbf{y}(-1)^{1/2}$. Equating the real and complex parts of $\mathbf{A}(\mathbf{x}+\mathbf{y}(-1)^{1/2})=(\mathbf{x}+\mathbf{y}(-1)^{1/2})(a+b(-1)^{1/2})$ implies $\mathbf{A}\mathbf{x}=\mathbf{x}a-\mathbf{y}b$ and $\mathbf{A}\mathbf{y}=\mathbf{y}a+\mathbf{x}b$. Premultiplying these two expressions by $\mathbf{y}'$ and $\mathbf{x}'$ and then subtracting yields $0=\mathbf{y}'\mathbf{x}a-\mathbf{y}'\mathbf{y}b-\mathbf{x}'\mathbf{y}a-\mathbf{x}'\mathbf{x}b=-b(\mathbf{x}'\mathbf{x}+\mathbf{y}'\mathbf{y})$, which implies $b=0$; thus λ is real.

Making use of the equation $\mathbf{A}\mathbf{c}_i=\mathbf{c}_i\lambda_i$, we have $\lambda_i=\mathbf{c}_i'\mathbf{A}\mathbf{c}_i/\mathbf{c}_i'\mathbf{c}_i=c_i'\mathbf{A}\mathbf{c}_i$. Therefore:

THEOREM 32. *If* $\mathbf{A}$ *is positive definite, all of its characteristic roots are positive. If* $\mathbf{A}$ *is positive semi-definite, all its characteristic roots are nonnegative.*

Any vector $\mathbf{x}$ can be written as a linear combination of the k characteristic vectors $\mathbf{x}=\Sigma_i\mathbf{c}_i z_i=\mathbf{C}\mathbf{z}$, where $\mathbf{C}$ is a $(k\times k)$ matrix whose columns are the eigenvectors of $\mathbf{A}$. The quadratic form $\mathbf{x}'\mathbf{A}\mathbf{x}$ can then be written as

$$\mathbf{x}'\mathbf{A}\mathbf{x}=\mathbf{z}'\mathbf{C}'\mathbf{A}\mathbf{C}\mathbf{z}=\mathbf{z}'\boldsymbol{\Lambda}\mathbf{z},$$

where $\boldsymbol{\Lambda}$ is a diagonal matrix with the eigenvalues λ_i on the diagonal. The condition $\mathbf{C}'\mathbf{A}\mathbf{C}=\boldsymbol{\Lambda}$ follows from the fact that $\mathbf{c}_j'\mathbf{A}\mathbf{c}_i=\mathbf{c}_j'\mathbf{c}_i\lambda_i$. It is also easy to show that $\mathbf{C}'\mathbf{C}=\mathbf{I}_k$. Collecting these results together, we have the following theorem:

THEOREM 33. *For any* $k\times k$ *real symmetric matrix* $\mathbf{A}$ *there exists a* $k\times k$ *real matrix* $\mathbf{C}$ *such that* $\mathbf{C}'\mathbf{A}\mathbf{C}=\boldsymbol{\Lambda}$ *where* $\boldsymbol{\Lambda}$ *is a diagonal matrix and* $\mathbf{C}$ *is orthogonal,* $\mathbf{C}'\mathbf{C}=\mathbf{I}_k$.

If the columns of $\mathbf{C}$ are used as basis vectors then the quadratic surface

$\mathbf{x}'\mathbf{A}\mathbf{x}=r^2$ can be expressed in its *canonical form*

$$\mathbf{z}'\boldsymbol{\Lambda}\mathbf{z}=\sum_i \lambda_i z_i^2 = r^2$$

where $\mathbf{x}=\mathbf{C}\mathbf{z}$. If the eigenvalues λ_i are all positive, this is the equation of an ellipsoid centered at the origin with axes of length $r/\lambda_i^{1/2}$. An axis is taken by the transformation $\mathbf{x}=\mathbf{C}\mathbf{z}$ into a column of $\mathbf{C}$, that is, into an eigenvector.

The principal geometric result of this section can now be restated. If $\mathbf{A}$ is a square symmetric positive definite matrix, the quadratic surface $\mathbf{x}'\mathbf{A}\mathbf{x}=r^2$ is an ellipsoid located at the origin with axes equal to $\mathbf{c}_i r/(\lambda_i)^{1/2}$ where $\mathbf{c}_i$ is a characteristic vector of $\mathbf{A}$ and λ_i is the corresponding characteristic value.

The following result describes the projection of an ellipsoid onto an axis:

THEOREM 34 *The extreme values of the function* $\boldsymbol{\psi}'\mathbf{x}$ *evaluated over the surface* $\mathbf{x}'\mathbf{A}\mathbf{x}=r^2$ *(where* $\mathbf{A}$ *is invertible) occur at the points* $\mathbf{x}=\pm\mathbf{A}^{-1}\boldsymbol{\psi}\sqrt{r^2/\boldsymbol{\psi}'\mathbf{A}^{-1}\boldsymbol{\psi}}$ *. The function at these points takes on the values* $r\sqrt{\boldsymbol{\psi}'\mathbf{A}^{-1}\boldsymbol{\psi}}$ *. (A corollary is that the orthogonal projection of the surface* $\mathbf{x}'\mathbf{A}\mathbf{x}=r^2$ *onto the* ith *axis is the interval* $|x_i|\leqslant[\mathbf{A}^{-1}]_{ii}^{\frac{1}{2}}r$*.)*

Proof: Maximization of $\boldsymbol{\psi}'\mathbf{x}$ subject to the constraint $\mathbf{x}'\mathbf{A}\mathbf{x}=r^2$ leads to the Lagrangian derivatives $\mathbf{0}=\boldsymbol{\psi}+\lambda\mathbf{A}\mathbf{x}$ where λ is the Lagrange multiplier. Solving for $\mathbf{x}$ yields $\mathbf{x}=-\lambda^{-1}\mathbf{A}^{-1}\boldsymbol{\psi}$, which can be used to determine λ: $r^2=\mathbf{x}'\mathbf{A}\mathbf{x}=\boldsymbol{\psi}'\mathbf{A}^{-1}\boldsymbol{\psi}\lambda^{-2}$. Thus $\lambda^2=\boldsymbol{\psi}'\mathbf{A}^{-1}\boldsymbol{\psi}/r^2$, and the two $\mathbf{x}$ vectors are $\mathbf{x}=\pm\mathbf{A}^{-1}\boldsymbol{\psi}r\sqrt{(\boldsymbol{\psi}'\mathbf{A}^{-1}\boldsymbol{\psi})^{-1}}$.

Conjugate Axes

The coordinate system of the eigenvectors of a symmetric positive definite matrix $\mathbf{A}$ is not necessarily the only coordinate system in which $\mathbf{x}'\mathbf{A}\mathbf{x}=r^2$ assumes the canonical form of an ellipsoid. The transformation $\mathbf{z}=\mathbf{P}^{-1}\mathbf{x}$ takes $\mathbf{x}'\mathbf{A}\mathbf{x}$ into $\mathbf{z}'\mathbf{P}'\mathbf{A}\mathbf{P}\mathbf{z}$, which is the canonical form of an ellipsoid if $\mathbf{P}'\mathbf{A}\mathbf{P}$ is a positive-diagonal matrix. Indicating a column of $\mathbf{P}$ by $\mathbf{P}_i$, the matrix $\mathbf{P}'\mathbf{A}\mathbf{P}$ is a positive-diagonal matrix if $\mathbf{P}_i'\mathbf{A}\mathbf{P}_j=0$ for $i\neq j$. Such a $\mathbf{P}$ can be constructed in the following way. Choose any vector $\mathbf{P}_1$ to be the first column. For $\mathbf{P}_2$ choose any vector satisfying $\mathbf{P}_2'\mathbf{A}\mathbf{P}_1=0$, which from the discussion above requires that $\mathbf{P}_2$ be in the hyperplane tangent to the ellipsoid at $\mathbf{P}_1$. Next find a $\mathbf{P}_3$ in the intersection of the tangent hyperplanes at $\mathbf{P}_1$ and $\mathbf{P}_2$, etc. Such a sequence of $\mathbf{P}_i$ vectors is called a *set of conjugate axes* of the ellipsoid $\mathbf{x}'\mathbf{A}\mathbf{x}=r^2$, and is illustrated in Figure A2.

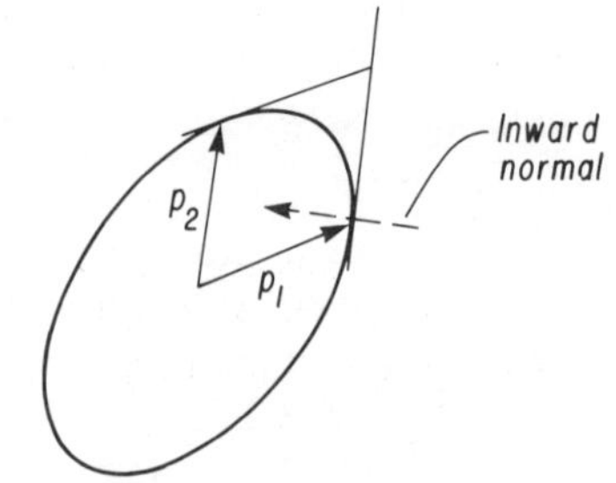

Fig. A.2 Conjugate axes.

Common Conjugate Axes

A pair of ellipsoids, $\mathbf{x}'\mathbf{A}\mathbf{x}=r_1^2$ and $\mathbf{x}'\mathbf{B}\mathbf{x}=r_2^2$ where **A** and **B** are positive definite, has a set of common conjugate axes, illustrated in Figure A3. This can be expressed algebraically in Theorem 35.

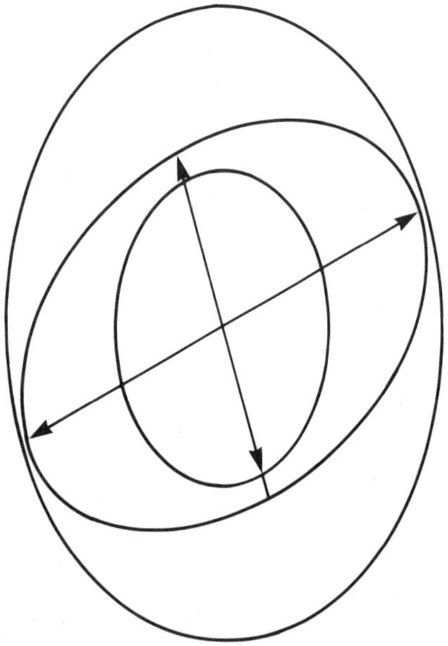

Fig. A.3 Common conjugate axes.

THEOREM 35. *Given a pair of real* $(m\times m)$ *symmetric matrices* **A** *and* **B**, *there exists an* $(m\times m)$ *nonsingular matrix* **P** *such that* $\mathbf{P}'\mathbf{B}\mathbf{P}$ *is a diagonal matrix, and* $\mathbf{P}'\mathbf{A}\mathbf{P}$ *is a diagonal matrix with ones and zeroes on the diagonal.*

Proof: Using Theorem 33, find a matrix $\mathbf{C}$ such that $\mathbf{C'AC}$ is a diagonal matrix $\boldsymbol{\Lambda}$ with any zero diagonal elements in the lower-right corner,

$$\boldsymbol{\Lambda}=\begin{bmatrix}\boldsymbol{\Lambda}_1 & \mathbf{0}\\ \mathbf{0} & \mathbf{0}\end{bmatrix}.$$

Let $\mathbf{E}$ be $\mathbf{C'BC}$, partition so that $\mathbf{E}_{22}$ corresponds to the full block of zeroes of $\mathbf{C'AC}$. Let $(\mathbf{E}_{22}^{-1}\mathbf{E}_{21})$be a matrix such that $\mathbf{0}=\mathbf{E}_{21}-\mathbf{E}_{22}(\mathbf{E}_{22}^{-1}\mathbf{E}_{21})$, and let

$$\mathbf{F}=\begin{bmatrix}\mathbf{I} & \mathbf{0}\\ -\mathbf{E}_{22}^{-1}\mathbf{E}_{21} & \mathbf{I}\end{bmatrix}$$

and note that $\mathbf{F}'\boldsymbol{\Lambda}\mathbf{F}=\boldsymbol{\Lambda}$ and

$$\mathbf{F'EF}=\begin{bmatrix}\mathbf{E}_{11}-\mathbf{E}_{12}\mathbf{E}_{22}^{-1}\mathbf{E}_{21} & \mathbf{0}\\ \mathbf{0} & \mathbf{E}_{22}\end{bmatrix}.$$

Then find a matrix $\mathbf{R}_1$ such that $\mathbf{R}_1'\boldsymbol{\Lambda}_1^{-1/2}(\mathbf{E}_{11}-\mathbf{E}_{12}\mathbf{E}_{22}^{-1}\mathbf{E}_{21})\boldsymbol{\Lambda}_1^{-1/2}\mathbf{R}_1=\mathbf{D}_1$, a diagonal matrix, and $\mathbf{R}_1'\mathbf{R}_1=\mathbf{I}$; also find a matrix $\mathbf{R}_2$ such that $\mathbf{R}_2'\mathbf{E}_{22}\mathbf{R}_2=\mathbf{D}_2$, a diagonal matrix, and $\mathbf{R}_2'\mathbf{R}_2=\mathbf{I}$. Then let

$$\mathbf{P}=\mathbf{CF}\begin{bmatrix}\boldsymbol{\Lambda}_1^{-1/2}\mathbf{R}_1 & \mathbf{0}\\ \mathbf{0} & \mathbf{R}_2\end{bmatrix}.$$

If $\mathbf{A}$ is invertible the diagonal elements of $\mathbf{D}$ can be shown to be the roots of the polynomial $|\mathbf{B}-\lambda\mathbf{A}|=0$, the columns of $\mathbf{P}$ satisfy the eigenvector equation $(\mathbf{B}-d_i\mathbf{A})\mathbf{P}_i=\mathbf{0}$, and the matrix $\mathbf{P}$ is unique if there are no multiple roots $d_i\neq d_j$, for $i\neq j$.

APPENDIX **2**

PROBABILITY DISTRIBUTIONS

The probability distributions used in this book are reported in this appendix. Relevant properties are listed. Johnson and Kotz (1972) and Raiffa and Schlaifer (1961) are useful sources of additional details and proofs.

Definitions

A multivariate random variable $\mathbf{x}$ is said to have the probability density function (p.d.f.) $f(\mathbf{x})$ if the probability that $\mathbf{x}$ is in any region R is

$$P(\mathbf{x} \in R) = \int_R f(\mathbf{t})\, d\mathbf{t}$$

where the symbols stand for the integral of f over the region R. If the symbol R is suppressed, the integral by convention extends over the whole domain of definition of f. Thus, for example $\int f(\mathbf{x})\, d\mathbf{x} = 1$ indicates that the integral of a p.d.f. over its domain is equal to one.

The *marginal* p.d.f. of the subvector $\mathbf{x}_I$, where $\mathbf{x}' = (\mathbf{x}_I' \mathbf{x}_J')$ is

$$g(\mathbf{x}_I) = \int_{\mathbf{x}_J} f(\mathbf{x})\, d\mathbf{x}_J$$

where the symbols stand for integration over the domain of $\mathbf{x}_J$. The *conditional* p.d.f. of the subvector $\mathbf{x}_I$ given $\mathbf{x}_J$ is

$$h(\mathbf{x}_I | \mathbf{x}_J) = \frac{f(\mathbf{x})}{g(\mathbf{x}_J)}.$$

Whereas here we have selected the symbols f, g, and h to distinguish three different densities, henceforth we, without confusion, write $f(\mathbf{x})$, $f(\mathbf{x}_I)$, and $f(\mathbf{x}_I | \mathbf{x}_J)$.

The mean or expectation of an element x_i is

$E(x_i)=\int x_i f(\mathbf{x})\,d\mathbf{x}$. The mean vector of a multivariate random variable is $E\mathbf{x}=\{E(x_i)\}=\int \mathbf{x}f(\mathbf{x})\,d\mathbf{x}$. The variance-covariance matrix of the random vector $\mathbf{x}$ is

$$V(\mathbf{x})=E(\mathbf{x}-E(\mathbf{x}))(\mathbf{x}-E(\mathbf{x}))'.$$

Beta Distribution

The beta p.d.f. is

$$f_\beta(p|r,n)=\left[B(r,n-r)\right]^{-1}p^{r-1}(1-p)^{n-r-1},\quad 0\leqslant p\leqslant 1$$

where $B(r,n-r)=(r-1)!(n-r-1)!/(n-1)!$, and $r,n>0$. The first two moments of a beta random variable are

$$E(p)=\frac{r}{n}$$

$$V(p)=\frac{r(n-r)}{n^2(n+1)}.$$

Multivariate Normal Distributions

The nondegenerate r-variate normal p.d.f. is

$$f_N^r(\mathbf{x}|\mathbf{u},\Sigma)=(2\pi)^{-r/2}|\boldsymbol{\Sigma}|^{-1/2}\exp\left[-\tfrac{1}{2}(\mathbf{x}-\boldsymbol{\mu})'\boldsymbol{\Sigma}^{-1}(\mathbf{x}-\boldsymbol{\mu})\right]$$

where $\mathbf{x}$ and $\boldsymbol{\mu}$ are $r\times 1$ vectors and $\boldsymbol{\Sigma}$ is an $r\times r$ symmetric positive definite matrix. The mean and variance are

$$E(\mathbf{x})=\boldsymbol{\mu}$$

$$V(\mathbf{x})=\boldsymbol{\Sigma}.$$

Partition $\mathbf{x}$, $\boldsymbol{\mu}$, and $\boldsymbol{\Sigma}$ conformably:

$$\mathbf{x}'=(\mathbf{x}_I'\mathbf{x}_J'),\boldsymbol{\mu}=(\boldsymbol{\mu}_I'\boldsymbol{\mu}_J)\qquad\text{and}$$

$$\boldsymbol{\Sigma}=\begin{bmatrix}\boldsymbol{\Sigma}_{II} & \boldsymbol{\Sigma}_{IJ}\\ \boldsymbol{\Sigma}_{JI} & \boldsymbol{\Sigma}_{JJ}\end{bmatrix}$$

with $\mathbf{x}_I$ having q elements and $\mathbf{x}_J$ having $r-q$ elements. The marginal and conditional distributions are

$$f(\mathbf{x}_I)=f_N^q(\mathbf{x}_I|\,\boldsymbol{\mu}u_I,\boldsymbol{\Sigma}_{II})$$

$$f(\mathbf{x}_I|\mathbf{x}_J)=f_N^q\left(\mathbf{x}_I|\,\boldsymbol{\mu}_I+\boldsymbol{\Sigma}_{IJ}\boldsymbol{\Sigma}_{JJ}^{-1}(\mathbf{x}_J-\boldsymbol{\mu}_J),\,\boldsymbol{\Sigma}_{II}-\boldsymbol{\Sigma}_{IJ}\boldsymbol{\Sigma}_{JJ}^{-1}\boldsymbol{\Sigma}_{JI}\right).$$

Notation. $\mathbf{x}\sim N(\boldsymbol{\mu},\boldsymbol{\Sigma})$ stands for "$\mathbf{x}$ is normally distributed with mean $\boldsymbol{\mu}$ and variance $\boldsymbol{\Sigma}$."

THEOREM. *If* $\mathbf{x} \sim N(\boldsymbol{\mu}, \boldsymbol{\Sigma})$ *with* $\boldsymbol{\Sigma}$ $r \times r$ *positive definite and if* $\mathbf{A}$ *is* $m \times r$ *with rank* $m \leqslant r$, *then*

$$\mathbf{z} = (\mathbf{a} + \mathbf{A}\mathbf{x}) \sim N(\mathbf{a} + \mathbf{A}\boldsymbol{\mu}, \mathbf{A}\boldsymbol{\Sigma}\mathbf{A}').$$

A *degenerate* normal distribution is an r-dimensional distribution that concentrates the probability on a subspace of dimension $q < r$. The covariance matrix $\boldsymbol{\Sigma}$ of a degenerate normal is singular. The density function is then defined only conditionally on the subspace in which the random variable certainly must lie.

An *improper* normal distribution has a density in the form of a normal distribution but does not integrate to one because $\boldsymbol{\Sigma}^{-1}$ is singular. After a transformation if necessary, an improper normal can be written as the product of a lower dimensional proper normal times an improper uniform distribution on the other variables:

$$f(\mathbf{x}_I, \mathbf{x}_J) \propto f_N(\mathbf{x}_I).$$

By convention, the marginal and conditional distributions of $\mathbf{x}_I$ are taken to be the proper distribution $f_N(\mathbf{x}_I)$.

Gamma Distribution

The gamma p.d.f. is

$$f_\gamma(h|s^2, \nu) = \left[\left(\frac{\nu}{2} - 1\right)!\right]^{-1} \left[\frac{\nu s^2}{2}\right]^{\nu/2} h^{\frac{\nu}{2} - 1} \exp\left[-\frac{1}{2}\nu s^2 h\right]$$

for $h > 0$, $\nu > 0$, $s^2 > 0$. The moments of the gamma distribution are

$$E(h) = s^{-2}$$

$$V(h) = \left(\frac{\nu s^4}{2}\right)^{-1}.$$

Notation. $h \sim \Gamma(s^2, \nu)$ stands for "h is gamma distributed with parameters s^2 and ν."

Multivariate Student Distribution

The nondengenerate r-variate Student p.d.f. is the marginal of a normal-gamma p.d.f.

$$f_S^r(\mathbf{t}|\boldsymbol{\mu}, \mathbf{H}^{-1}, \nu) = \int_0^\infty f_N^r(\mathbf{t}|\boldsymbol{\mu}, (h\mathbf{N})^{-1}) f_\gamma(h|s^2, \nu)\, dh$$

$$= \frac{\nu^{(\nu/2)} \left(\frac{1}{2}\nu + \frac{1}{2}r - 1\right)! |\mathbf{H}|^{\frac{1}{2}}}{\pi^{(r/2)} \left(\frac{1}{2}\nu - 1\right)!} \left[\nu + (\mathbf{t} - \boldsymbol{\mu})'\mathbf{H}(\mathbf{t} - \boldsymbol{\mu})\right]^{-(\nu + r)/2}$$

where $\mathbf{H}=\mathbf{N}s^{-2}$, $\nu>0$, and $\mathbf{H}$ is symmetric positive definite. The first two moments are

$$E(\mathbf{t})=\boldsymbol{\mu} \qquad \nu>1$$

$$V(\mathbf{t})=\mathbf{H}^{-1}\frac{\nu}{\nu-2} \qquad \nu>2$$

Note that the Student distribution and also the normal distribution are written most conveniently in terms of the precision matrices $\mathbf{H}=\mathbf{V}^{-1}$ instead of the variance matrices. More importantly, prior to posterior analysis of location parameters involves formulas that are additive in the precision matrices. Bayesians such as Raiffa and Schlaifer (1961) for these reasons often express densities in terms of the precision parameters. In consideration of students who are more comfortable working with variances, I have adopted the schizophrenic notation above.

Partitioning as in the case of the normal distribution and letting $\mathbf{H}^{-1}=\mathbf{V}$, the marginal Student distribution is

$$f(\mathbf{t}_I)=f_S^q\left(\mathbf{t}_I|\,\boldsymbol{\mu}_I,(\mathbf{H}^{-1})_{II},\nu\right)$$

where

$$(\mathbf{H}^{-1})_{II}=\left(\mathbf{H}_{II}-\mathbf{H}_{IJ}\mathbf{H}_{JJ}^{-1}\mathbf{H}_{JI}\right)^{-1}=\mathbf{V}_{II}$$

and the conditional p.d.f. is

$$f(\mathbf{t}_I|\mathbf{t}_J)=f_S^q\left(\mathbf{t}_I|\,\boldsymbol{\mu}_I^*,(\mathbf{H}_{II})^{-1}s^2,\nu+r-q\right)$$

where

$$\boldsymbol{\mu}_I^*=\boldsymbol{\mu}_I+\mathbf{V}_{IJ}\mathbf{V}_{JJ}^{-1}(\mathbf{t}_J-\boldsymbol{\mu}_J)=\boldsymbol{\mu}_I-\mathbf{H}_{II}^{-1}\mathbf{H}_{IJ}(\mathbf{t}_J-\boldsymbol{\mu}_J)$$

$$(\mathbf{H}_{II})^{-1}=\mathbf{V}_{II}-\mathbf{V}_{IJ}\mathbf{V}_{JJ}^{-1}\mathbf{V}_{JI}$$

$$s^2=\frac{\nu+(\mathbf{t}_J-\boldsymbol{\mu}_J)'\mathbf{V}_{JJ}^{-1}(\mathbf{t}_J-\boldsymbol{\mu}_J)}{\nu+r-q}$$

Notation. $\mathbf{t}\sim S(\boldsymbol{\mu},\mathbf{V},\nu)$ stands for "$\mathbf{t}$ has a Student distribution with parameters $\boldsymbol{\mu}$, $\mathbf{V}$, and ν."

Wishart Distribution

A $k\times k$ symmetric positive definite matrix $\boldsymbol{\Omega}$ is said to have a Wishart distribution if its density function is

$$f_W^k(\boldsymbol{\Omega}|\mathbf{S},\nu)=c|\boldsymbol{\Omega}|^{(\nu-k-1)/2}\exp\left[-\tfrac{1}{2}\operatorname{tr}\boldsymbol{\Omega}\mathbf{S}\right]$$

where

$$c^{-1}=|\mathbf{S}|^{\nu/2}2^{\nu k/2}\pi^{k(k-1)/4}\prod_{i=1}^{k}\Gamma\left[\frac{\nu+1-i}{2}\right]$$

for $k<\nu$ and $\mathbf{S}$ a $(k\times k)$ symmetric positive-definite matrix. Properties of a Wishart distribution are discussed in Zellner (1971, pp. 389–394). The one property used in this book is the following relationship between the Wishart and Student distributions.

THEOREM. *If conditional on* $\boldsymbol{\Omega}$, *the random vector* $\mathbf{t}$ *is normally distributed with mean* $\boldsymbol{\mu}$ *and covariance matrix* $(T\boldsymbol{\Omega})^{-1}$ *and if the matrix* $\boldsymbol{\Omega}$ *has a Wishart distribution with parameters* $\mathbf{S}$ *and* ν, *then marginally* $\mathbf{t}$ *has a multivariate Student distribution with parameters* $\boldsymbol{\mu}$, $\mathbf{S}/T$ *and* $\nu+1-k$.

Proof: The joint distribution of $\mathbf{t}$ and $\boldsymbol{\Omega}$ can be written as

$$f(\mathbf{t},\boldsymbol{\Omega})\propto|\boldsymbol{\Omega}|^{1/2}\exp\left[-\frac{T}{2}(\mathbf{t}-\boldsymbol{\mu})'\boldsymbol{\Omega}(\mathbf{t}-\boldsymbol{\mu})\right]$$

$$|\boldsymbol{\Omega}|^{(\nu-k-1)/2}\exp\left[-\frac{1}{2}\operatorname{tr}\boldsymbol{\Omega}\mathbf{S}\right]$$

$$=|\boldsymbol{\Omega}|^{(\nu-k+1-1)/2}\exp\left[-\frac{1}{2}\operatorname{tr}\boldsymbol{\Omega}\mathbf{W}\right]$$

where $\mathbf{W}=\mathbf{S}+T(\mathbf{t}-\boldsymbol{\mu})(\mathbf{t}-\boldsymbol{\mu})'$. The last line in this expression has the form of a Wishart distribution, and the matrix $\boldsymbol{\Omega}$ can be integrated from the expression merely by inserting the appropriate normalizing constant:

$$f(\mathbf{t})\propto|\mathbf{W}|^{-(\nu+1)/2}$$

$$=\left(|\mathbf{S}|\left(1+T(\mathbf{t}-\boldsymbol{\mu})'\mathbf{S}^{-1}(\mathbf{t}-\boldsymbol{\mu})\right)\right)^{-(\nu+1)/2}$$

$$\propto f_s^k\left(\mathbf{t}|\,\boldsymbol{\mu},\mathbf{S}/T,\nu+1-k\right).$$

APPENDIX 3

PROOF OF THEOREMS 5.5 AND 5.8[1]

Proof of Theorem 5.5

The variables are first scaled so that the prior precision is $\mathbf{D}^* = d\mathbf{I}$. We let C_n be the set of all combinations of the first k integers taken n at a time. Define $H(I,J)$ as the minor of $\mathbf{H}$ formed by deleting the rows $i \in I$ and columns $j \in J$ with $I, J \in C_n$ for some $n \leqslant k$. Furthermore, let $H(Im,Jn)$ be the minor formed by deleting the rows I and m, and the columns J and n.

Expanding the characteristic polynomial as in Gantmacher (1959, p. 70) yields

$$|\mathbf{H}+d\mathbf{I}| = \sum_{j=0}^{k} p_j d^j \quad \text{with} \quad p_j = \sum_{I \in C_j} H(I,I) = \sum_{I \in C_j} |\mathbf{H}_I|,$$

where $|\mathbf{H}_I| = H(I,I)$ and where we let $|\mathbf{H}_I| = 1$ for $I = C_k$.

We can derive a similar expansion for the adjoint, which takes the form

$$|\mathbf{H}+d\mathbf{I}|(\mathbf{H}+d\mathbf{I})^{-1} = \sum_{j=0}^{k-1} \mathbf{B}_j d^j.$$

We derive a formula for $\mathbf{B}_j$ by collecting all terms that involve d^j. In our notation, the cofactor of the (m,n)th element of $\mathbf{H}+d\mathbf{I}$ is

$$(-1)^{m+n}(H+dI)(m,n).$$

In the expansion of this determinant, d^j occurs in any term that contains exactly j diagonal elements:

$$\prod_{i \in I} (h_{ii}+d),$$

[1]This material is taken from Leamer and Chamberlain (1976).

where $I \in C_j$, $j \leqslant k-1$, and I contains neither m nor n. This product of j diagonal elements is multiplied by $(H+dI)(Im,In)$, which can be written as the sum of $H(Im,In)$ plus terms that contain d. The latter are thereby excluded from consideration, since they would create higher powers of d. Thus if we set $H(Im,In)=0$ for m or $n \in I$, the (m,n)th element of $\mathbf{B}_j$ is

$$(-1)^{m+n}\Sigma_{I \in C_j} H(Im,In).$$

Then defining the "deleted" inverse $\mathbf{H}_I^{-1}$ to be the $k \times k$ matrix whose (m,n)th element is

$$(-1)^{m+n}H(Im,In)/|\mathbf{H}_I|, \qquad \text{we have} \quad \mathbf{B}_j = \sum_{I \in C_j} |\mathbf{H}_I|\mathbf{H}_I^{-1}.$$

Now note that $\mathbf{b}_I = \mathbf{H}_I^{-1}\mathbf{H}\mathbf{b}$ is the restricted least-squares point with $\beta_i = 0$ for $i \in I$. Thus we can write the conditional posterior mean as

$$\mathbf{b}^{**} = (\mathbf{H}+d\mathbf{I})^{-1}\mathbf{H}\mathbf{b} = \Sigma_{j=0}^{k} d^j \sum_{I \in C_j} |\mathbf{H}_I|\mathbf{b}_I/|\mathbf{H}+d\mathbf{I}| \tag{1}$$

with

$$|\mathbf{H}+d\mathbf{I}| = \Sigma_{j=0}^{k} d^j \sum_{I \in C_j} |\mathbf{H}_I|. \tag{2}$$

The result (5.15) involves a straightforward transformation of $\mathbf{H}$ in these formulas to allow for the arbitrary diagonal matrix $\mathbf{D}$ in the place of $d\mathbf{I}$.

Proof of Theorem 5.8

The proof of theorem involves a slight rewriting of equations (1) and (2).

APPENDIX **4**

ASSORTED PROBLEMS

Problems. Chapter 2, Sections 2.1–2.2

1. Determine a probability for each of the following events, and explain what is meant by probability in each case.

(a) A one in the next roll of a die.
(b) R is the first letter of the first name of the 168th entry in the Los Angeles phone book.
(c) R is the first letter of the last name of the eighth president of the United States.
(d) R is the first letter of the last name of the president of the United States in the year 2000.

2. A coin is to be flipped. If it lands heads up, the following statement is made: "Nixon's weight is negative." Otherwise, the statement is made: "Nixon's weight is positive." Which of the following statements are true?

(a) The probability that a true statement will be made is one half.
(b) The probability that a true statement will be made is either one or zero, depending on which statement is made.
(c) If we get a head and say "Nixon's weight is positive," the probability that this is true is one-half.

3. What is the probability of an event A if the odds against A are (a) 2 to 1 (b) 3 to 1 (c) 3 to 2

4. In a three-horse race the odds against the favorite are even (one to one); against the second horse, two to one; and against the "sleeper," three to one. Find stakes that make you a sure winner.

5. It has been shown that coherence implies the conditional probability axiom $P(A|B)P(B) = P(A \cap B)$ when A is a proper subset of B. Prove this for any events A and B.

6. A coin is flipped. If it lands heads up a ball is drawn from an urn containing one red and three black balls. Otherwise, a ball is drawn from an urn containing three red and one black balls. Given that a red ball results from this experiment, what is the probability that it came from the first urn?

7. A bookie offers even odds that the Yankees will win the World Series. He also announces that the Yankees are two-to-one favorites to have more home runs. Given that they have more home runs, they are two-to-one favorites to win the series. Construct a set of bets that make the bookie a sure loser.

Problems. Chapter 2, Sections 2.3–2.5

1. Fill in the following table where r_1 and n_1 are beta prior parameters and the sample consists of one success in five trials.

r_1	n_1	95% interval	$E(\tilde{p})$	r_2	n_2	95% interval	$E(\tilde{p})$
0	0						
1	2						
2	4						
5	10						
10	20						

2. Formulate approximate prior distributions on the binomial parameter p if the random variable X takes on the value one for the event: a randomly selected United States citizen

   a) is male.
   b) has more than 48 chromosomes.
   c) has three teeth.
   d) has more than 82,496 hairs on his body.

   Suppose for these events we observe 1 success ($X = 1$) in a sample of size 5. Compare the probability of a one on the first trial with the probability of a one on the sixth trial. Give explanations for your inferences.

3. Using a beta prior on p, give an example when the posterior variance exceeds the prior variance.

4. (a) If your opinions about p are described by a beta distribution with parameters r and n, what is the probability of getting a success on the next trial?
   (b) Starting with a beta prior with parameters r_1 and n_1 and observing r successes in n trials, what is the probability of another success?
   (c) Suppose you have observed n successes in n trials, with the "noninformative" prior $n_1 = r_1 = 0$. What is the probability of another success, or is it defined?
   (d) Formulate a beta prior for p = proportion of coins that land heads up. What is the probability of another head if no heads are observed, $n = 1, 10, 100, 1000$?

5. (a) Suppose that X and Y are both Bernoulli random variables (r.v.) with the same parameter p. (i) Given that p is a random variable in the sense that it has a prior distribution, can X and Y be independent? (ii) What do we mean when we say that the number of successes in n *independent* Bernoulli trials has a binomial distribution? (Call this *conditional independence*.)
   (b) If X is a Bernoulli r.v. with parameter p_x and Y is a Bernoulli r.v. with parameter p_y, what restriction on the joint prior on (p_x, p_y) implies X and Y (marginally) independent?

6. A sample of size 10 with mean 50 was taken from a normal population with unknown mean $\tilde{\mu}$ and variance 100. A normal prior for $\tilde{\mu}$ was formed with mean m_1 and variance v_1. Fill in the following table.

		Prior				Posterior	
m_1	v_1	95% interval	$P(\tilde{\mu}>48)$	m_2	v_2	95% interval	$P(\tilde{\mu}>48)$
50	1000						
50	1						
40	1000						
40	1						

7. Consider a population of size N consisting of pN ones and $(1-p)N$ twos. How does one make inferences based on a sample of size n about the mean $\mu = p + 2(1-p) = 2 - p$ when the sample (with replacement) results in s two's and f one's?

8. Formulate approximate prior distributions for the following means:
   (a) Average IQ of Harvard students.
   (b) Average weight in pounds of Harvard students.
   (c) Average length in tenths of an inch of wooden foot-long rulers.
   (d) Average weight in marspounds of Martians.

   Suppose we sample these populations, and in a random sample of size 5 we calculate a mean of $m = 121$ and a standard deviation of $s = 5$. What are the posterior probabilities that you assign to the events
   (a) $120 \leqslant \tilde{\mu} \leqslant 122$.
   (b) $\tilde{\mu} \leqslant 121$.

9. Discuss whether it is more appropriate when reporting results of a scientific experiment to provide the data, the sufficient statistics, or the posterior p.d.f.

10. Suppose that we begin with a normal prior for μ with mean m_1 and variance v_1. Two samples from a normal distribution are taken with means m_a and m_b and with the sample sizes n_a and n_b. Given σ^2, compute:

    (a) The "posterior" distribution for μ given m_a alone.
    (b) The posterior distribution for μ given m_a and m_b by using the result of part (a) as a prior for sample b.
    (c) The posterior for μ given a sample mean $m = (m_a n_a + m_b n_b)/(n_a + n_b)$ with sample size $n_a + n_b$. Compare your answers to (b) and (c).

11. The following are given: a normal population with uncertain mean μ and known variance α^2, a random sample of size n, and a normal prior distribution for μ with mean m_1 and variance v_1. Let $m_2 = E(\mu|m)$ where m is the sample mean.

    (a) Write down the posterior mean m_2 given the sample mean, m.
    (b) Compute the expected value of m_2 given μ but not m; given m and μ.
    (c) Compute the expected value of m_2 given neither m nor μ.
    (d) Compute the posterior variance v_2 of μ given the sample mean m.
    (e) Compute the expected variance given μ but not m; given m but not μ.
    (f) Compute the distribution of m_2 given neither m nor μ.

12. (a) One thousand measurements with a ruler of the width of a finger yielded a mean of 1.9682 and a standard deviation $s=.06$ centimeters. What is a 95% interval for the width of the finger, assuming a diffuse prior where relevant?
(b) A single measurement with a micrometer yielded a value of 1.96105. Reconcile (a) and (b).

13. A random sample of size n is to be taken from a normal population with *known* mean and the *unknown* variance σ^2. Beginning with a prior for $h=\sigma^{-2}$ in the gamma family, compute the posterior distribution for h.

Problems. Chapter 3

1. Find the mean and variance of the constrained estimator (3.15).

2. Show that the F statistic for testing $\mathbf{R}\boldsymbol{\beta}=\mathbf{r}$ is

$$F=\frac{(\mathbf{Rb}-\mathbf{r})'\left(\mathbf{R}(\mathbf{X}'\mathbf{X})^{-1}\mathbf{R}'\right)^{-1}(\mathbf{Rb}-\mathbf{r})}{ps^2}$$

where p is the rank of $\mathbf{R}$.

3. Find the posterior distribution of the residual vector $\mathbf{u}$, $f(\mathbf{u}|\mathbf{Y},\mathbf{X})$, given a conjugate prior for $\boldsymbol{\beta},\sigma^2$.

4. Given the regression equation $\mathbf{Y}=\mathbf{X}\boldsymbol{\beta}+\mathbf{u}$, with $\mathbf{u}$ normally distributed with mean vector $\mathbf{0}$ and covariance matrix $\sigma^2\mathbf{V}$, with $\mathbf{V}\neq\mathbf{I}$, show that the generalized least-squares estimator

$$\mathbf{b}(\mathbf{V})=(\mathbf{X}'\mathbf{V}^{-1}\mathbf{X})^{-1}\mathbf{X}'\mathbf{V}^{-1}\mathbf{Y}$$

is the maximum likelihood estimator, given $\mathbf{V}$. Show also that it is the best linear unbiased estimator of $\boldsymbol{\beta}$.

5. Let a regression equation be $\mathbf{Y}=\mathbf{X}\boldsymbol{\beta}+\mathbf{Z}\boldsymbol{\gamma}+\mathbf{u}$, where $\mathbf{X}$ and $\mathbf{Z}$ are observable matrices, $\mathbf{u}$ is unobservable and distributed normally with mean $\mathbf{0}_T$, and variance is $\sigma^2\mathbf{I}_T$. Let $\mathbf{M}_z=\mathbf{I}_T-\mathbf{Z}(\mathbf{Z}'\mathbf{Z})^{-1}\mathbf{Z}'$.

(a) Show that $\mathbf{M}_z\mathbf{M}_z=\mathbf{M}_z$
(b) Show that the least-squares estimate of $\boldsymbol{\beta}$ is $(\mathbf{X}'\mathbf{M}_z\mathbf{X})^{-1}\mathbf{X}'\mathbf{M}_z\mathbf{Y}$
(c) If $\boldsymbol{\gamma}$ has a prior distribution with mean $\mathbf{0}$ and variance $v\mathbf{I}$, what is the variance of $\boldsymbol{\varepsilon}=\mathbf{Z}\boldsymbol{\gamma}+\mathbf{u}$? Find the generalized least-squares

estimator of $\boldsymbol{\beta}$ given the regression equation $\mathbf{Y}=\mathbf{X}\boldsymbol{\beta}+\boldsymbol{\varepsilon}$. Show that this estimator converges to the least-squares estimator of $\boldsymbol{\beta}$ reported in part (b) as v goes to infinity.

6. Assume that $\mathbf{Y}=\mathbf{X}\boldsymbol{\beta}+\mathbf{u}$, $\mathbf{u}\sim N(0,\sigma^2\mathbf{I}_T)$, and

$$\mathbf{X}'\mathbf{X}=\frac{1}{3}\begin{pmatrix} 1 & -1 \\ -1 & 4 \end{pmatrix} \qquad \mathbf{Y}'\mathbf{Y}=8$$

$$\mathbf{X}'\mathbf{Y}=\begin{pmatrix} 1 \\ 1 \end{pmatrix}$$

$$\sigma^2=1$$

(a) Compute the least-squares estimate of $\boldsymbol{\beta}$.
(b) Compute a 95% confidence interval for β_1.
(c) Test the hypothesis $\beta_1=0$ versus the alternative $\beta_1\neq 0$.
(d) Compute a 95% confidence interval for $\beta_1+\beta_2$.
(e) Test the hypothesis $\beta_1+\beta_2=0$ versus the alternative $\beta_1+\beta_2\neq 0$.
(f) Compute the least-squares estimate of β_1, and β_2 given $\beta_1+\beta_2=0$.

7. Repeat 6, but assume that σ^2 is unknown and that $\mathbf{Y}'\mathbf{Y}=8$ and $T=12$.

8. Using the data in problem 6 and a normal prior for $\boldsymbol{\beta}$ with mean $\mathbf{0}$ and variance $\mathbf{I}$, determine the posterior distribution of $\boldsymbol{\beta}$. What are 95% posterior intervals for β_1, β_2, and $\beta_1+\beta_2$?

9. Using the data in problem 6 and the prior in problem 8 with variance defined up to a scale factor λ, find the equation for the curve décolletage.

10. Assuming σ^2 is unknown as in problem 7, compute a posterior distribution using a normal-gamma prior for $(\boldsymbol{\beta},\sigma^2)$ with parameters

$$\mathbf{b}^*=(0,0)$$

$$\mathbf{N}^*=\begin{pmatrix} 1 & 0 \\ 0 & 1 \end{pmatrix}$$

$$\nu^*=10$$

$$s^{*2}=1.$$

What are 95% posterior intervals for β_1, β_2, and $\beta_1+\beta_2$?

Problems. Chapter 4

1. It is known that x is normally distributed with unknown mean μ and variance $\sigma^2 = 1$.

(a) Construct a .05 level test of $H_0: \mu = 0$ versus $H_1: \mu = 1$.
(b) Write down the function $P(H_0|x)/P(H_1|x)$.
(c) If $P(H_0)/P(H_1) = 1$, what is the significance level of the following test: accept H_0 if $P(H_0|x)/P(H_1|x) > 1$; otherwise reject H_0.

2. Suppose that $\bar{x} \sim N(\mu, \sigma^2/n)$ where $\sigma^2 = 1$ and n is the sample size.

(a) Construct a .05 level test of $H_0: \mu = 0$ versus $H_1: \mu \neq 0$, given $n = 1$, 10, and 100. For each value of n, draw approximately the error characteristic curve, $P(\text{error}|\,\mu)$.
(b) Construct a .05 level test of $H_0': |\mu| \leqslant .1$ versus $H_1': |\mu| > .1$, for $n = 1$, 10, 100. For each value of n, draw approximately the error characteristic curve, $P(\text{error}|\,\mu)$.
(c) Suppose that you meant to test H_0' against H_1' but instead constructed a .05 level test for H_0 versus H_1. What is the error characteristic curve for $n = 1, 10, 100$. What is the actual significance level of the test?

3. Make use of the data reported in problem 6, Chapter 3 to do the following problem.

(a) Let the prior allocate positive probability to the hypothesis $\beta_1 = 0$, and let it otherwise be (appropriately) diffuse. Compute for $T = 10$, $T = 100$, and $T = 1000$, $P(\beta_1 = 0|\mathbf{Y})/P(\beta_1 \neq 0|\mathbf{Y})$.
(b) Compute the classical t value for testing $\beta_1 = 0$ for $T = 10$, $T = 100$, and $T = 1000$. Is the hypothesis $\beta_1 = 0$ accepted or rejected?
(c) Compute $P(\beta_1 = 0|\mathbf{Y})/P(\beta_2 = 0|\mathbf{Y})$, using as a prior the uniform distribution over the lines $\beta_1 = 0$ and $\beta_2 = 0$.
(d) Suppose in testing $\beta_1 = 0$ versus $\beta_2 = 0$ the first hypothesis is accepted if it yields an equation with the higher R^2. Let $g(\beta_1, \beta_2)$ be the probability of making error of accepting the wrong model. Write down an expression for $g(\beta_1, \beta_2)$ and explain why it depends on β_1 and β_2. What is the maximum value of $g(\beta_1, \beta_2)$? What is the type I error probability if $\beta_1 = 0$ is the null hypothesis and $\beta_2 = 0$ is the alternative.

Problems. Chapter 5

1. Find the set of constrained least-squares points given constraints of the form $\mathbf{MR}\boldsymbol{\beta}=\mathbf{Mr}$, where $\mathbf{R}$ is a given $p\times k$ matrix with rank p, $p\leqslant k$, and $\mathbf{r}$ is a given $p\times 1$ vector, and where $\mathbf{M}$ is an arbitrary $p\times p$ matrix.

2. Using the notation of this chapter, let the least-squares estimator of $\psi'\boldsymbol{\beta}$ be $\psi'\mathbf{b}$, and let a constrained least-squares estimator be $\psi'\hat{\boldsymbol{\beta}}(\mathbf{R})$ where $\mathbf{R}$ is a row vector and where $\hat{\boldsymbol{\beta}}(\mathbf{R})=\mathbf{b}-(\mathbf{X}'\mathbf{X})^{-1}\times\mathbf{R}'(\mathbf{R}(\mathbf{X}'\mathbf{X})^{-1}\mathbf{R}')^{-1}\mathbf{R}')^{-1}\mathbf{Rb}$. Show that $\psi'\hat{\boldsymbol{\beta}}(\mathbf{R})$ has a smaller mean-squared error than $\psi'\mathbf{b}$ if and only if the "true" squared t is less than one, $\tau^2=(\mathbf{R}\boldsymbol{\beta})^2/\sigma^2\mathbf{R}(\mathbf{X}'\mathbf{X})^{-1}\mathbf{R}'$.

3. Given the normal linear regression model with known variance and a multi-variate normal prior distribution for the coefficient vector, show that tangencies between a posterior ellipsoid and either a prior or a likelihood ellipsoid lie on the information contract curve.

4. Comment wisely on the following statements that have been made about the linear regression model $\mathbf{Y}=\mathbf{x}\beta+\mathbf{z}\gamma+\mathbf{u}$.

   (a) In the nonexperimental sciences the explanatory variables $\mathbf{x}$ and $\mathbf{z}$ are not subject to control, and they may be highly collinear. Sometimes the variables are collinear because they are drawn from a "collinear" population; alternatively, $\mathbf{x}$ and $\mathbf{z}$ may be independently distributed, but by chance are correlated in this particular sample. It is important to distinguish these two possibilities. A useful test of the "collinearity problem" is a test of the hypothesis that $\mathbf{x}$ and $\mathbf{z}$ are distributed independently.

   (b) Collinearity between the explanatory variables is less of a problem if the variables are negatively correlated than if they are positively correlated.

5. Given the following data, find the (four) principal component regression estimates and comment on the desirability of estimation subject to principal component restrictions.

$$\mathbf{X}'\mathbf{X}=\begin{bmatrix}3 & -1 & 0\\ -1 & 3 & 0\\ 0 & 0 & 1\end{bmatrix}\qquad \mathbf{X}'\mathbf{Y}=\begin{bmatrix}8\\ 8\\ 8\end{bmatrix}$$

6. Making use of the data given in problem 6, Chapter 3,

(a) Draw the ellipsoid of constrained estimates.
(b) Assuming the number of observations T is equal to 10, compute the pretest estimator (5.10) of β_1, with a test level of .05 and the hypothesis $\beta_2=0$.
(c) Can a Stein–James estimate of $\boldsymbol{\beta}$ be computed? Is the least-squares estimate admissible?
(d) Making use of a spherical prior, graph the information contract curve, and find the rotation invariant average regressions.
(e) Find the principal component regression estimates of $\boldsymbol{\beta}$.
(f) Compute the conditional 95% confidence interval for β_1 given β_2.
(g) If the prior distribution is spherical, is there a collinearity problem?
(h) For each coefficient, compute the measures of collinearity c_1, and c_2.

7. T observations of a two-variable regression model yielded the following moments

$$\mathbf{Y'Y}=2 \qquad T=12$$

$$\mathbf{Y'X}=\begin{pmatrix}1\\1\end{pmatrix} \qquad \mathbf{X'X}=\begin{pmatrix}1 & -1\\ -1 & 1\end{pmatrix}$$

Assume that the matrix $\mathbf{X}$ is fixed.

(a) Find a 2×1 vector $\boldsymbol{\lambda}$ such that $\mathbf{X}\boldsymbol{\lambda}=0$.
(b) Find a vector $\boldsymbol{\beta}$ that is observationally equivalent to the vector (3,2).
(c) Is the function $\beta_1+\beta_2$ identified?
(d) Draw a graph indicating the likelihood contours.
(e) Find an identifying restriction.
(f) If the prior is spherical, find a linear combination of β_1 and β_2 about which the experiment is personally uninformative.
(g) Find a linear combination about which the experiment is publicly informative.

BIBLIOGRAPHY

Alam, K. and J. R. Thompson (1964), "Estimation of the mean of a multivariate normal distribution," Indiana Technical Report.

Allen, David M. (1971), "Mean square error of prediction as a criterion for selecting variables," *Technometrics* **13**, 469–481.

Almon, S. (1965), "The distributed lag between capital appropriations and expenditures," *Econometrica* **33**, 178–196.

Alpert, M. and H. Raiffa (1969), "A progress report on the training of probability assessors," mimeo.

Ames, Edward and Stanley Reiter (1961), "Distributions of correlation coefficients in economic time series," *J. Amer. Stat. Ass.* **56**, 637–656.

Anderson, T. W. (1951), "Estimating linear restrictions on regression coefficients for multivariate normal distributions," *Ann. Math. Stat.* **22**, 327–351.

Anderson, T. W. (1958), *An Introduction to Multivariate Statistical Analysis*, Wiley, New York.

Anderson, T. W. (1962), "The choice of the degree of a polynomial regression as a multiple decision problem," *Ann. Math. Statist.* **33**, 255–272.

Ando, A. and G. M. Kaufman (1964), "Bayesian analysis of reduced form systems," *Alfred P. Sloan School of Management*, MIT, Working Paper No. 98-64.

Ando, A. and G. M. Kaufman (1965), "Bayesian analysis of the independent multi-normal process—neither mean nor precision known," *J. Amer. Statist. Ass.* **60**, 347–358.

Ando, A. and G. M. Kaufman (1966), "Evaluation of an ad hoc procedure for estimating parameters of some linear models," *Rev. Econ. Statist.* **XLVIII**, 334–340.

Anscombe, F. J. and J. W. Tukey (1963), "The examination and analysis of residuals," *Technometrics* **5**, 141.

Aoki, M. (1967), *Optimization of Stochastic Systems*, Academic, New York.

Asch, S. E. (1952), *Social Psychology*, Prentice-Hall, Englewood Cliffs, N.J.

Atkinson, R. C., G. H. Bower, and E. J. Crothers (1965), *An Introduction to Mathematical Learning Theory*, Wiley, New York.

Atkinson, A. C. and D. R. Cox (1974), "Planning experiments for discriminating between models," *J. Roy. Statis. Soc. B* **36**, 321–335.

Bancroft, T. A. (1944), "On biases in estimation due to the use of preliminary tests of significance," *Ann. Math. Statist.* **15**, 190–204.

Bancroft, T. A. (1964), "Analysis and inference for incompletely specified models involving the use of preliminary tests of significance," *Biometrics* **20**, 427–442.

Baranchik, A. (1964), "Multiple regression and estimation of the mean of the multivariate normal distribution," *Technical Report #51*, Department of Statistics, Stanford University.

Baranchik, A. J. (1970), "A family of minimax estimators of the mean of a multivariate normal distribution," *Ann. Math. Statist.* **41**, 642–645.

Barnard, G. A. (1958), "Thomas Bayes—a biographical note (with a reproduction on 'An essay towards solving a problem in the doctrine of chances,' by Thomas Bayes)," *Biometrika* **45**, 293–315.

Barnard, G. A., G. M. Jenkins and C. B. Winsten (1962), "Likelihood inference and times series," *J. Roy. Statist. Soc. A* **125**, 321–352.

Barnard, G. M. (1969), Summary remarks In N. L. Johnson and H. Smith Jr., Eds., *New Developments in Survey Sampling*, Wiley-Interscience, New York.

Barnett, V. (1973), *Comparative Statistical Inference*, Wiley, New York.

Balestra, P. and M. Nerlove (1966), "Pooling cross section and time series data in the estimation of a dynamic model: the demand for natural gas," *Econometrica* **34**, 585–612.

Bayes, Rev. T. (1763), "An essay toward solving a problem in the doctrine of chances," *Phil. Trans. Roy. Soc.* (London) **53**, 370-418.

Berkson, J. (1938), "Some difficulties of interpretation encountered in the application of the chi-square test," *J. Amer. Statis. Ass.* **33**, 526–536.

Berkson, J. (1958), "Smoking and lung cancer: some observations of two recent reports," *J. Amer. Statist. Assoc.* **53**, 28–38.

Berkson, J. (1963), "Smoking and Lung Cancer," *Amer. Statistician* **17**, 15–22.

Bernard, C. (1927), *An Introduction to the Study of Experimental Method*, MacMillan, New York.

Birnbaum, A. (1962), "On the foundation of statistical inference," *J. Amer. Statist. Ass.* **57**, 269–306.

Blattberg, R. C. and N. J. Gonedes (1975), "A comparison of the stable and Student distributions as statistical models for stock prices," *J. Business* **47**, 244–280.

Bock, M. E., T. A. Yancey and G. G. Judge (1973), "The statistical consequences of preliminary test estimators in regression," *J. Amer. Statist. Ass.* **68**, 109–116.

Bock, M. E. (1975), "Minimax estimators of the mean of a multivariate distribution," *Ann. Stat.* **3**, 209–218.

Bose, R. C. (1944), "The fundamental theorem of linear estimation," *Proc. 31st Indian Sci. Congress*, 2–3.

Box, G. E. P. (1966), "Use and abuse of regression," *Technometrics* **8**, 625–629.

Box, G. E. P. and D. R. Cox (1964), "An analysis of transformations," *J. Roy. Stat. Soc. B* **26**, 211–243.

Box, G. E. P. and N. R. Draper (1965), "The Bayesian estimation of common parameters from several responses," *Biometrika* **52**, 355–365.

Box, G. E. P. and W. J. Hill (1967), "Discrimination among mechanistic models," *Technometrics* **9**, 57–71.

Box, G. E. P. and G. Jenkins (1970), *Time Series Analysis*, Holden Day, San Francisco.

Box, G. E. P. and G. C. Tiao (1962), "A further look at robustness via Bayes' theorem," *Biometrika* **49**, 419–432.

Box, G. E. P. and G. C. Tiao (1964), "A Bayesian approach to the importance of assumptions applied to the comparison of variances," *Biometrika* **51**, 153–167.

Box, G. E. P. and G. C. Tiao (1968), "A Bayesian approach to some outlier problems," *Biometrika* **55**, 119–129.

Box, G. E. P. and G. C. Tiao (1973), *Bayesian Inference in Statistical Analysis*, Addison Wesley, Reading, Massachusetts.

Brodbeck, M. (ed.) (1968) *Readings in the Philosophy of the Social Sciences*, MacMillan, New York.

Brown, R. L., J. Durbin and J. M. Evans (1975), "Techniques for testing the constancy of regression relationships over time," *J. Roy. Stat. Soc. B*, 149–192.

Carnap, R. (1962) *Logical Foundations of Probability*, The University of Chicago Press, Chicago.

Casson, M. C. (1974), "Generalised Errors in Variables Regression," *Rev. Econ. Stud.*, **41**, 347–52.

Chamberlain, G. (1975), "Unobservables in econometric models," unpublished Ph.D. dissertation, Harvard University.

Chamberlain, G. and E. Leamer (1976), "Matrix weighted averages and posterior bounds," *J. Roy. Statist. Soc. B* **38**, 73–84.

Chapman, L. J. and J. P. Chapman (1967), "Genesis of popular but erroneous psychodiagnostic observations," *J. Abnor. Psy.* **72**, 193–204.

Chetty, V. K. (1968a), "Bayesian analysis of Haavelmo's models," *Econometrica* **36**, 582–602.

Chetty, V. K. (1968b), "On pooling time series and cross-section data," *Econometrica* **36**, 279–290.

Chipman, J. S. (1964), "On least squares with insufficient observations," *J. Amer. Statist. Assoc.* **59**, 1078–1111.

Chow, G. C. (1960), "Tests of equality between sets of coefficients in two linear regressions," *Econometrica* **28**, 591–605.

Christ, C. (1966), *Econometric Models and Methods*, Wiley, New York.

Cohen, A. (1965), "Estimates of linear combinations of the parameters in the mean vector of a multivariate distribution," *Ann. Math. Stat.* **36**, 78–87.

Cooley, T. F. and E. C. Prescott (1973a), "Tests of an adaptive regression model," *Rev. Econ. Stat.* **55**, 248–256.

Cooley, T. F. and E. C. Prescott (1973b), "The adaptive regression model," *Intern. Econ. Rev.* **14**, 364–371.

Copas, J. B. (1972), "The likelihood surface in the linear functional relationship problem," *J. Roy. Stat. Soc. B* **34**, 274–278.

Cornfield, J. E. (1969), "The Bayesian outlook and its applications," *Biometrics* **25**, 617–657.

Cornfield, J. (1975), "A statistician's apology (presidential address)," *J. Amer. Statis. Ass.* **70**, 7–14.

Cover, T. M. (1969), "Hypothesis testing with finite statistics," *Ann. Math. Stat.* **40**, 828–835.

Cox, D. R. (1958), "Some problems connected with statistical inference," *Ann. Math. Stat.* **29**, 352–372.

Cragg, J. G. and B. G. Malkiel (1968), "The consensus and accuracy of some predictions of the growth of corporate earnings," *J. Finance* **23**, 67–84.

Cyert, R. M. and M. H. DeGroot (1970), "Bayesian analysis and duopoly theory," *J. Polit. Econ.* **5**, 1168–1184.

Cyert, R. M. and M. H. DeGroot (1971), "Interfirm learning and the kinked demand curve," *J. Econ. Theory* **3**, 272–287.

Dagenais, M. G. (1972), "Choosing among alternative linear regression models: a general Bayesian solution with non-informative priors," Institute Deconomic Appliquee, Montreal.

Dawid, A. P., M. Stone and J. V. Zidek (1973), "Marginalization paradoxes in Bayesian and structural inference," *J. Roy. Stat. Soc. B* **34**, 189–233.

Day, N. E. (1969), "Estimating the components of a mixture of normal distributions," *Biometrika* **56**, 463–474.

deFinetti, B. (1937), "Foresight: its logical laws, its subjective sources," reprinted in H. E. Kyburg, Jr. and H. G. Smokler (Eds.) (1964) *Studies in Subjective Probability*, Wiley, New York.

deFinetti, B. (1974) *Theory of Probability, Vol. 1*, Wiley, New York.

DeGroot, M. H. (1970) *Optimal Statistical Decisions*, McGraw-Hill, New York.

DeGroot, M. H. (1974), "Reaching a consensus," *J. Amer. Stat. Assoc.* **69**, 118–121.

Dempster, A. (1973), "Alternatives to least squares in multiple regression," In Kabe, D. G. and R. P. Gupta, Eds., *Multivariate Statistical Inference*, North-Holland, Amsterdam, pp. 25–40.

Dhrymes, P. et. al. (1972), "Criteria for evaluation of econometric models," *Ann. Econ. Soc. Meas.* **1**, 259–290.

Dickey, J. M. (1967a), "Expansion of t-densities and related complete integrals," *Ann. Math. Statist.* **38**, 503–510.

Dickey, J. M. (1967b), "Matric-variate generalizations of the multivariate t-distribution and the inverted multivariate t-distribution," *Ann. Math. Stat.* **38**, 511–518.

Dickey, J. M. (1971), "The weighted likelihood ratio, linear hypotheses on normal location parameters," *Ann. Math Stat.* **42**, 204–223.

Dickey, J. M. (1973), "Scientific reporting and personal problems: Student's hypothesis," *J. Roy. Stat. Soc.* B **35**, 285–305.

Dickey, J. M. (1975), "Bayesian alternatives to the F-test and least-squares estimates in the normal linear model," In S. E. Fienberg and A. Zellner (Eds.)

Bayesian Studies in Econometrics and Statistics, North-Holland, Amsterdam.

Dickey, J. M. and Fischer (1975), "Coherent consensus precedes certainty," *Technical Report No. 28*, Stat. Sci. Division, State University of New York at Buffalo.

Dickey, J. M. and P. Freeman (1975), "Population-distributed personal probabilities," *J. Amer. Statist. Ass.* **70**, 362–364.

Dickey, J. M. and B. P. Leintz (1970), "The weighted likelihood ratio, sharp hypotheses about chances, the order of a Markov chain," *Ann. Math. Stat.* **41**, 214–226.

Doyle, A. C. (1888) *A Study in Scarlet* reprinted in Baring-Gould, W. S. (1967) *The Annotated Sherlock Holmes*, Clarkson N. Potter, Inc., New York.

Drèze, J. (1962), "The Bayesian approach to simultaneous equations estimation," Research Memorandum No. 67, Technological Institute, Northwestern University.

Drèze, J. (1970), "Econometrics and decision theory," *Econometrica* **40**, 1–17.

Drèze, J. (1975), "Bayesian theory of identification in simultaneous equations models," In S. E. Fienberg and A. Zellner, Eds., *Studies in Bayesian Econometrics and Statistics*, North-Holland, Amsterdam.

Drèze, J. (1976), "Limited information estimation from a Bayesian viewpoint," *Econometrica* **44**, 1045–1073.

Dunn, O. J. (Sept. 1959), "Confidence intervals for the means of dependent normally distributed variables," *J. Amer. Stat. Ass.* **54**, 613–621.

Durbin, J. (1953), "A note on regression when there is extraneous information about one of the coefficients," *J. Amer. Stat. Ass.* **48**, 799–808.

Dwyer, P. S. (1967), "Some applications of matrix derivatives in multivariate analysis," *J. Amer. Stat. Ass.* **62**, 607–625.

Edwards, J. B. (1969), "The relationship between the *F*-test and $\bar{R}^2$," *Amer. Stat.*, **23**, 28.

Edwards, W. (1968), "Conservatism in human information processing," In B. Kleinmutz, Ed., *Formal Representation of Human Judgement*, Wiley, New York.

Edwards, W., H. Lindman and L. J. Savage (1963), "Bayesian statistical inference for psychological research," *Psychol. Rev.* **70**, p. 193–242.

Efron, B. (1971), "Does an observed sequence of numbers follow a simple rule? (Another look at Bode's law)" with comments by I. J. Good, I. D. J. Bross, A. Stuart, J. M. A. Danby, C. Blyth, J. Pratt, (1971) *J. Amer. Stat. Ass.* **66**, 552–568.

Efron, B. and C. Morris (1973), "Stein's estimation rule and its competitors—an empirical Bayes approach," *J. Amer. Stat. Ass.* **68**, 117–130.

Efron, B. and C. Morris (1975), "Data analysis using Stein's estimator and its generalization," *J. Amer. Stat. Ass.* **70**, 311–319.

Eisenberg, E. and D. Gale (1975), "Consensus of subjective probabilities: the pari-mutual method," *Ann. of Math. Stat.* **30**, 165–168.

Ellsberg, D. (1961), "Risk, ambiguity and the Savage axioms," *Quar. J. Econ.* **75**, 643–669.

Ericson, W. A. (1969a), "A note on the posterior mean of a population mean," *J. Roy. Stat. Soc. B* **31**, 332–334.

Ericson, W. A. (1969b), "Subjective Bayesian models in sampling finite populations (with discussion)," *J. Roy. Stat. Soc. B* **31**, 195–233.

Estes, W. K. (1972), "Research and theory on the learning of probabilities," *J. Amer. Statist. Assoc.* **67**, 81–102.

Farrar, D. E. and R. R. Glauber (1967), "Multicollinearity in regression analysis: the problem revisited," *Rev. Econ. Stat.*, **XLIX**, 92–107.

Feldstein, M. (1973), "Multicollinearity and the mean square error criterion," *Econometrica* **41**, 337–346.

Feldstein, M. (1974), "Errors in variables: a consistent estimator with smaller MSE in finite samples," *J. Amer. Stat. Ass.* **69**, 990–996.

Fellner, W. (1965) *Probability and Profit*, Richard D. Irwin, Inc., Homewood, Illinois.

Ferguson, T. S. (1967) *Mathematical Statistics: A Decision Theoretic Approach* Academic, New York.

Festinger, L. (1950), "Informal social communication," *Psychol. Rev.* **57**, 271–282.

Fienberg, S. E. and A. Zellner (Eds.) (1975) *Studies in Bayesian Econometrics and Statistics*, North-Holland, Amsterdam.

Fisher, L. (1972), "The behavior of Bayesians in bunches," *Amer. Stat.* **26**, 19–20.

Fisher, R. A. (1960) *The Design of Experiments*, Hafner, New York.

Fisher, W. D. (1962), "Estimation in the linear decision model," *Intern. Econ. Rev.* **3**, 1–29.

Fischoff, B. (1975), "The Silly Certainty of Hindsight," *Psychol. Today*, **32**, 72–77.

Florens, J. P., M. Mouchart, and J. F. Richard (1974), "Bayesian inference in error in variables models," *J. Multivariate Anal.* **4**, 419–452.

Foster, M. H. and M. L. Martin (Eds.) (1966) *Probability, Confirmation and Simplicity*, The Odyssey Press, New York.

Fraser, D. A. S. (1968) *The Structure of Inference*, Wiley, New York.

Frisch, R. (1934) *Statistical Confluence Analysis by Means of Complete Regression Systems*, University Institute of Economics, Oslo, Norway.

Furnival, George M. (1971), "All possible regressions with less computation," *Technometrics* **13**, 403–412.

Gantmacher, F. R. (1960), *The Theory of Matrices*, Chelsea, New York.

Garside, M. J. (1965), "The best subset in multiple regression analysis," *Appl. Statist.* **14**, 196–201.

Gaver, K. M. and M. S. Geisel (1974), "Discriminating among alternative models: Bayesian and non-Bayesian methods," In P. Zarembka (Ed.), *Frontiers of Econometrics*, Academic, New York.

Geary, R. C. (1949), "Determination of linear relations between systematic parts of variables with errors of observation the variances of which are unknown," *Econometrica* **17**.

Geisel, M. S. (1970), "Comparing and choosing among parametric statistical models: a Bayesian analysis with microeconomic applications," unpublished Ph.D. dissertation, University of Chicago.

Geisser, S. (1966), "Predictive discrimination," In P. Krishnaiah (Ed.) *Multivariate Analysis*, Academic, New York, 149–163.

Geisser, S. (1975), "The predictive sample reuse method with applications," *J. Amer. Stat. Ass.* **70**, 320–327.

Geisser, S. and J. Cornfield (1963), "Posterior distributions for multivariate normal parameters," *J. Roy. Stat. Soc. B* **25**, 368–376.

Gillies, D. A. (1973) *An Objective Theory of Probability*, Methune and Co., London.

Godambe, V. P. (1974) Review of deFinetti, "Probability, induction and statistics," *J. Amer. Stat. Ass.* **69**, 578–580.

Godambe, V. P. and D. A. Sprott (1971) *Foundations of Statistical Inference*, Holt Rinehart and Winston of Canada, Toronto.

Godambe, V. P. and M. E. Thompson (1971), "Bayes, fiducial and frequency aspects of statistical inference in regression analysis in survey sampling," *J. Roy. Stat. Soc. B* **31**, 361–376.

Goldberger, A. S. (1964) *Econometric Theory*, Wiley, New York.

Goldberger, A. S. (1972), "Maximum likelihood estimation of regression models containing unobservable variables," *Int. Econ. Rev.* **13**, 1–15.

Goldberger, A. S. (1974), "Unobservable variables in econometrics," In P. Zarembka, (Ed.), *Frontiers of Econometrics*, Academic, New York.

Goldstein, M. (1974), "Approximate Bayesian inference with incompletely specified prior distributions," *Biometrika* **61**, 629–631.

Goldstein, M. and A. F. M. Smith (1974), "Ridge type estimation for regression analysis," *J. Roy. Stat. Soc. B* **36**, 284–291.

Good, I. J. (1950) *Probability and the Weighing of Evidence*, Hafner, New York.

Good, I. J. (1965) *The Estimation of Probabilities*, MIT Press, Cambridge, Mass.

Good, I. J. (1966), "A derivation of the probabilistic explication of information," *J. Roy. Stat. Soc. B* **28** 578–581.

Good, I. J. (1968), "Corroboration, explanation, evolving probability, simplicity, and a sharpened razor," *Brit. J. Phil. Sci.* **19**, 123–143.

Good, I. J. (1969), "A subjective evaluation of Bode's law and an 'objective' test for approximate numerical rationality," with discussion by H. D. Hartley, I. J. Bross, H. A. David, M. Zelen, R. E. Bargmann, F. J. Anscombe, M. Davis and R. L. Anderson, *J. Amer. Stat. Ass.* **64**, 23–66.

Good, I. J. (1971), "46656 varieties of Bayesians," letters to the editor, *Amer. Stat.* **25**, 62–63.

Good, I. J. (1972), "Christian Wolff, Titius, Bode, and Fibonacci," letters to the editor, *Amer. Stat.* **26**, 48–49.

Good, I. J. (1974), "A correction concerning complexity," *Brit. J. Phil. Sci.* **25**, 289.

Good, I. J. (1975), "Explicativity, corroboration, and the relative odds of hypotheses," *Synthese* **30**, 39–73.

Good, I. J. (1976), "Dynamic probability, computer chess, and the measurement of knowledge," In E. W. Elcock and D. Michie (Eds.), *Machine Representations of Knowledge*, Reidel, Hingham, Mass.

Gossett, W. S. (1908), "The probable error of a mean," *Biometrika* **6**, p. 1.

Granger, C. W. J. and P. Newbold (1974), "Spurious regressions in econometrics," *J. Econometrics* **2**, 111–120.

Granger, C. W. J. and P. Newbold (1976) *Forecasting Economic Times Series*, Academic Press, New York.

Graybill, F. (1961) *An Introduction to Linear Statistical Models*, McGraw-Hill, New York.

Graybill, F. A. (1969) *Introduction to Matrices with Applications in Statistics*, Wadsworth, Belmont, Calif.

Griliches, Z. (1961), "A note on serial correlation bias in estimates of distributed lags," *Econometrica* **29**, 65–73.

Griliches, Z. (1974), "Errors in variables and other unobservables," *Econometrica* **42**, 979–998.

Haitovsky, Y. (1969), "A note on the maximization of $\bar{R}^2$," *Amer. Stat.* **23**, 20–21.

Hampton, J. M., P. G. Moore and H. Thomas (1973), "Subjective probability and its measurement," *J. Roy. Stat. Soc. A* **136**, 43–63.

Harmon, H. H. (1960) *Modern Factor Analysis*, University of Chicago Press, Chicago.

Hartigan, J. A. (1964), "Invariant prior distributions," *Ann. Math. Stat.* **35**, 836–845.

Hartigan, J. A. (1969), "Linear Bayesian methods," *J. Roy. Stat. Soc. B* **31**, 446-454.

Hartigan, J. A. (1970), "Exact confidence intervals in regression problems with independent symmetric errors," *Ann. Math. Stat.* **41**, 1992–1998.

Hauser, R. M. and A. S. Goldberger (1971), "Treatment of unobservable variables in path analysis," In H. L. Costner, (Ed.), *Sociological Methodology*, Jossey-Bass, San Francisco.

Heath, D. and W. Sudderth (1976), "DeFinetti's theorem on exchangeable variables," *Amer. Stat.*, **30**, 188–89.

Hellman, M. E. and T. M. Cover (1970), "Learning with finite memory," *Ann. Math. Statist.* **41**, 765–782.

Hildreth, C. and J. P. Houck (1968), "Some estimators for a linear model with random coefficients," *J. Amer. Stat. Ass.* **63**, 584–595.

Hill, B. M. (1965), "Inference about variance components in the one-way model," *J. Amer. Stat. Ass.* **60**, 806–825.

Hill, B. M. (1969), "Foundations for the theory of least squares," *J. Roy. Stat. Soc. B* **31**, 89–97.

Hocking, Ian (1976) *The Emergence of Probability: A Philosophical Study of Early Ideas About Probability, Induction and Statistical Inference*, Cambridge University Press, Cambridge, England.

Hocking, R. R. (1976), "The analysis and selection of variables in linear regression," *Biometrics* **32**, 1–49.

Hoerl, A. E. and R. W. Kennard (1970a), "Ridge regression: biased estimation for nonorthogonal problems," *Technometrics* **12**, 55–67.

Hoerl, A. E. and R. W. Kennard (1970b), "Ridge regression: applications to nonorthogonal problems," *Technometrics* **12**, 69–82.

Hogarth, R. M. (1975), "Cognitive processes and the assessment of subjective probability distributions," *J. Amer. Stat. Assoc.* **70**, 271–294.

Hogg, R. V. (1974), "Adaptive robust procedures: a partial review and some suggestions for future applications and theory," *J. Amer. Stat. Ass.* **69**, 909–927.

Holton, G. (1962), "Models for understanding the growth and excellence of scientific research," In S. E. Gaubard and G. Holton, (Eds.) *Excellence and Leadership in a Democracy*, Columbia University Press, New York.

Holland, J. D. (1962), "The Reverend Thomas Bayes, F. R. S. (1702–61)," *J. Roy. Stat. Soc. A* **125**, 451–461.

Hooper, J. W. (1962), "Partial Trace Correlations," *Econometrica*, **30**, 324–331.

Hotelling, H. (1936), "Relations Between Two Sets of Variates," *Biometrika*, **28**, 321–77.

Hotelling, H. (1940), "The selection of variates for use in prediction with some comments on the general problem of nuisance parameters," *Ann. Math. Stat.* **11**, 271–283.

Howe, M. J. A. (1970) *Introduction to Human Memory*, Harper and Row, New York.

Huber, P. J. (1972), "Robust statistics: a review," *Ann. of Math. Stat.* **43**, 1041–1067.

Huntsberger, D. V. (1955), "A generalization of a preliminary testing procedure for pooling data," *Ann. Math. Stat.* **26**, 734–743.

James, W. and C. Stein (1961), "Estimation with quadratic loss," *Proceedings of the Fourth Berkeley Symposium on Mathematical Statistics and Probability*, University of California Press, Berkeley, pp. 361–379.

Jaynes, E. T. (1968), "Prior probabilities," *IEEE Trans. Systems Science and Cybernetics*, SSC-4, p. 227–241.

Jeffreys, H. (1957) *Scientific Inference* (2nd ed.), Cambridge University Press, Cambridge, England.

Jeffreys, H. (1961) *Theory of Probability* (3rd ed.), Oxford University Press, London.

Johnson, N. and S. Kotz (1972) *Continuous Multivariate Distributions*, Wiley, New York.

Johnston, J. (1973) *Econometric Methods*, McGraw-Hill, New York.

Jones, L. Y. (1974), "Sexual relations among married couples," (interview of C. F. Westoff), *Princeton Alumni Weekly* **75**. Reprinted from *People Weekly* 1974.

Jöreskog, K. G. (1963) *Statistical Estimation in Factor Analysis: A New Technique and its Foundation*, Almquist and Wiksell, Stockholm.

Jöreskog, K. G. (1966), "Testing a simple hypothesis in factor analysis," *Psychometrika* **31**, 165–178.

Jöreskog, K. G. (1970), "A general method for analysis of covariance structures," *Biometrika* **57**, 239–251.

Jöreskog, K. G. and A. S. Goldberger (1975), "Estimation of a model with multiple indicators and multiple causes of a single latent variable," *J. Amer. Stat. Ass.* **70**, 631–639.

Judge, G. G., M. E. Bock and T. A. Yancey (1974), "On post data model evaluation," *Rev. Econ. Stat.* **56**, 245–253.

Kabe, D. G. and R. P. Gupta (1973) *Multivariate Statistical Inference*, North-Holland, New York.

Kadane, J. B. (1975), "The role of identification in Bayesian theory," In S. E. Feinberg and A. Zellner, Eds., *Studies in Bayesian Econometrics and Statistics*, North-Holland, Amsterdam.

Kalbfleisch, J. G. and D. A. Sprott (1970), "Application of likelihood methods to models involving large numbers of parameters," *J. Roy. Stat. Soc. B* **32**, 175–194.

Kalman, R. E. (1960), "A new approach to linear filtering and prediction theory," Transactions of ASME, Series D, *J. Basic Eng.* **82**, 35–45.

Kalman, R. E. and R. S. Bucy (1961), "New results in linear filtering and prediction theory," Transactions of ASME, Series D, *J. Basic Eng.*, **83**, 95–108.

Kendall, M. G. and A. Stuart (1961) *The Advanced Theory of Statistics*, Griffin, London.

Kennard, Robert (1971), "A note on the C_p statistic," *Technometrics* **13**, 899–900.

Kennedy, W. J. and T. A. Bancroft (1971), "Model building for predicting in regression based upon repeated significance tests," *Ann. Math. Stat.* **42**, 1273–1284.

Kerridge, D. (1967), "Errors of prediction in multiple regression with stochastic regressor variables," *Technometrics* **9**, 309–311.

Keynes, J. M. (1921) *A Treatise on Probability*, Harper and Row, New York.

Kiefer, J. and J. Wolfowitz (1956), "Consistency of maximum likelihood estimator in the presence of infinitely many nuisance parameters," *Ann. Math. Stat.* **27**, 887–906.

Kimeldorf, G. S. (1965), "Applications of Bayesian statistics to actuarial graduation," unpublished Ph.D. dissertation, University of Michigan.

Klein, G. E. (1968), "Selection regression programs," *Rev. Econ. Statist.* **L**, 288–290.

Kmenta, J. (1971) *Elements of Econometrics*, MacMillan, New York.

Kmenta, J. and R. F. Gilbert (1970), "Estimation for seemingly unrelated regressions with autoregressive disturbances," *J. Amer. Stat. Ass.* **65**, 186–197.

Knight, F. H. (1921) *Risk, Uncertainty and Profit*, Houghton Mifflin, New York.

Koopmans, T. (1937), *Linear Regression Analysis of Economic Time Series*, Netherlands Econometric Institute, Haarlem-de Erwen F. Bohn N. V.

Koopmans, T. C. and O. Reiersol (1950), "The identification of structural characteristics," *Ann. Math. Stat.* **21**, 165–181.

Koyck, L. (1954) *Distributed Lags and Investment Analysis*, North-Holland, Amsterdam.

Krupp, Sherman Roy, Ed. (1966) *The Structure of Economic Science*, Prentice-Hall, Englewood Cliffs, N.J.

Kuhn, T. S. (1962) *The Structure of Scientific Revolutions*, University of Chicago Press, Chicago.

Kuhn, T. S. (1969), "Logic of discovery or psychology of research?" In Lakatos, I. and A. Musgrave, Eds., *Criticism and the Growth of Knowledge*, Cambridge University Press, Cambridge, England.

Kyburg, H. E. and H. E. Smokler, Eds. (1964) *Studies in Subjective Probability*, Wiley, New York.

Lakatos, I. and A. Musgrave, Eds. (1969) *Criticism and the Growth of Knowledge*, Cambridge University Press, Cambridge, England.

LaMotte, L. R. and R. R. Hocking (1970), "Computational efficiency in the selection of regression variables," *Technometrics* **12**, 83–93.

Larson, H. J. and T. A. Bancroft (1963a), "Biases in prediction by regression for certain incompletely specified models," *Biometrika* **50**, 391–402.

Larson, H. J. and T. A. Bancroft (1963b), "Sequential model building for prediction in regression analysis," *Ann. Math. Statist.* **34**, 462–479.

LaValle, I. H. (1970) *An Introduction to Probability Inference and Decision*, Holt, Rinehart and Winston, New York.

Lawley, D. N. and A. E. Maxwell (1964) *Factor Analysis as a Statistical Method*, Butterworth, London.

Lawley, D. N. and A. E. Maxwell (1973), "Regression and factor analysis," *Biometrika* **60**, 331–338.

Lazarsfeld, P. F. and N. W. Henry, Eds., (1966) *Readings in Mathematical Social Science*, The MIT Press, Cambridge, Mass.

Leamer, E. E. (1970), "Inference with non-experimental data: a Bayesian view," unpublished Ph.D. dissertation, University of Michigan.

Leamer, E. E. (1971), "Another note on posterior means," mimeo.

Leamer, E. E. (1972), "A class of informative priors and distributed lag analysis," *Econometrica* **40**, 1059–1081.

Leamer, E. E. (1973), "Multicollinearity: a Bayesian interpretation," *Rev. Econ. Stat.* **55**, 371–80.

Leamer, E. E. (1974), "False models and post-data model construction," *J. Amer. Stat. Ass.* **69**, 122–131.

Leamer, E. E. (1975a), "A result on the sign of restricted least squares estimates," *J. Econometrics* **3**, 387–390.

Leamer, E. E. (1975b), "'Explaining your results' as access-biased memory," *J. Amer. Stat. Ass.* **70**, 88–93.

Leamer, E. E. (1976), "The simple geometry of least squares confidence regions," mimeo.

Leamer, E. E. (1977), "Regression-selection strategies and revealed priors," *J. Amer. Stat. Soc.*, under review.

Leamer, E. E. and G. Chamberlain (1976), "A Bayesian interpretation of pre-testing," *J. Roy. Stat. Soc. B* **38**, 85–94.

Lempers, F. B. (1971), *Posterior Probabilities of Alternative Linear Models*, Rotterdam University Press, Rotterdam.

Lee, W. (1971) *Decision Theory and Human Behavior*, Wiley, New York.

Lindley, D. V. (1947), "Regression lines and the linear functional relationship," *J. Roy. Stat. Soc.* (supplement) **9**, 218–244.

Lindley, D. V. (1957), "A statistical paradox," *Biometrika* **44**, 187–192.

Lindley, D. V. (1965) *Introduction to Probability and Statistics from a Bayesian Viewpoint, Part 2, Inference*, Cambridge University Press, Cambridge, England.

Lindley, D. V. (1968), "The choice of variables in multiple regression," *J. Roy. Stat. Soc. B* **31**, 31–66.

Lindley, D. V. (1971a) *Bayesian Statistics, A Review*, Regional Conference Series in Applied Mathematics, S.I.A.M.

Lindley, D. V. (1971b) *Making Decisions*, Wiley, New York.

Lindley, D. V. and A. F. M. Smith (1972), "Bayes estimates for the linear model," *J. Roy. Stat. Soc.* **34**, 1–18.

Lindley, D. V. and G. M. El-Sayyad (1968), "The Bayesian estimation of a linear functional relationship," *J. Roy. Stat. Soc. B* **30**, 190–202.

Lindsey, J. K. (1974), "Construction and comparison of statistical models," *J. Roy. Stat. Soc. B* **36**, 418–425.

Liu, T. C. (1960), "Underidentification, structural estimation, and forecasting," *Econometrica* **28**, 855–865.

Liviatan, N. (1961), "Errors in variables and Engle curve analysis," *Econometrica* **29**, 336–362.

Lovell, M. C. (1975), "Data grubbing," mimeo.

Luce, R. D. and H. Raiffa (1958) *Games and Decisions*, Wiley, New York.

Madansky, A. (1959), "The fitting of straight lines when both variables are subject to error," *J. Amer. Stat. Ass.* **54**, 173–205.

Maddala, G. S. (1971), "The use of variance components models in pooling cross section and time series data," *Econometrica* **39**, 341–358.

Malinvaud, E. (1970) *Statistical Methods of Econometrics*, North-Holland, Amsterdam.

Marquardt, Donald W. (1970), "Generalized inverses, ridge regressions, biased linear estimation, and nonlinear estimation," *Technometrics* **12**, 591–612.

Massy, W. F. (1965), "Principle components regression in exploratory statistical research," *J. Amer. Stat. Soc.* **60**, 234–256.

Maxwell, A. E. (1961), "Recent trends in factor analysis," *J. Roy. Stat. Soc. A* **124**, 49–59.

Mayer, T. (1975), "Selecting economic hypotheses by goodness of fit," *Econ. J.*, **85**, 877–82.

Mellor, D. H. (1971) *The Matter of Chance*, Cambridge University Press, Cambridge, England.

Merton, R. K. (1957), "Priorities in scientific discovery," *Amer. Sociol. Rev.* **22**, 635–659.

Meyer, D. L. and R. Collier, Eds., (1970) *Bayesian Statistics*, F. E. Peacock, Itasca, Ill.

Moran, P. A. P. (1971), "Estimating structural and functional relationships," *J. Multivariate Anal.* **1**, 232–255.

Mulkay, M. J., G. N. Gilbert and S. Woolgar (1975), "Problem areas and research networks in science," *Sociology* **9**, 188–203.

Nagel, E. (1939) *Principles of the Theory of Probability*, University of Chicago Press, Chicago.

Narula, S. C. (1974), "Predictive mean square error and stochastic regressor variables, *Appl. Stat.* **23**, 11–17.

National Bureau of Economic Research (1973) *Ann. Econ. Soc. Meas.* **2**, October.

Nerlove, M. and K. F. Wallis (1966), "Use of the Durbin-Watson statistic in inappropriate situations," *Econometrica* **34**, 235–238.

Neyman, J. and E. S. Scott (1948), "Consistent estimates based on partially consistent observations," *Econometrica* **16**, 1–32.

Neyman, J. and E. S. Scott (1951), "On certain methods of estimating the linear structural relation," *Ann. Math. Stat.* **22**, 352–361; and 1952 correction, **23**, p. 115.

Novick, M. R. (1969), "Multiparameter Bayesian indifference procedures," (with discussion) *J. Roy. Stat. Soc. B* **31**, 29–64.

Oskamp, S. (1965), "Overconfidence in case-study judgements," *J. Consulting Psychol.* **29**, 261–265.

Park, C. N. and A. L. Dudycha (1974), "A cross validation approach to sample size determination for regression models," *J. Amer. Stat. Ass.* **69**, 214–218.

Polanyi, M. (1964) *Personal Knowledge*, Harper and Row, New York.

Polya, G. (1963) (2nd ed.) (1st ed. 1954) *Patterns of Plausible Inference*, Princeton University Press, Princeton, N.J.

Popper, K. R. (1972) *Objective Knowledge, An Evolutionary Approach*, Clarendon Press, Oxford.

Pratt, J. W. (1965), "Bayesian interpretation of standard inference statements (with discussion)," *J. Roy. Stat. Soc. B* **27**, 169–203.

Pratt, J. W. (1967), "Testing hypotheses in linear models: some comments and Bayesian alternatives," Talk for the Econometric Society, mimeo.

Pratt, J. W. (1970), "A note on posterior means," mimeo.

Pratt, J. W., H. Raiffa and R. Schlaifer (1964), "The foundations of decision under uncertainty: an elementary exposition," *J. Amer. Stat. Ass.* **59**, 353–375.

Pratt, J. W., H. Raiffa and R. Schlaifer (1965), *Introduction to Statistical Decision Theory*, McGraw-Hill, New York.

Prescott, E. (1972), "The multiperiod control problem under uncertainty," *Econometrica* **40**, 1043–1058.

Pruitt, D. G. (1971), "Choice shifts in group discussion: an introductory review," *J. Personal. Soc. Psychol.* **20**, 339–360.

Quandt, R. E. (1958), "The estimation of the parameters of a linear regression system obeying two separate regimes," *J. Amer. Stat. Ass. B* **53**, 873–880.

Raduchel, W. J. (1971), "Multicollinearity once again," *Harvard Institute of Economic Research* Paper No. 205.

Raiffa, H. (1961), "Risk, ambiguity and the Savage axioms: comment," *Quart. J. Econ.* **75**, 690–694.

Raiffa, H. (1968) *Decision Analysis*, Addison-Wesley, Reading, Mass.

Raiffa, H. and R. Schlaifer (1961) *Applied Statistical Decision Theory*, Harvard University Press, Cambridge, Mass.

Ramsey, F. P. (1926), "Truth and probability," In Kyburg and Smokler (Eds.) *Studies in Subjective Probability*, Wiley, New York.

Ramsey, J. B. (1969), "Tests for specification errors in classical linear least-squares regression analysis," *J. Roy. Stat. Soc. B* **31**, 350–371.

Ramsey, J. B. (1974), "Classical Model Selection through Specification Error Tests," in P. Zarembka, Ed., *Frontiers in Econometrics*, Academic Press, New York.

Rao, C. R. (1965) *Linear Statistical Inference and Its Applications*, Wiley, New York.

Reichenbach, H. (1949) *Theory of Probability*, University of California Press, Berkeley.

Reichenbach, H. (1958) *The Rise of Scientific Philosophy*, University of California Press, Berkeley.

Reiersol, O. (1945), "Confluence analysis by means of instrumental sets of variables," *Arkiv for Mathematik, Astronomi och Fysik* **32**, 1–119.

Reiersol, O. (1950), "Identifiability of a linear relation between variables which are subject to error," *Econometrica* **18**, 375–389.

Robbins, H. (1964), "The empirical Bayes approach to statistical problems." *Ann. Math. Stat.* **35**, 1–20.

Roberts, H. V. (1965), "Probabilistic prediction," *J. Amer. Stat. Assoc.* **60**, 50–62.

Roberts, H. V. (1966) *Statistical Inference and Decision*, University of Chicago Press, Illinois.

Robertson, C. A. (1974), "Large-sample theory for the linear structural relation," *Biometrika* **61**, 353–359.

Robinson, P. M. (1974), "Identification, estimation and large-sample theory for regressions containing unobservable variables," *Int. Econ. Rev.* **15**, 680–692.

Rosenberg, B. (1972), "The estimation of stationary stochastic regression parameters re-examined," *J. Amer. Stat. Ass.* **67**, 650–654.

Rosenberg, B. (1973), "Linear regression with randomly dispersed parameters," *Biometrika* **60**, 65–72.

Rothenberg, T. (1963), "A Bayesian analysis of simultaneous equations systems," Econometric Institute Report 6315, Rotterdam, Holland.

Rothenberg, T. (1970), "The Bayesian approach and alternatives in econometrics—II," with comments by G. M. Kaufman and J. W. Pratt In M. D. Intriligator (Ed.), *Frontiers of Quantitative Economics*.

Savage, L. J. (1954) *The Foundation of Statistics*, Wiley, New York.

Savage, L. J. (1962a), "Subjective probability and statistical practice," In L. J. Savage et al. *The Foundations of Statistical Inference*, Wiley, New York, pp. 9–35.

Savage, L. J. et al. (1962b) *The Foundations of Statistical Inference*, Wiley, New York.

Schatzoff, M., R. Tsao and S. Fienberg (1968), "Efficient calculations of all possible regressions," *Technometrics* **10**, 769–780.

Scheffe′, H. (1959) *The Analysis of Variance*, Wiley, New York.

Schlaifer, R. (1959) *Probability and Statistics for Business Decisions*, McGraw-Hill, New York.

Schmidt, S. A. (1969) *Measuring Uncertainty: An Elementary Introduction to Bayesian Statistics*, Addison-Wesley, Reading, Mass.

Sclove, S. L. (1968), "Improved estimators for coefficients in linear regressions," *J. Amer. Stat. Ass.* **63**, 596–606.

Sclove, S. L., C. Morris and R. Radhakrishnan (1972), "Nonoptimality of preliminary test estimators for multinormal mean," *Ann. Math. Stat.* **43**, 1481–1490.

Shiller, R. (1973), "A distributed lag estimator derived from smoothness priors," *Econometrica* **41**, 775–788.

Silvey, S. D. (1969), "Multicollinearity and imprecise estimation," *J. Roy. Statis. Soc. B*, **31**, 539–552.

Simon, H. A. and H. Guetzkow (1966), "Mechanisms involved in pressures toward uniformity in groups," in Lazarsfeld, P. F. and N. W. Henry (Eds.) *Readings in Mathematical Social Science*, MIT Press, Cambridge, Mass.

Sims, C. A. (1971), "Discrete approximation to continuous time distributed lags in econometrics," *Econometrica* **39**, 545–563.

Skinner, B. F. (1948), "Superstition in the pigeon," *J. Exper. Psychol.* **38**, 168–172.

Slovic, P. (1966), "Value as a determiner of subjective probability," *IEEE Transactions on Human Factors in Electronics* HFE-7, 223–224.

Slovic, P. (1972), "Psychological study of human judgement: implications for decision making," *J. Finance* **27**, 779–800.

Smith, A. F. M. (1973), "Bayes estimates in one-way and two-way models," *Biometrika* **60**, 319–329.

Solari, M. E. (1969), "The maximum likelihood solution of the problem of estimating a linear functional relationship, *J. Roy. Stat. Soc. B* **31**, 611–613.

Solow, R. M. (1960), "On a family of lag distributions," *Econometrica* **28**, 393–406.

Stein, C. M. (1956), "Inadmissibility of the usual estimator of the mean of a multivariate normal distribution," *Proceedings of the Third Berkeley Symposium on Mathematical Statistics and Probability*, pp. 197–206.

Stein, C. M. (1962), "Confidence sets for the mean of a multivariate normal distribution," *J. Roy. Stat. Soc. B* **24**, 265–269.

Stone, M. (1961), "The opinion pool," *Ann. Math. Stat.* **32**, 1339–1342.

Stone, M. (1974), "Cross-validatory choice and assessment of statistical predictions," *J. Roy. Stat. Soc. B* **36**, 111–147.

Stone, M. and B. G. F. Springer (1965), "A paradox involving quasi prior distributions," *Biometrika* **52**, 623–627.

Strawderman, W. E. and A. Cohen (1971), "Admissibility of estimators of the mean vector of a multivariate normal distribution with quadratic loss," *Ann. Math. Stat.* **42**, 270–296.

Suppes, P. (1960), "Some open problems in the foundations of subjective probability," *Information and Decision Processes* by Machol (Ed.), McGraw-Hill, New York.

Suppes, P. (1968), "Concept formation and Bayesian decisions," In J. Hintikka and P. Suppes (Eds.) *Aspects of Inductive Logic*, North-Holland, Amsterdam, pp. 21–48.

Suppes, P. (1974), "The measurement of belief (and discussion)," *J. Roy. Stat. Soc. B* **36**, 160–191.

Swamy, P. A. V. B. (1970), "Efficient inference in a random coefficient regression model," *Econometrica* **38**, 311–323.

Swamy, P. A. V. B. (1971) *Statistical Inference in a Random Coefficient Regression Model*, Springer Verlag, Berlin–Heidelberg–New York.

Taylor, L. D. (1967), "Estimation by minimizing the sum of absolute errors and the Pareto distribution," mimeo.

Theil, H. (1954) *Linear Aggregation of Economic Relations*, North-Holland, Amsterdam.

Theil, H. (1957), "Specification errors and the estimation of economic relationships," *Rev. Int. Stat. Instit.* **25**, 41–51.

Theil, H. (1963), "On the use of incomplete prior information in regression analysis," *J. Amer. Stat. Ass.* **58**, 401–414.

Theil, H. (1961) *Economic Forecasts and Policy*, North-Holland, Amsterdam.

Theil, H. (1971) *Principles of Econometrics*, Wiley, New York.

Theil, H. and A. S. Goldberger (1961), "On pure and mixed statistical estimation in economics," *Int. Econ. Rev.* **2**, 65–78.

Thornber, H. (1966), "Applications of decision theory to econometrics," unpublished Ph.D. dissertation, U. of Chicago.

Tiao, G. C. and M. M. Ali (1971), "Analysis of correlated random effects: linear model with two random components," *Biometrika* **58**, 37–51.

Tiao, G. C. and N. R. Draper (1968), "Bayesian analysis of linear models with two random components with special reference to the balanced incomplete block design," *Biometrika* **55**, 101–117.

Tiao, G. C. and W. Y. Tan (1965), "Bayesian analysis of random-effect models in the analysis of variance. I. Posterior distribution of variance components," *Biometrika* **52**, 37–53.

Tiao, G. C. and A. Zellner (1964), "Bayes' theorem and the use of prior knowledge in regression analysis," *Biometrika* **51**, 219–230.

Toro-Vizcarrondo, C. and T. D. Wallace (1968), "A test of the mean square error criterion for restrictions in linear regressions," *J. Amer. Stat. Ass.* **63**, 558–572.

Turnovsky, S. J. (1969), "A Bayesian approach to the theory of expectations," *J. Econ. Theory* **1**, 220–227.

Tversky, A. and D. Kahneman (1971b), "The judgement of frequency and probability by availability of instances," *Oregon Research Institute Research Bulletin* **11**, p. 6.

Tversky, A. and D. Kahneman (1971a), "Belief in the law of small numbers," *Psychol. Bull.* **76**, 105–110.

Tversky, A. and D. Kahneman (1974a), "Assessing uncertainty," *J. Roy. Stat. Soc.* **36**, 148–159.

Tversky, A. (1974b), "Judgment under uncertainty: heuristics and biases," *Science* **186**, 1124–1131.

Wallace, T. D. (1964), "Efficiencies for stepwise regressions," *J. Amer. Stat. Ass.* **59**, 1179–1182.

Wallace, T. D. and V. G. Ashar (1972), "Sequential methods of model construction," *Rev. Econ. Stat.* **54**, 172–178.

Wallace, T. D. and A. Hussain (1969), "The use of error components models in combining cross section with time series data," *Econometrica* **37**, 55–72.

Warner, S. L. (1975), "Advocate scoring for unbiased information," *J. Amer. Stat. Assoc.* **70**, 15–22.

Watson, S. R. (1974), "On Bayesian inference with incompletely specified prior distributions," *Biometrika* **61**, 193–196.

Whittaker, E. T. and G. Robinson (1940), 3rd edn. *The Calculus of Observations*, Blackie and Son, London and Glasgow.

Wilks, S. S. (1962) *Mathematical Statistics*, Wiley, New York.

Winkler, R. L. (1967a), "The assessment of prior distributions in Bayesian analysis," *J. Amer. Stat. Ass.* **62**, 776–800.

Winkler, R. L. (1967b), "The quantification of judgement: some methodological suggestions," *J. Amer. Stat. Ass.* **62**, 1105–1120.

Winkler, R. L. (1972) *An Introduction to Bayesian Inference and Decision*, Holt, Rinehart and Winston, New York.

Wold, S. (1956), "Causal inference from observational data: a review of ends and means," *J. Roy. Stat. Soc. A* **119**, 28–61.

Wright, R. L. (1968), "Unidentifiability, Bayesian inference and the linear functional relation with errors in both variables," unpublished Ph.D. dissertation, University of Michigan.

Yule, G. U. (1926), "Why do we sometimes get nonsense correlation between time series?—A study in sampling and the nature of time series," *J. Roy. Stat. Soc.* **89**, 1–64.

Zellner, A. (1962), "An efficient method of estimating seemingly unrelated regressions and tests for aggregation bias," *J. Amer. Stat. Ass.* **57**, 348–368.

Zellner, A. (1963), "Estimators for seemingly unrelated regression equations: some exact finite sample results," *J. Amer. Stat. Ass.* **58**, 977–992.

Zellner, A. (1970a), "The Bayesian approach and alternatives in econometrics—I," with comments by G. M. Kaufman and J. W. Pratt In M. D. Intriligator (Ed.) *Frontiers of Quantitative Economics*, North-Holland, Amsterdam.

Zellner, A. (1970b), "Estimation of regression relationships containing unobservable independent variables," *Int. Econ. Rev.* **11**, 441–454.

Zellner, A. (1971) *An Introduction to Bayesian Inference in Econometrics*, Wiley, New York.

Zellner, A. (1975), "Bayesian analysis of regression error terms," *J. Amer. Stat. Ass.* **70**, 139–144.

Zellner, A. (1976), "Bayesian analysis of the regression model with multivariate Student-*t* error terms," *J. Amer. Stat. Ass.* **71**, 400–405.

Zellner, A. and V. K. Chetty (1965), "Prediction and decision problems in regression models from the Bayesian point of view," *J. Amer. Stat. Ass.* **60**, 608–616.

Zellner, A. and M. Geisel (1968), "Sensitivity to control to uncertainty and form of the criterion function," In D. G. Watts (Ed.) *The Future of Statistics*, Academic, New York.

Zellner, A. and M. Geisel (1970), "Analysis of distributed lag models with application to consumption function estimation," *Econometrica* **38**, 865–888.

Zellner, A., J. Kmenta and J. Dreze (1966), "Specification and estimation of Cobb-Douglas production function models," *Econometrica* **34**, 784–795.

Zellner, A. and F. Palm (1974), "Time series analysis and simultaneous equation models," *J. Econometrics* **2**, 17–54.

Zellner, A. and G. C. Tiao (1964), "Bayesian analysis of the regression model with autocorrelated errors," *J. Amer. Stat. Ass.* **59**, 763–778.

Zellner, A. and W. Vandaele (1975), "Bayes-Stein estimators for *k*-means, regression and simultaneous equation models." In S. E. Fienberg and A. Zellner (Eds.) *Studies in Bayesian Econometrics and Statistics*, North-Holland, Amsterdam.

Zimon, J. (1968) *Public Knowledge*, Cambridge University Press, London.

INDEX

Applied Probability and Statistics (Continued)

GROSS and CLARK · Survival Distributions: Reliability Applications in the Biomedical Sciences

GROSS and HARRIS · Fundamentals of Queueing Theory

GUTTMAN, WILKS, and HUNTER · Introductory Engineering Statistics, *Second Edition*

HAHN and SHAPIRO · Statistical Models in Engineering

HALD · Statistical Tables and Formulas

HALD · Statistical Theory with Engineering Applications

HARTIGAN · Clustering Algorithms

HILDEBRAND, LAING, and ROSENTHAL · Prediction Analysis of Cross Classifications

HOEL · Elementary Statistics, *Fourth Edition*

HOLLANDER and WOLFE · Nonparametric Statistical Methods

HUANG · Regression and Econometric Methods

JAGERS · Branching Processes with Biological Applications

JESSEN · Statistical Survey Techniques

JOHNSON and KOTZ · Distributions in Statistics
Discrete Distributions
Continuous Univariate Distributions-1
Continuous Univariate Distributions-2
Continuous Multivariate Distributions

JOHNSON and KOTZ · Urn Models and Their Application: An Approach to Modern Discrete Probability Theory

JOHNSON and LEONE · Statistics and Experimental Design in Engineering and the Physical Sciences, Volumes I and II, *Second Edition*

KEENEY and RAIFFA · Decisions with Multiple Objectives

LANCASTER · An Introduction to Medical Statistics

LEAMER · Specification Searches: Ad Hoc Inference with Nonexperimental Data

McNEIL · Interactive Data Analysis

MANN, SCHAFER, and SINGPURWALLA · Methods for Statistical Analysis of Reliability and Life Data

MEYER · Data Analysis for Scientists and Engineers

OTNES and ENOCHSON · Digital Time Series Analysis

PRENTER · Splines and Variational Methods

RAO and MITRA · Generalized Inverse of Matrices and Its Applications

SARD and WEINTRAUB · A Book of Splines

SEARLE · Linear Models

THOMAS · An Introduction to Applied Probability and Random Processes

WHITTLE · Optimization under Constraints

WILLIAMS · A Sampler on Sampling

WONNACOTT and WONNACOTT · Econometrics

WONNACOTT and WONNACOTT · Introductory Statistics, *Third Edition*

WONNACOTT and WONNACOTT · Introductory Statistics for Business and Economics, *Second Edition*

YOUDEN · Statistical Methods for Chemists

ZELLNER · An Introduction to Bayesian Inference in Econometrics